Science, Technology and Medicine in Modern History
General Editor: John V. Pickstone, Centre for the History of Science, Technology and Medicine, University of Manchester, England (www.man.ac.uk/CHSTM)

One purpose of historical writing is to illuminate the present. At the start of the third millennium, science, technology and medicine are enormously important, yet their development is little studied.

The reasons for this failure are as obvious as they are regrettable. Education in many countries, not least in Britain, draws deep divisions between the sciences and the humanities. Men and women who have been trained in science have too often been trained away from history, or from any sustained reflection on how societies work. Those educated in historical or social studies have usually learned so little of science that they remain thereafter suspicious, overawed, or both.

Such a diagnosis is by no means novel, nor is it particularly original to suggest that good historical studies of science may be peculiarly important for understanding our present. Indeed this series could be seen as extending research undertaken over the last half-century. But much of that work has treated science, technology and medicine separately; this series aims to draw them together, partly because the three activities have become ever more intertwined. This breadth of focus and the stress on the relationships of knowledge and practice are particularly appropriate in a series which will concentrate on modern history and on industrial societies. Furthermore, while much of the existing historical scholarship is on American topics, this series aims to be international, encouraging studies on European material. The intention is to present science, technology and medicine as aspects of modern culture, analysing their economic, social and political aspects, but not neglecting the expert content which tends to distance them from other aspects of history. The books will investigate the uses and consequences of technical knowledge, and how it was shaped within particular economic, social and political structures.

Such analyses should contribute to discussions of present dilemmas and to assessments of policy. 'Science' no longer appears to us as a triumphant agent of Enlightenment, breaking the shackles of tradition, enabling command over nature. But neither is it to be seen as merely oppressive and dangerous. Judgement requires information and careful analysis, just as intelligent policy-making requires a community of discourse between men and women trained in technical specialities and those who are not.

This series is intended to supply analysis and to stimulate debate. Opinions will vary between authors; we claim only that the books are based on searching historical study of topics which are important, not least because they cut across conventional academic boundaries. They should appeal not just to historians, nor just to scientists, engineers and doctors, but to all who share the view that science, technology and medicine are far too important to be left out of history.

Titles include:

Julie Anderson, Francis Neary and John V. Pickstone
SURGEONS, MANUFACTURERS AND PATIENTS
A Transatlantic History of Total Hip Replacement

Roberta E. Bivins
ACUPUNCTURE, EXPERTISE AND CROSS-CULTURAL MEDICINE

Linda Bryder
WOMEN'S BODIES AND MEDICAL SCIENCE
An Enquiry into Cervical Cancer

Roger Cooter
SURGERY AND SOCIETY IN PEACE AND WAR
Orthopaedics and the Organization of Modern Medicine, 1880–1948

Jean-Paul Gaudillière and Ilana Löwy (*editors*)
THE INVISIBLE INDUSTRIALIST
Manufacture and the Construction of Scientific Knowledge

Christoph Gradmann and Jonathan Simon (*editors*)
EVALUATING AND STANDARDIZING THERAPEUTIC AGENTS, 1890–1950

Alex Mold and Virginia Berridge
HEALTH AND SOCIETY SINCE THE 1960S
Voluntary Action and Illegal Drugs

Ayesha Nathoo
HEARTS EXPOSED
Transplants and the Media in 1960s Britain

Neil Pemberton and Michael Worboys
MAD DOGS AND ENGLISHMEN
Rabies in Britain, 1830–2000

Cay-Rüdiger Prüll, Andreas-Holger Maehle and Robert Francis Halliwell
A SHORT HISTORY OF THE DRUG RECEPTOR CONCEPT

Thomas Schlich
SURGERY, SCIENCE AND INDUSTRY
A Revolution in Fracture Care, 1950s–1990s

Eve Seguin (*editor*)
INFECTIOUS PROCESSES
Knowledge, Discourse and the Politics of Prions

Crosbie Smith and Jon Agar (*editors*)
MAKING SPACE FOR SCIENCE
Territorial Themes in the Shaping of Knowledge

Stephanie J. Snow
OPERATIONS WITHOUT PAIN
The Practice and Science of Anaesthesia in Victorian Britain

Carsten Timmermann and Julie Anderson (*editors*)
DEVICES AND DESIGNS
Medical Technologies in Historical Perspective

Science, Technology and Medicine in Modern History
Series Standing Order ISBN 978–0–333–71492–8 hardcover
Series Standing Order ISBN 978–0–333–80340–0 paperback
(*outside North America only*)

You can receive future titles in this series as they are published by placing a standing order.
Please contact your bookseller or, in case of difficulty, write to us at the address below with
your name and address, the title of the series and one of the ISBNs quoted above.

Customer Services Department, Macmillan Distribution Ltd, Houndmills, Basingstoke,
Hampshire RG21 6XS, England

Evaluating and Standardizing Therapeutic Agents, 1890–1950

Edited by

Christoph Gradmann
Professor in History of Medicine, University of Oslo

and

Jonathan Simon
Maître de conférences, Université de Lyon

First published 2010 by
PALGRAVE MACMILLAN

Palgrave Macmillan in the UK is an imprint of Macmillan Publishers Limited, registered in England, company number 785998, of Houndmills, Basingstoke, Hampshire RG21 6XS.

Palgrave Macmillan in the US is a division of St Martin's Press LLC, 175 Fifth Avenue, New York, NY 10010.

Palgrave Macmillan is the global academic imprint of the above companies and has companies and representatives throughout the world.

Palgrave® and Macmillan® are registered trademarks in the United States, the United Kingdom, Europe and other countries

ISBN 978–0–230–20281–8   hardback

This book is printed on paper suitable for recycling and made from fully managed and sustained forest sources. Logging, pulping and manufacturing processes are expected to conform to the environmental regulations of the country of origin.

A catalogue record for this book is available from the British Library.

A catalog record for this book is available from the Library of Congress.

10   9   8   7   6   5   4   3   2   1
19 18 17 16 15 14 13 12 11 10

Contents

List of Figures

List of Tables

Acknowledgements

The editors would like to thank the German Research Foundation (DFG, Gr 2116/1-1) for financing the seminar held in Heidelberg in 2005 at which most of these papers were originally delivered. We would also like to thank the anonymous reviewer for useful comments as well as the team at Palgrave-Macmillan for helping us through the production process, notably Michael Strang and Ruth Ireland. We would also like to thank John Pickstone for accepting this contribution to the series and for his careful reading of the articles. The final stages of preparing the manuscript and the elaboration of the index of names were graciously supported by the European Science Foundation's Research Network Drugs (http://drughistory. eu/) and the University of Oslo.

Notes on Contributors

Christian Bonah (MD, Ph.D.) is Professor of the history of life and health sciences at the University of Strasbourg and currently holds a research professorship with the Institut Universitaire de France (IUF). He has published on social and cultural history of the medical sciences including *L'expérimentation humaine. Discours et pratiques en France, 1900–1940* (Paris: Les Belles Lettres, 2007). He has recently co-edited *Harmonizing Drugs: Standards in 20th Century Pharmaceutical History* (Paris: Glyphe, 2009) and *Histoire et médicament au XIXe et XXe siècle* (Glyphe, 2005) as well as a forthcoming collaboration with D. Cantor and M. Doerries; *Meat, Medicine and Human Health in the Twentieth Century* (Pickering and Chatto, 2010).

Alberto Cambrosio is Professor of Social Studies of Medicine at McGill University. His area of expertise lies at the crossroads of medical sociology and the sociology of science and technology. His work centres on biomedical practices and innovation, in particular at the clinical–laboratory interface, with a focus on the application of modern biological techniques to the diagnosis and the therapy of cancer, the development of cancer clinical trials as a new style of practice, and the role of visual imagery in the development of immunology. His book *Biomedical Platforms* (MIT Press, 2003), co-authored with Peter Keating, analyzes the transformation of medicine into biomedicine and the re-specification of diagnostic, nosological and therapeutic practices.

Gabriel Gachelin is an honorary Chef de Service at the *Institut Pasteur* (immunology) and a former Team Leader in immunology at the Tsukuba Life Science Centre (RIKEN), Japan. He presently works on the history of serotherapy in France and on comparative Franco-Brazilian studies on the history of tropical diseases, medical geography and medical entomology. He has coordinated *Les organismes modèles dans la recherché médicale* (Presses Universitaires de France, 2006) and was co-editor of two special issues of *Parassitologia* on the *History of Tropical Diseases in Brazil* (2005) and on *Medical Entomology* (2008).

Jean-Paul Gaudillière is a historian of science and medicine. He teaches at the Ecole des Hautes Etudes en Sciences Sociales in Paris and is co-director of CERMES (Centre de Recherche Médecine, Sciences, Santé et Société). He has worked on the transformation of biological and medical research in the twentieth century. He is currently writing a history of biological therapies. He has published *Inventer la biomédecine* (La Découverte, 2002) and *La médecine et les sciences. XIXème-XXème siècles* (La Découverte, 2006). He recently edited a special issue of *Studies in History and Philosophy of the Biological and*

Biomedical Sciences on 'Drug Trajectories' and of *History and Technology* on 'How Pharmaceuticals Became Patentable in the Twentieth Century'.

Christoph Gradmann is Professor in the History of Medicine at the University of Oslo, Norway. He received his doctorate as a historian from the University of Hannover in 1992 and has held positions at the University of Heidelberg and the Max Planck Institute for the History of Science in Berlin. His research interests cover the history of infectious diseases in modernity, the history of antibiotic resistance, biography as a genre and the history of standardization in twentieth-century medicine. He is the author of *"Historische Belletristik". Populäre historische Biographien in der Weimarer Republik* (Frankfurt, 1993). He was guest editor of *Medizinhistorisches Journal* 40 (2005), issue 2 *Die Deutsche Forschungsgemeinschaft (DFG) und die medizinisch-biologische Forschung im 20. Jahrhundert* and (with Volker Hess) *Science in Context*, 21 (2008), issue 2 *Vaccines as Medical, Industrial, and Administrative Objects*. His most recent book publication is *Laboratory Disease: Robert Koch's Medical Bacteriology* (Baltimore, 2009).

Anne Hardy is Professor of the History of Modern Medicine at the Wellcome Trust Centre for the History of Medicine, University College London. Her research interests are in the modern period, more especially in the history of disease, environment and nutrition. Her publications include *The Epidemic Streets: Infectious Disease and the Rise of Preventive Medicine, 1856–1900* (Oxford, 1993), and *Health and Medicine in Britain since 1860* (London, 2001). Her most recent publication is *Of Medicine and Men: Biographies and Ideas in European Social Medicine between the World Wars* (Frankfurt, 2009) co-edited with Iris Borowy, to which she also contributed an essay on Thorvald Madsen.

Anne I. Hardy is Adviser for science communication at the Goethe University Frankfurt. She was formerly a postdoctoral fellow at the University of Heidelberg, where she worked in the 'evaluations – project' funded by the Deutsche Forschungsgemeinschaft. She is the author of *Ärzte, Ingenieure und städtische Gesundheit. Medizinische Theorien in der Hygienebewegung des 19. Jahrhunderts* (2005) and *Lise Meitner* (2002) (in collaboration with Lore Sexl).

Axel C. Hüntelmann is a postdoctoral research fellow at the Bielefeld Graduate School in history and sociology where he works on the cultural history of growth 1770–1970. In recent years he has worked on the German Imperial Health Office and other bio-political institutions between 1876 and 1933. He is preparing a biography of the immunologist Paul Ehrlich, and he has studied serum and vaccine regulation in the German Empire and in France. Recent publications include 'Das Diphtherie-Serum und der Fall Langerhans', in: *MedGG* 24 (2006), p. 71–104; *Hygiene im Namen des Staates. Das Reichsgesundheitsamt 1876–1933*, Göttingen 2008; 'Dynamics of Wertbestimmung', in: *Science in Context* 21 (2008). Together with Johannes Vossen and Herwig Czech, he edited a collection of articles on German Health Offices: *Gesundheit und Staat. Studien zur Geschichte der Gesundheitsämter 1870–1950*, Husum 2006.

Mariama Kaba is HES Professor (Haute Ecole Spécialisée – University of Applied Sciences) at the Haute école de travail social et de la santé in Lausanne, Switzerland. She teaches history of social work and history of medicine and health, and her research topics include disability, childhood, and social institutions, with a gender perspective. She is currently finishing her doctoral thesis on the history of disability in French-speaking Switzerland (19th century – early 20th century) at the University of Lausanne.

Ulrike Lindner is a senior lecturer at the University of Bielefeld. She specializes in the comparative history of health policy, in gender and health and more recently, in comparative imperial history as well as global history. She is the author of *Gesundheitspolitik in der Nachkriegszeit. Großbritannien und Deutschland im Vergleich* (Munich 2004) and co-editor with Merith Niehuss of *Ärztinnen –Patientinnen Frauen im deutschen und britischen Gesundheitswesen des 20. Jahrhunderts* (Köln 2002). Recently, she has finished a book manuscript on *Colonial Encounters: Germany and Britain as European Imperial Powers in Africa Before WW1*.

Pauline M. H. Mazumdar is Professor Emeritus of the History of Medicine at the Institute for History and Philosophy of Science and Technology, University of Toronto, Canada. She is the author of *Eugenics, Human Genetics and Human Failings* (London: Routledge, 1992) on the British Eugenics Society; *Species and Specificity: An Interpretation of the History of Immunology* (Cambridge, 1995). She is presently working on a monograph on the history of standardization under the League of Nations, under the working title of *Standardisation and Collective Security*.

Cay-Rüdiger Prüll is Lecturer at the Medical Faculty of the University of Freiburg in Germany. Since 2009, he has also been an assistant at the Institute for the History of Medicine at the University of Heidelberg. His work focuses on the social and cultural history of nineteenth- and twentieth-century medicine, especially the history of pathology and pharmacology as well as military medicine and the relationship of medicine to the public in Western Germany after 1945. His publications include: 'Part of a Scientific Masterplan? – Paul Ehrlich (1854–1915) and the Origins of his Receptor Concept', *Medical History* 47 (2003), pp. 332–56; 'Caught between the Old and the New – Walther Straub (1874–1944), the Question of Drug Receptors and the Rise of Modern Pharmacology', *Bulletin of the History of Medicine* 80 (2006), pp.465–89.

Jonathan Simon is *Maître de conférences* at Université Lyon 1 (Université de Lyon) where he teaches history of science and history of medicine. He has worked on the relationship between chemistry and pharmacy as well as the development of serotherapy at the end of the nineteenth century. Recent publications include *Chemistry, the Impure Science* (Imperial College Press, 2008, with Bernadette Bensaude-Vincent), *Chemistry, Pharmacy and Revolution,* (Ashgate, 2005) and 'Emil Behring's Medical Culture: From Disinfection to Serotherapy' *Medical History,* 2007.

Michael Worboys is Director of the Centre for the History of Science, Technology and Medicine and Wellcome Unit for the History of Medicine at the University of Manchester. He has written on the history of colonial science and medicine, the history of bacteriology and laboratory medicine, and the history of infectious diseases. He is the co-author, with Neil Pemberton, of *Mad Dogs and Englishmen: Rabies in Britain, 1830–2000* (Palgrave Macmillan, 2007). He is completing a project on the history of fungal infections and is beginning a new study of the 'manufacture' of the modern pedigree dog.

Introduction: Evaluating and Standardizing Therapeutic Agents, 1890–1950

Christoph Gradmann and Jonathan Simon

It seems evident today that vaccines and other medicines, whether they are intended to be injected, swallowed, or even applied topically, should meet stringent criteria of quality. Indeed, such pharmaceutical products are embedded in specific regulations that have been successively put into place since WWII. The general regulatory consensus is that they should not be unnecessarily harmful,[1] and they should contain specified amounts of the active ingredients that a qualified physician has prescribed, combined with only those excipients necessary to facilitate administration and that have also been shown to be innocuous. Furthermore, and this takes us to the heart of the present volume, the dose of the active principle in each unit of consumption – be it pill, phial or suppository – should be fixed and known. The idea behind this requirement is that the effectiveness of the medicine, something one cannot see just by looking at a liquid or a pill, or establish by any simple test, should be precisely known and guaranteed. This, of course, implies that the 'activity' of the active principle can be precisely measured in appropriate, well-defined units. Indeed, the pharmaceutical ideal is that the precise physiological effect of the administered product should be pre-dictable and under complete control, an ideal that is, of course, impossible to realize given the variety that exists among human beings (and other living organisms) and their divergent individual responses to both disease and medical treatment.[2]

These quality standards are such an integral part of modern medical culture, at least in the industrialized and post-industrial world, that it might come as a surprise to many how recent this quality control regime is in the context of the long history of pharmacy. For centuries, each individual apothecary was responsible for guaranteeing the quality of the medicines he sold in his pharmacy in a context where he prepared the majority of these himself. The declared aim of this guild-based system of professional responsibility was to protect the apothecary's customers against poisonous substances, although, as a corollary, the amounts of the ingredients that went into any prescription were precisely dictated by the pharmacopoeia.

1

As a general rule, the contents of the pharmacopoeia, which often determined what could be legally dispensed by an apothecary, were fixed by groups of elite doctors and pharmacists who transcribed the combinations of ingredients – with quantities specified to the nearest grain – that formed the contemporary standards.[3] Nevertheless, however precise, these quantities were arrived at through traditional empirical considerations, and the dosages were certainly not based on anything like the modern concept of a pure active ingredient that constitutes the point of reference today. Finally, to close this summary of traditional ways of regulating medicines, we should note that the apothecary mostly dealt with plant and animal extracts, as well as an increasing number of chemical preparations starting in the eighteenth century. Few of the products that graced the shelves of a pharmacy in the eighteenth century would, however, survive into its twentieth-century descendant, except in purified, if not chemically synthesized form.

This principle of the personal responsibility of the apothecary for the quality of his wares changed at the end of the nineteenth century, partly in response to new ways of thinking about organizing and legislating over public health in Europe, partly in response to innovations in therapeutics, and partly due to the influence that industrial methods of production had on the pharmaceutical market. Indeed, the regime in which the apothecary bore the exclusive responsibility for all that he sold was based on the idea that he would likewise personally produce all these medicines. Ideally he would gather the leaves or roots that constituted the raw materials, and perform the operations that transformed them into the final medicament, although this ideal was rarely realized, with raw materials often bought off suppliers. In the nineteenth century, the finished medicines were increasingly obtained by the apothecary in a ready-to-sell form without a precise indication of their contents. Indeed, if a pharmaceutical entrepreneur (usually an apothecary himself) wanted to make money from an innovative product, he had good commercial reasons for keeping the ingredients secret. In the absence of patents that might protect the producer, revealing this information to the retail apothecary carried the risk of losing his business, as there was nothing to stop the retailer producing the medicine himself. Evidently, without a good idea of what the product contained, the apothecary would be unable to test its quality in the sense of ensuring that it corresponded with any advertised ingredients. Such secret remedies, specialities, or patent medicines, as they were variously known, while present in the seventeenth and eighteenth centuries, became an increasingly prominent feature of nineteenth-century pharmacy. Governments either struggled to keep up with contemporary practice by deploying new legislation or effectively turned a blind eye to this burgeoning market, which often contributed significantly to a country's overseas sales. Thus, as was the case in France, a nascent pharmaceutical industry flourished in a legal grey area in which informal rules more than any formal regulation dictated acceptable conduct.[4] Nevertheless, one of the primary concerns of

governments remained the traditional public health goal of keeping poisonous products off the market. Meanwhile, the question of the efficacy of the
medicines that were available – a veritable obsession today – does not appear
to have been a priority. This situation changed at the end of the nineteenth
and beginning of the twentieth centuries.

The development of *Wertbestimmung* was a key element in a new movement in public health in which government institutions started taking an
interest in the therapeutic value, the '*Wert*' (hence the use of the German
word *Wertbestimmung* – 'determination of value') of new therapeutic agents,
particularly vaccines and sera. These new processes of evaluation were initially constructed around therapeutic agents that while of animal origin were
not animal extracts in the traditional sense of the term. The product that lies
at the root of this new 'scientific' approach to quality control was the serum
used for treating diphtheria, which was first introduced between 1892 and
1894. This medicine, given by injection, and used principally on children,
consisted of serum separated out of blood taken from horses immunized
against diphtheria. This medicine was produced and 'policed' by relatively
large-scale producers – both private and public enterprises – and public
health institutions, with the particular configuration depending heavily on
the local – usually national – context. Indeed, some questions that arise from
an examination of the introduction and initial large-scale manufacture of
this serum are the reasons behind the different forms of production (private,
public, or charitable institute) and the implications that these forms had on
the system as well as the process for evaluating and establishing standards
for the serum. It seems that the antitoxin for treating diphtheria, with its
associated methods for establishing the quality of the serum served as a
model for the development, application and handling of other medicines in
the twentieth century, and indeed, the issue of this relationship is discussed
in several of the chapters in the present volume. Thus, the regimes of quality
control that form the focus for this collection of papers were aimed at fixing
the therapeutic value or worth of a medicine in different terms than those
habitually used for the retail apothecary trade. It was no longer enough to
know what the ingredients were and how much of each was in the medicament in question. Instead, there was now a perceived need to know – or to
fix – a product's activity by performing animal experiments and, in the case
of novel therapies, by establishing animal models, thereby giving an indication, in principle, of the effect it would have on the patient's organism.

Wertbestimmung and industrial culture

It is evident from the timing, as well as from the subject, that this shift
to a new scientific mode of evaluating therapeutic agents was related to a
whole range of standardization techniques associated with the Industrial
Revolution, which swept across Europe in the nineteenth century. While

this wider movement has interested historians of science and technology, it has not, however, been explored with the same intensity in the history of medicine and pharmacology.

One aspect of standardization that has emerged clearly from work in the history and philosophy of science is the role of standards in exchanges between scientists, or, more importantly, disciplinary communities of scientists. Like 'objectivity', standardization became an important goal as fields grew to a scale where it was no longer possible to know all the actors personally. Doubts concerning the reliability of 'anonymous' collaborators were allayed by alternative 'technologies of trust', notably the use of machines and a preference for quantitative measurements. Indeed, units of measurement and, in particular, numbers served as a means for regulating these domains where personal contact and judgement of character no longer sufficed. Ted Porter has argued that we can interpret this movement as the rise of the 'accounting ideal' in the sciences, with numbers serving to validate scientific expertise, particularly in contested political environments.[5] Thus, nineteenth-century science became an increasingly anonymous world in which machines were asked to ensure the standards that had been constituted by the gentleman of the Republic of Science in the wake of the Scientific Revolution.[6]

In the context of the history of medicine in the nineteenth century, Volker Hess has suggested how the use of the thermometer provided an 'objective' measure which facilitated exchanges between doctors and their patients. A numerical temperature could be seen as a welcome innovation on both sides in light of the changing nature of the doctor–patient relationship. With the introduction of widespread health insurance, both doctors and their new working-class patients had to deal with people from distant social classes in a different context from that of traditional charitable hospital care. In this new situation, where the working poor were legitimate clients who could make demands on the doctor, the numbers provided by the thermometer could serve both to make these demands (requests for sick leave among others) and to legitimate medical decisions on 'objective' grounds.

One way to objectively determine the quantity of a pharmaceutical in numerical terms is the 'chemical approach'; purify the chemical compounds that enter into the substance and determine the exact weight of each one. Indeed, the requirements of industrial production and its associated regime of quality control have led to this kind of technical response that is now ubiquitous in the pharmaceutical sector. If you look on (or inside) any package of phials for injection or pills (of generic aspirin, for example),[7] you can find the quantity of the pure active ingredient per unit, per pill or per package; 'Active Ingredient: Acetylsalicylic acid, 300mg'. This kind of information is so transparent, that most consumers never stop to think about what it means or why it is there? It is the producer's assurance (usually guaranteed, in principle at least, by the state) that the pill you are about to ingest (to continue with the same example) contains precisely this amount of the physiologically active substance that, it is hoped, is capable of alleviating, if not curing your medical complaint. Implicit is the idea that this active ingredient was synthesized or purified in a completely 'unpolluted' environment, and a precisely defined amount of this known chemical compound was put into the pill in question (and it is the only 'active' substance present). This context of chemically pure, synthesized medicaments dates from the end of the nineteenth century, and the model for the kind of information that accompanies it is the chemical industry rather than the apothecary. The extract of willow bark, while maybe just as effective as aspirin, has the disadvantage of not containing a single 'pure' chemical compound.

Nevertheless what counts for the patient is not so much the amount of active ingredient as the dosage – how many pills to take per hour, per day, or per week. If the dosage is determined in terms of grams or milligrams of the active ingredient, then the calculation is relatively straightforward, but how many patients consume pills in terms of the content of the active ingredients? What counts for most people (many doctors included) are the limits – minimal and maximal – of the dosage in terms of the units administered (pills, volume of injection, etc.), as these are considered to hold the key to maximizing the curative effect of the medicine while minimizing the associated risks (overdose, side-effects, etc.) In terms of a simple dose-response model in which each single compound works its physiological effect within minimal and maximal limits, the synthesized 'pure' chemical compound is the grail at the end of every pharmaceutical company's research and development quest. Such synthetic (even if based on natural products) pharmaceuticals evidently possess another key advantage for pharmaceutical companies as they can be protected by patents (either the molecules themselves, or the means of producing them), which are considerably easier to defend for 'pure' chemical compounds. Nevertheless, it is not only the issue of intellectual property rights that has pushed the industry towards the chemically pure active ingredient as the foundation of a rational drug-based therapy. Synthesizing and combining the ingredients of a therapy in a production plant gives a greater degree of control over the product, and ensures the identity and uniformity of the units that are delivered to the consumer.

Given the importance attributed to the dosage, one might suppose that there is a single, universal physiological response to the ingestion of a pure chemical compound, salicylic acid or other, by a human being of a given weight. On this assumption, the move from the weighing of chemically pure compounds to *Wertbestimmung* depends on finding the means to determine this response, putting a numerical value on the physiological activity. Thus, *Wertbestimmung* involved the development of standardized techniques and approaches that resulted in assigning numbers to phials or pills. This practice started in the area of sera and vaccines, which, precisely because their active ingredients were not pure chemical compounds, could not simply be weighed, and the numbers produced by these techniques ensured their legitimate place in the modern pharmaceutical sphere. Over the course of the twentieth century, these techniques were progressively transferred to the realm of the chemically pure medicines in order to translate weight of chemicals into expected physiological effects. While it is still the weight of the active ingredient that is found on the packaging, it is these measures of physiological effectiveness that justify the all-important dosages that are prescribed. But even these numbers have retained a certain independence from the actual effect of the therapy or vaccine on any particular patient due to inherent variation in the patient population.[8] Nevertheless, they have undoubtedly served to make the products themselves more uniform, and have become an essential ingredient in developing any modern medicine. As we shall see, these evaluations also had considerable significance in the economic sphere, as their absence could block the entry of a product onto the medical market, and their presence could serve as an invaluable marketing tool in a fiercely competitive domain.

Standard units also provide a site for the meeting of technology and science, evoking a mix of economic, scientific, and other interests. Thus, studies on the determination of electrical units at the end of the nineteenth century have revealed the interpenetration of 'basic' science, applied technology, and economic interests. As Bruce Hunt in his study of Maxwell and Fleeming Jenkin, as well as Simon Schaffer, and Norton Wise and Crosbie Smith have amply demonstrated, electricity – and in particular the measurement of resistance – was a field where the scientific community met the world of applied engineering, as the fixing of standards was needed to ensure the quality – or even the simple functioning – of the Transatlantic as well as other underground cables.[9]

Wertbestimmungsverfahren as a 'boundary object'

The considerations outlined above also support the idea that the practices for evaluation of a range of novel biological therapies that came to be known as *Wertbestimmungsverfahren* can be thought of as dynamic communicative tools available to different groups of people dealing with such medicines. A medical researcher trying to define a lethal dose in a guinea pig, a clinician attempting to cure a child suffering from diphtheria and an industrialist trying to produce and sell a medicine, even though all pursuing rather different ends, could integrate their efforts via a procedure that consisted in a quantifying measurement of a medicine's curative power and which resulted in its standardization as an industrial product. In this sense, *Wertbestimmung* can reasonably be considered as a form of boundary object, and some of the contributions in this volume do indeed adopt this approach.[10] The standardization of the efficacy of a given therapy or vaccine also served to communicate knowledge between medical researchers, clinicians, industrial producers and health administrators as well as mediate their various interests with respect to the product. Boundary objects in this sense are loosely defined and achieve stability by means of active interchange between the various groups of actors who handle them. As we can see in the long-term effects of the system that was put in place in the case of the diphtheria antitoxin, these interactions can serve to fuel and integrate dynamic historical processes. In the present volume, the papers by Christian Bonah, Jean-Paul Gaudillière, Anne I. Hardy and Pauline M. H. Mazumdar illustrate how this form of standardization in medicine continued to serve as a model for other more or less similar products well into the twentieth century. Besides highlighting the multiplicity of actors and dynamic historical processes, analysing vaccines and sera as boundary objects also allows us to emphasize several different but interconnected features of 'scientific' medicine in relation to standardization starting from around 1880. Substances like the diphtheria antitoxin were at once novel objects of knowledge, industrially produced medicines and healthcare products subjected to innovative forms of control and regulation by local and national authorities. Indeed, while the issue of standardization is implicit in all of these aspects, the standardization of novel biological therapies in its specific form was in this sense dependent on the hybrid character of such medicines. On the one hand, diphtheria antitoxin was a 'natural' biological substance produced by the functioning of a living animal, while on the other hand it was the product of a sophisticated process of research, development and industrial production that could not readily be integrated into the traditional repertoire of the local pharmacist. The need to evaluate the quality of such medicines of biological origin thus triggered the development of a specific procedure for testing their efficacy, known as '*Wertbestimmung*'. It is the aim of this book, therefore, to understand this concept of *Wertbestimmung* in its specificity, as well as showing how the concept that had been developed in a particular set of circumstances came to serve as an influential – although far from monolithic – model for future developments in the medical sciences and pharmacy across the twentieth century.

While *Wertbestimmung* cannot be reduced to standardization or boundary work, the literature on these themes helps by raising a number of issues and proposing a set of analytical tools. Thus, as with standardization, while the professed aim of *Wertbestimmung* is to guarantee the quality of a product, it also performed other complementary functions. This is perhaps easiest to see in the well-documented case of diphtheria serum in Germany. Here, an independent Institute responsible for quality control required the endless repetition of the same technical gestures on the part of its staff in order to pass or fail batches of serum sent in for approval from all over the German empire. Indeed, the testing procedures were so repetitive that Ehrlich demanded the possibility to do original research as compensation for having to perform these routine tests.[11] Furthermore, these tests served to discipline and standardize the production and quality control of the firms producing the serum, as they had to keep up with the exigencies of the testing institute in order to stay in business. Third, the whole system functioned to standardize the production and distribution of biological medicines throughout Germany, eliminating any independent producers, and automatically penalizing any approach that did not fit with the testing regime put in place. To sum up, we want to insist on the fact that various forms of standardization were concomitantly involved in what came to be called *Wertbestimmung* and that they were dependent on one another. These include standardization in industrial production, standardization of the measurement of efficacy using animal models, the use of standardized test poisons or antitoxins in this context, and – gradually – the application of statistical methods in the clinical evaluation of these products. With the benefit of hindsight, we can now see the significant historical destiny of this tendency, with the statistical treatment of well-defined cohorts replacing the traditional reliance on a physician's clinical experience when making judgements about the quality of a medicine.[12]

Wertbestimmung and the history of the pharmaceutical industry

We have already hinted at the intimate relationship between the regime of *Wertbestimmung* and the early history of the pharmaceutical industry. Indeed, the second half of the nineteenth century witnessed a rapid expansion of the Western pharmaceutical industry, in part pushed by new therapies like the sera, and in part (probably a much larger part) pulled by the development of a consumer culture that embraced medicines.[13] This early history of the pharmaceutical industry is somewhat paradoxical, as the industry experienced considerable growth even before the discovery of the modern 'miracle' drugs that effectively placed a great deal more therapeutic power in the hands of doctors.[14]

Nevertheless, several historians of pharmacy have remarked that the end of the nineteenth century was a period when scientists entered the pharmaceutical industry in relatively large numbers. As was the case for the émigré Dr Gottlieb in Sinclair Lewis's novel *Arrowsmith*, who reluctantly accepted the offer to work for Dawson T. Hunziker and Co., Inc. of Pittsburgh, both sides stood to gain from associations between academic scientists and the pharmaceutical industry.[15] The former obtained relief from time-consuming academic duties, better facilities, more money, and a certain amount of freedom in their research, while the pharmaceutical company stood to benefit from its investment in such research as much in terms of its image as a 'scientific' enterprise as from any new products that might emerge. The growing distinction between 'ethical' companies that engaged in respectable scientific practice (and were permitted to advertise in the official medical press) and the rest of the sector only served to accentuate this tendency. The pharmaceutical companies also hoped to emulate the success story provided by the German synthetic chemical industry that employed large numbers of organic chemists producing artificial dyestuffs with the potential of generating huge profits.[16]

Of course, the topics that have been introduced in this brief introduction do not exhaust the relevant historical context for the introduction of *Wertbestimmung* at the end of the nineteenth century. Thus, one factor that we have not discussed here is the financing of health care and the transformations it experienced in this period. The fact that health insurance companies rather than individuals were starting to pay for therapies meant a dramatic change in the power of negotiation over the price of medicines as well as new restraints on their availability and new mechanisms for their distribution. Furthermore, the existence of a hygiene movement that had achieved considerable success in combating infectious disease by improving living conditions rather than providing treatment affected public expectations in terms of public health initiatives in this period. Indeed, *Wertbestimmung* can be approached from many perspectives, and, however distinctive, the practice clearly constitutes a part of a wider movement of modernity affecting science, technology and medicine. Thus, for example, as Paul Weindling has pointed out in the case of the treatment for diphtheria, the introduction of serum therapy along with the scientific laboratory diagnosis of diphtheria served to raise the image of hospitals in the eyes of a bourgeois clientele, which was now tempted to entrust its children to institutions that could offer exclusive expertise in a 'high-tech' medicine for a widespread and very dangerous disease.[17]

While the collection of papers that follows explores some but not all of these perspectives, we hope that they open up enough avenues to provide a solid basis for future research in this and related areas of the history of medicine.

The structure of the book

Because of the importance of the work around the evaluation of the therapeutic quality of this serum, and particularly the contributions to an ever more sophisticated system of evaluation by Paul Ehrlich and his collaborators in Imperial Germany, we have dedicated the first part of the book exclusively to the question of diphtheria antitoxin in different contexts.

The book opens, therefore, with a series of papers that concern the evaluation of diphtheria serum in Germany and France. As an introduction to the German context, Cay-Rüdiger Prüll offers a biography of Paul Ehrlich that situates his work in the area of evaluation, exploring it in institutional and scientific terms. This first paper also shows how the research into fixing and theorizing the curative power of serum was tied into Ehrlich's professional trajectory. The next two papers, by Axel C. Hüntelmann and Anne I. Hardy go into more detail concerning the practices of evaluating the sera in Germany, starting with the treatment for diphtheria but eventually including sera for tetanus and erysipelas, among others. Following this, there are two papers that treat the situation in France, which developed quite differently from that in Germany, particularly in terms of the evaluation of the therapeutic quality of the diphtheria and other sera. While Gabriel Gachelin gives the general outlines of the history of serum production in France in the context of the history of the Institute's human and animal vaccines, Jonathan Simon discusses the importance of serum quality for the Pasteur Institute's reputation, if not its very survival. Serum production was not, however, limited to France and Germany, and, in the following paper, Mariama Kaba explores the production and place of diphtheria serum in Switzerland, thereby taking the history beyond the two dominant microbiological powers of nineteenth-century Europe. Pauline M. H. Mazumdar opens up the perspective of this section centred on diphtheria serum even wider by considering the moves to establish international standards, and their relationship to the earlier evaluation work on the diphtheria serum. As a conclusion to this section, Anne Hardy has written a synthetic paper that draws out the major points raised by the first seven papers, relating the different case studies to the history of the State Serum Institute in Denmark.

The second part of the book concerns other medical microbiological products, and one plant extract. Thus, this second section shows how the techniques, or more importantly the way of thinking about *Wertbestimmung*, developed in the context of the sera were applied more or less successfully in other domains. While the therapeutic techniques vary widely, from Wright's process of auto-immunization to the use of digitaline in acute heart disease, the papers share a concern with the modes and meanings of evaluation and the challenge of standardizing these materials for use on humans and animals. First of all, Michael Worboys explores the importance of setting standards in Almroth Wright's auto-immune therapy, then Jean-Paul Gaudillière examines the problems posed by the need to standardize hormones in the absence of a 'reaction' model familiar from the toxin-based sera. Next, Christian Bonah explores how attempts were made to apply the model of *Wertbestimmung* for the sera in the case of a physiologically potent plant extract, digitalis. Here, the problem was in a sense the opposite of that facing the serum producers, in that such a powerful agent provoked more concern about an excessive effect rather than an ineffective product. Ulrike Lindner discusses the issue of standards and the quality of the vaccine in a comparative context, treating the more recent case of the introduction of the polio vaccine in Britain and West Germany. In his conclusion, Alberto Cambrosio teases out some central themes involved in the subject of *Wertbestimmung*, and some lessons that we might be able to draw for contemporary medical science.

Notes

1. For life-threatening conditions, like cancer and HIV, the active ingredients may well be harmful, but then the associated risks are only considered justified in light of sufficient individual benefit with respect to the risk. There is also a requirement that the side-effects be clearly indicated.
2. Allergic reactions remain unpredictable, and are usually only discovered subsequent to some accident. Furthermore, the rise of strains of tuberculosis and other microbial diseases resistant to common antibiotics points to the uncertain nature of treatment outcomes, even when one uses the most 'efficacious' of modern drugs.
3. In France, a grain was a weight unit of about 53 mg, but was slightly more in England, where there were 20 grains in a scruple, 3 scruples in a dram and 8 drams in an ounce.
4. For more on this transformation in France, see Sophie Chauveau, *L'Invention Pharmaceutique: La pharmacie française entre l'Etat et la société au XXe siècle*, Paris: Sanofi-Synthélabo, 1999, and Matthew Ramsey, 'Academic Medicine and Medical Industrialism: The Regulation of Secret Remedies in Nineteenth-Century France' in Mordechai Feingold and Ann F. La Berge (eds), *French Medical Culture in the Nineteenth Century*, Clio Medica 25, Amsterdam: Rodopi, 1994, pp. 25–78. For relevant developments in Germany, see G. Huhle-Kreutzer, *Die Entwicklung arzneilicher Produktionsstätten aus Apothekenlaboratorien. Dargestellt an ausgewählten Beispielen*. Stuttgart: Deutscher Apotheker Verlag, 1989.
5. Ted Porter, *Trust in Numbers: The Pursuit of Objectivity in Science and Public Life*, Princeton: Princeton University Press, 1995, and for a more specific treatment of the importance of standards in Germany, see David Cahan, *An Institute for an Empire: The Physikalisch-Technische Reichsanstalt, 1871–1918*. Cambridge, New Rochelle, and New York: Cambridge University Press, 1989.
6. Steven Shapin, *A Social History of Truth: Civility and Science in Seventeenth-Century England*, Chicago: University of Chicago Press, 1994, where he argues for the origin of scientific standards of truth in such gentlemanly interactions, and Lorraine Daston and Peter Galison, 'The Image of Objectivity', *Representations*, 1992, 81–128 for a discussion of the rise of 'mechanical' objectivity.
7. We choose this example deliberately, as acetylsalicylic acid was first synthesized by Bayer chemists in 1897, before being launched as a brand-name product 'Aspirin' in 1899. This model for the development of synthetic pharmaceuticals dates from the same period as the *Wertbestimmung*.
8. The issue of the variability of patients and treatment regimens is raised by a number of authors in the present volume. Different people react differently to the same compound and even the same person can react differently at different times. These differences include such things as variation in clinical efficacy and allergic reactions.
9. Bruce J. Hunt, 'The Ohm is where the Art is: British Telegraph Engineers and the Development of Electrical Standards', *Osiris*, 1994, 9: 48–63; Simon Schaffer 'Late Victorian Metrology and Its Instrumentation: A Manufactory of Ohms' in Robert Bud and Susan Cozzens (eds), *Invisible Connections, Instruments, Institutions and Science*, SPIE Optical Engineering Press, 1992: 23–56, and Norton Wise and Crosbie Smith, *Energy and Empire: A Biographical Study of Lord Kelvin*, Cambridge: Cambridge University Press 1989.
10. S. L. Star and J. R. Griesemer, 'Institutional Ecology, Translations and Boundary Objects: Amateurs and Professionals in Berkeley's Museum of Vertebrate Zoology, 1907–1939' *Social Studies of Science*, 1989, 19: 387–420.

11. See the contribution by Cay Rüdiger Prüll.
12. Harry Marks, *The Progress of Experiment*, Cambridge: Cambridge University Press, 1997.
13. This culture of consumption is argued for in Olivier Faure, *Les Francais et leur médecine au XIXe siècle*, Paris: Belin, 1993.
14. For the history of the pharmaceutical industry in Germany, see Wolfgang Wimmer, '*Wir Haben fast immer was Neues: Gesundheitswesen und Innovationen der Pharma-Industrie in Deutschland, 1880–1935*', Berlin: Duncker and Humblot, 1994, while for France, see Sophie Chauveau, *L'Invention Pharmaceutique : La pharmacie française entre l'Etat et la société au XXe siècle*, Paris: Sanofi-Synthélabo, 1999.
15. Sinclair Lewis, *Arrowsmith*, New York: Harcourt Brace, 1925, Chapter 13.
16. See, Jonathan Liebenau, *Medical Science and Medical Industry: The Formation of the American Pharmaceutical Industry*, London: Macmillan, 1987, and John P. Swann, *Academic Scientists and the Pharmaceutical Industry: Cooperative Research in Twentieth-Century America*, Baltimore: The Johns Hopkins University Press, 1988.
17. Paul Weindling, 'From Medical Research to Clinical Practice: Serum Therapy for Diphtheria in the 1890s' in J. V. Pickstone (ed), *Medical Innovations in Historical Perspective*, Basingstoke: Palgrave Macmillan, 1992: 72–83.

1
Paul Ehrlich's Standardization of Serum: *Wertbestimmung* and Its Meaning for Twentieth-Century Biomedicine

Cay-Rüdiger Prüll

As is amply illustrated in other chapters in this volume, Paul Ehrlich (1854–1915) was a key figure in the introduction of new methods to standardize drugs at the end of the nineteenth century. Starting in the 1890s, the Berlin bacteriologist and immunologist created a method to measure the efficacy of therapeutic sera. This method had a wide impact on scientific medicine in the twentieth century, but for Ehrlich it constituted just one step, albeit a very important one, in his intellectual and scientific career. This paper argues for the importance of Ehrlich's method for medicine in general, and, more particularly that it was representative of developments in nineteenth-century scientific medicine that prepared the ground for the rise of biomedicine in the twentieth century. In this respect, Ehrlich's standardization method participated in an ensemble of fundamental cultural changes that underwrote scientific medicine.[1]

In what follows, I will first deal with Ehrlich's life and the factors which led to the development of his standardization method. Secondly, I will present a description and analysis of the construction of the method and its context. Thirdly, I will focus on the influence of serum standardization on Ehrlich's academic career, on medicine and on the cultural entanglement of the discipline. Ehrlich's method not only led to a new kind of professional position for its creator, but it also promoted a new theory of drug action, a new style of scientific medical work, a new relationship between physician and patient and a new relationship between medicine, politics and industry.

Paul Ehrlich's life and his academic pathway to standardization

Paul Ehrlich studied medicine between 1872 and 1878 in Breslau, Strasbourg, Freiburg and Leipzig. He was mainly influenced by studies at the universities of Strasbourg and Breslau. The well-known anatomist Wilhelm Waldeyer

(1836–1921) inspired him in the Strasbourg years between 1872 and 1874, while Ehrlich was educated in the morphological tradition of the nineteenth century when preparing histological specimens.[2] He then ventured into the field of staining human tissues and cells in order to explore their functions, becoming a pioneer in these staining techniques, work he continued during his studies in Breslau.[3] There he was influenced by his cousin, Carl Weigert (1845–1904), but above all by the pathologist Julius Cohnheim (1839–84).[4] In these formative years, Ehrlich developed two basic scientific ideas that would shape his future work in medicine. Firstly, he realized the importance of staining for medicine in general and histology in particular.[5] Secondly, he came to believe that the process of staining depended on an easily reversible chemical reaction between dye-stuff and cell.[6] In 1878, Ehrlich defended a dissertation entitled 'Contributions to a Theory and Practice of Histological Staining'[7] and obtained a job as first physician with Theodor Frerichs (1819–85), the well-known Professor of Internal Medicine and head of the Medical Clinic at the Charité-Hospital in Berlin.[8] Frerichs believed he could improve the treatment of his patients by integrating laboratory research and diagnostics with clinical work. Thus, he supported Ehrlich's staining experiments,[9] allowing him to combine his work in the wards with his work in the laboratory. Ehrlich carried out anatomo-pathological examinations of tissue specimens taken from the post-mortem room to learn more about the causes of death,[10] but primarily he worked on clinical pathology, examining patients' bodily fluids using both dyes and animal experiments.[11] The most outstanding example for this method was Ehrlich's experimentation on blood, in which, following animal experiments, different dyes were tested on the blood of both healthy and sick patients, allowing him to analyze different kinds of white and red blood cells. This experimental work was then applied to the diagnosis and therapy of both blood diseases and infectious diseases, making Ehrlich a pioneer in modern haematology.[12]

Ehrlich's situation changed rapidly after the death of Frerichs in 1885. The Medical Clinic of the Charité Hospital was now renamed the 'First Medical Clinic' and handed over to Ernst von Leyden (1832–1910), who had been Professor of Internal Medicine in Berlin since 1876 and was head of the so-called Preparatory (*Propädeutische*) Clinic of the Charité Hospital. The chair of the Preparatory Clinic, now vacant, was given to Carl Gerhardt (1833–1902), who until this time had been Professor of Internal Medicine in Würzburg. The Preparatory Clinic itself was renamed the 'Second Medical Clinic'.[13] Ehrlich now worked under Gerhardt, who was mainly interested in clinical work in the wards and not in the laboratories attached to them. Ehrlich caught tuberculosis in 1888, resigned his post at the clinic, and went to Egypt to try and recover.[14]

Ehrlich returned to Berlin in 1889.[15] While the first stage of his academic research, which ended with his voyage to Egypt, had been devoted to staining and its application to practical medicine, the second stage, which lasted until around 1905, was devoted to theoretical immunological work.[16] Upon his return from Egypt, Ehrlich found himself unemployed and it was only with the financial support of his father-in-law that he managed to set up a small laboratory in Berlin. His new work was in large part inspired by Robert Koch (1843–1910) and contemporary ideas about anti-bacterial treatments.[17] As early as 1882, Ehrlich had developed an effective method for staining Koch's newly discovered tubercle bacillus,[18] and he also succeeded in immunizing mice against the plant poisons, Ricin and Abrin. In addition, he studied the transmission of immunity both by inheritance and via breast feeding, obtaining results that touched the basic processes of active and passive immunization.[19]

In 1890, only one year after his return, Koch offered his former assistant a post as clinical supervisor for scientific studies of tuberculosis at the City Hospital in the Moabit district of Berlin.[20] In collaboration with a colleague, Ehrlich explored the best-tolerated dosage of the tuberculin serum and carried out research on the therapeutic value of dyes like methylene blue. Here again, he combined histological sputum examinations using staining techniques with animal experimentation and therapeutic human experimentation.[21] In 1891, Robert Koch offered Ehrlich a laboratory in the newly founded Institute for Infectious Diseases (*Institut für Infektionskrankheiten*) where he collaborated with the group of bacteriological researchers working under Koch, which included Emil von Behring (1854–1917), Richard Pfeiffer (1858–1945), and August Wassermann (1866–1925). The discovery in 1890 by Behring and Shibasaburo Kitasato (1852–1931) of the phenomenon of antitoxins against diphtheria and tetanus considerably boosted Ehrlich's interest in immunology.[22] The organism's ability to form substances (today known as antibodies) to combat specific microbes opened up new treatment opportunities in the form of the therapeutic sera. Furthermore, Behring and Kitasato were able to show that bacterial toxins could be neutralized by antitoxins in the test tube. Thus, from 1891 onwards, Ehrlich worked chiefly on human immunology, including the standardization of sera and the principles and techniques of *Wertbestimmung*. When Ehrlich entered this new field of research via the diphtheria serum, he was probably not aware that it would bring his institutional clinical work to an end. He initially examined diphtheria serum to help resolve the production difficulties confronted by Behring in his collaboration with the Hoechst company (*Farbwerke Hoechst*, near Frankfurt/Main). Behring and Ehrlich agreed that the horse was well suited for the large-scale production of the serum,[23] but it proved difficult to routinely achieve a reliable therapeutic concentration of the product before 1894. As with tuberculin, the challenge was to standardize the dosage in terms of the serum's efficacy. Thus, Ehrlich was assigned the task of examining the exact quantitative relations between diphtheria toxin and antitoxin and developing a method for standardizing the use of the therapeutic serum.[24]

Paul Ehrlich's method of *Wertbestimmung* and the standardization of serum

Behring's problems with diphtheria serum were reflected in Ehrlich's first efforts to evaluate its efficacy. Starting in the summer of 1892, Ehrlich used goats to produce antitoxins, as this animal was very sensitive to infection with diphtheria and proved very robust with respect to the immunization procedures. Ehrlich and his co-workers were able to confirm his own and Behring's earlier findings, namely that there is a fixed proportion between the amount of toxin present and the amount of antitoxin needed to neutralize it. By boosting the antibody production of goats using suspensions of killed diphtheria bacilli, Ehrlich and his colleagues succeeded in obtaining useable doses of diphtheria antitoxin. This was possible with a minimal loss of animals as Ehrlich was able to use his earlier experience with plant poisons to correctly dose the killed bacilli. He also investigated the presence of antitoxins in the goats' milk, hoping this might provide a more convenient treatment. In 1893, Ehrlich tested his own serum on children with diphtheria in several Berlin city hospitals. These early experiments were generally successful, with no major side effects, although an increase in mortality among children treated at an advanced stage of the disease was a cause for concern. Furthermore, the appropriate dose for treating the children effectively could only be estimated for the serum produced at Koch's Institute, and these doses could not be applied either to sera coming from elsewhere in the German Reich or from outside Germany.[25]

The problems were now clearly posed, and Ehrlich aimed to solve them using immunological theory based on chemical laws applied to medicine and supported by mathematics. Ehrlich received backing for his serological work from the Prussian Ministry of Science and Education (*Ministerialrat im preußischen Kultusministerium*), thanks to the support of the prominent and powerful Councillor, Friedrich Althoff (1839–1908).[26] Previously, Ehrlich had never received any official post in Koch's Institute, presumably because of his being Jewish,[27] and when the 'control station' for therapeutic sera (*Controlstation für Heilsera*) opened in the Institute in 1895, it was entrusted to Koch's assistants August Wassermann and Hermann Kossel and not to Ehrlich, who was appointed deputy head of the department.[28] But in 1896, at Althoff's instigation, Ehrlich became head of the new Institute for Serum Research and Serum Testing (*Institut für Serumforschung und Serumprüfung*) located in Steglitz in the suburbs of Berlin.[29] The main function of the institute was the testing of sera, but it also enabled Ehrlich to focus on laboratory research into diphtheria.

Ehrlich was now able to work on his own method and theory of serum standardization and develop the set of practices that constituted the original version of *Wertbestimmung*. After only a year, Ehrlich presented his groundbreaking achievements in a now classic study, 'The Assay of the Activity of Diphtheria-curing Serum and Its Theoretical Basis'.[30] The starting point for his research was the practical problem that the dose of diphtheria serum administered to patients was often far too weak, so that repeated doses had to be given to the same patient. This widespread problem was a mystery since the phials of serum only left the manufacturers after having successfully passed a test of their efficacy. Ehrlich also experienced this apparent loss of strength with his own test serum, which served as the basis for checking the samples from the various companies. Normally, a specific amount of diphtheria toxin could be neutralized by 0.23 cubic centimetres of this test serum, but Ehrlich noticed that the potency of both toxin and serum decreased over time. The determination of the neutralization point started to lose its former meaning, presenting the challenge of developing new standards to evaluate the efficacy of diphtheria antitoxin samples.[31] From the beginning, this challenge was twofold. First, using mathematics, Ehrlich had to find a practical way of correlating the amounts of antitoxins and toxins that could neutralize one another. He also wanted to determine why the serum samples lost their potency, which would require rethinking his theory of immunization.

Ehrlich started with the pragmatic side of the research, checking and improving all the measures aimed at conserving the samples. To avoid the deleterious influences of water, oxygen, light and heat, Ehrlich adopted Behring's method of storing the serum as a dry powder and inserting the powder into vacuum tubes.[32] The second step was to standardize the serum, and for this Ehrlich turned once again to goats for the antitoxin, with immunization experiments conducted on guinea pigs. The method used at the time was to introduce a specific amount of toxin into a guinea pig and then inject a sufficient amount of test serum to neutralize the toxin so that the animal would survive at least four days. This method proved unsatisfactory as it depended too much on the researcher's subjective evaluation of the animal's condition. In his search for a more reliable method, Ehrlich focused on the death of an experimental animal as the most 'objective' criterion on which to base the evaluation of the efficacy of diphtheria sera. Reviewing his old experiments, he decided that the 'single lethal dose' of toxin would form a reliable starting point for his evaluations.[33] He defined this lethal dose as the amount of toxin that killed a 250 gram guinea pig within four days, and developed an elaborate method for precisely determining this quantity. He selected a batch of diphtheria serum capable of neutralizing one hundred lethal doses of toxin, and tested 11 samples of diphtheria toxin (each from a different origin) against the standardized serum, using guinea pigs. Then, he successively added poison from the sample under test until he arrived at the one hundred and first dose, which, based on his theory (assuming a model of serum-toxin neutralization), was the amount needed to kill the guinea pig within four days. This enabled him to find two threshold concentrations for each analyzed sample of diphtheria toxin. The first one was a completely neutralized solution of toxins, which caused no signs of disease when administered to a guinea pig (L_0), and the second one was the quantity of toxin that killed a 250 gram animal within 4 days (L_+). The difference between the first (neutral) and the final (lethal) solution was designated as the 'single lethal dose' (D = *einfache letale Dosis*, where $D = L_+ - L_0$).[34] Ehrlich's results were far from encouraging, as the concentrations of the solutions of the different diphtheria toxins varied markedly. The value obtained for L_0 differed between many of the toxins and therefore D could vary between five and ten doses of poison. Moreover, the solutions were not stable, but lost their toxicity after being stored for a certain length of time, although the number of antibody-binding units did not decrease. This meant that the toxic effect did not correspond with the capacity of the toxin to bind to the antitoxin. The only explanation for this phenomenon was that the toxins had undergone some changes that affected their toxicity.[35]

To try to explain these results, Ehrlich was obliged to reconsider the immunological theory he had developed when he was Frerichs's assistant, in which certain chemical 'side-chains' at the surface of the cell were able to bind specific toxins. Once occupied, these side-chains would then be unable to fulfil their normal physiological functions (to bind nutritive substances), and the cell would overcompensate by producing a large number of additional side-chains. These excess side-chains would be released into the blood stream, where they acted as antibodies or antitoxins, attaching to toxins and thereby preventing them from attacking the cells.[36] Ehrlich first developed this side-chain theory in 1885, when he was explaining staining phenomena in terms of the binding of oxygen to cells.[37] Although they had remained relevant as an underlying concept in his research, these 'side-chains' were mentioned in only one of his papers between 1885 and 1897, but he had nevertheless integrated them into his model of the immunological system.[38] This new research into the diphtheria antitoxin allowed him to mobilize this theory as an explanatory tool for practical serological purposes. Ehrlich supposed that the toxin consisted of two parts; a poisonous part, the so-called toxophore group (*toxophore Gruppe*) and a component responsible for binding to the antitoxin, the so-called haptophore group (*haptophore Gruppe*). He hypothesized that the toxophore group was not as stable as the haptophore one and, consequently, the toxophore groups gradually broke down, losing their toxic power. This explained the appearance of non-toxic toxins, which Ehrlich termed 'toxoids' (*Toxoide*), that while able to bind to antitoxins, no longer had any toxic effect on the body. Because they retained their capacity to bind the side-chains of the cell, toxoids were able to induce the production of antibodies, understood as an excess production of side-chains, which were released into the blood stream. This meant that the chemical processes of specific binding were combined with biological processes of regeneration. Ehrlich explained the binding mechanism in terms of an analogy, borrowing the idea of the 'lock-and-key' mechanism from the biochemist Emil Fischer (1852–1919) who had used it to describe the functioning of enzymes.[39]

Armed with this explanation for the diminishing toxicity of the diphtheria toxins, Ehrlich could develop a usable and effective method for standardizing diphtheria sera. Deploying complex mathematical equations, he approximately determined various relationships between toxins and toxoids in solutions of the diphtheria toxin, arriving at laws correlating the two components. These correlation laws enabled Ehrlich to give specific advice on how to determine the potency of diphtheria serum despite potential deterioration of toxin. A well-stored test serum (dry powder in a vacuum tube) should be dissolved in a specific well-defined mixture including sodium chloride and glycerine to give a test serum that could be used to determine the composition of the test toxin. The serum was administered to a guinea pig and toxin was then added until the animal died within four days after the application of the solutions (L_+). The serum in question was then checked against the L_+-value poison, with both being administered once again to a guinea pig; the serum was only considered effective if the animal lived for more than six days afterwards. If the animal died within four days of administration, the serum was not potent enough, and if it died on the fifth or sixth day, the potency was considered doubtful and the sample had to be sent back to the producer with the recommendation that they strengthen it. Ehrlich's protocol for this complex test procedure was full of detailed information and advice concerning the production, storage and repeated testing of the test sera and the test toxins.[40]

Standardization; *Wertbestimmung* and its consequences for Ehrlich and modern medicine

Ehrlich's 1897 paper and the published method for standardizing diphtheria serum were crucial in ensuring that the serum was an effective product, and in 1918 the method was adopted by the Biological Standardization Commission of the League of Nations. Although it was subsequently improved by different researchers who succeeded in reducing the rather excessive number of research animals used in the protocol, Ehrlich's method remained influential right up until the early 1930s.[41] Sir Henry Hallett Dale (1875–1968), one of the leading pharmacologists of the first half of the twentieth century, spent the end of 1903 and the beginning of 1904 with Ehrlich at his Institute for Experimental Research in Frankfurt am Main, and then used his knowledge of the standardization of diphtheria serum for the international standardization of insulin in the mid-1920s.

But Ehrlich's method was more than just a contribution to the narrow field of serum standardization and therapy; through it, he made at least three significant contributions to scientific medicine more generally. Ehrlich provided a new theory of drug binding, a new institution for research into practical pharmacology, and a vector for the acceptance of the techniques of scientific medicine in Western society. The next three sections of this paper will focus on these contributions, underlining the consequences of diphtheria standardization on Ehrlich's subsequent career, which continued until his death in 1915.

The receptor concept, chemotherapy and biomedicine

Ehrlich wanted a better understanding of how sera worked within the animal and human body, as he believed that their effective use as therapeutic agents was hampered by this lack of knowledge. In his 1897 paper, Ehrlich not only applied the idea of 'side-chains' to immunology, but also expanded his vague and provisional idea of side-chains into a fully fledged 'side-chain theory'. In turn, this theory became the basis of the immunological investigations that dominated his research between 1897 and approximately 1905.[42] During these years, Ehrlich expanded the side-chain-theory into a receptor-theory, with side-chains being renamed 'receptors' in 1900. In 1908, Ehrlich was awarded the Nobel Prize for his chemical immunological work together with Elie Metchnikoff (1845–1916).[43]

This concept of receptors is remarkable for several reasons. One is that the concept – derived from serum standardization – soon served as a theoretical tool for explaining drug binding and the action of drugs in general. At the beginning of his work on receptors, Ehrlich was chiefly interested in the patho-physiology of human cells and their interaction with microbes, believing that receptors bound only microbes and nutritive substances. But in 1907, he came to accept the binding of drugs to receptors.[44] This shift of ideas about drug binding resulted from his search for chemotherapeutic agents, which involved studying the effect of dyes on trypanosomes, the syphilis germ, starting in 1904. Furthermore, his new attitude was supported by the research of the Cambridge physiologist John Newport Langley (1852–1925), who was interested in the anatomy and function of the autonomic nervous system, and had independently developed his own 'receptor concept'.[45] The application of the receptor theory to chemotherapeutic problems formed the theoretical basis for his subsequent work on Salvarsan, the first chemotherapeutic substance for the treatment of syphilis, which became available in 1910. Up until his untimely death in 1915, Ehrlich continued to search for similar examples of a 'magic bullet', the concept that has made him one of the most prominent figures in the history of twentieth-century medicine.[46]

This extension of the receptor concept into the realm of drug action influenced pharmacological research more widely, even though 'side-chains' or 'receptors' were not visible. They were theoretical constructs supported mainly by Ehrlich's own explanations and representations, which were in competition with a 'physical theory' that explained drug binding by physical forces such as adhesion or surface-tension at the cell wall. The theory also competed with another research programme in pharmacology – studies on so-called transmitters, mediator substances that delivered information to receptors. In the 1930s and 1940s the community of pharmacologists found these mediating substances more promising than basic research on receptors. Indeed, research on receptors remained limited until the 1960s, when the concept finally provided a consistent explanation of the efficacy of drugs and – even more importantly – was used to develop new drugs. Today, receptors are indispensable in medical theory and practice.[47]

Another important point is that the receptor concept fitted in with a new approach, which came to dominate twentieth-century medicine, the investigation of the direct mode of action of drugs on cells and tissues. The morphologically oriented medicine of the nineteenth century gave way to the experimental investigation of disease *processes*, the application of physiology to medical problems, and above all the development of new therapies.[48] In a context where the aim was to find specific cures for specific diseases and the development of new synthetic drugs, it became increasingly important to investigate the precise ways in which drugs influenced the organism, the organs, the tissues and especially the cells of the body, and the receptor concept responded to these demands. Ehrlich was prescient in claiming that studies on receptors would 'open a new meaningful direction in biological research'.[49] He believed that receptors would explain not only immunological function but also the whole metabolism of the human body, placing him among the pioneers who applied biological laws to scientific medicine.

Ehrlich's career and a new institutional pattern of scientific medicine

As we suggested above, serum standardization also opened up a novel career path for Ehrlich, who did not fit the mould of the 'classic' German clinician. As early as his medical training, Ehrlich sought to be a 'clinical pathologist', hoping to apply the results of his staining experiments obtained in the laboratory to practical clinical medicine. In this sense, he followed the example of the pathologist Julius Cohnheim, who not only combined animal experimentation with autopsies, but who also worked in close contact with clinicians. In the context of German pathology, which was heavily influenced by the model of Rudolf Virchow (1821–1902) with an exclusive focus on static morphological work in the morgue, such an approach remained marginal.[50]

Working under Frerichs in Berlin, Ehrlich was able to organize his time in a way that fitted his own and Cohnheim's ideals. While Ehrlich called himself a 'clinician',[51] he had his own laboratory, with access to the wards for material and clinical experience. This arrangement ended when Carl Gerhardt replaced Frerichs; first, Ehrlich was restricted to his clinical bedside activities, before he finally lost his job and any contact with the patients. Ehrlich kept his medical career alive by turning to theoretical work on immunology and serum standardization. Describing his work on receptors in 1901, Ehrlich had this to say about his being cut off from the clinic: 'Because I myself am unable to perform such investigations on a greater number of patients, I thought it my duty to clarify the points of view and in this way to lay the base for work on a field whose importance for pathology and therapy will presumably be fully recognized only after many years'.[52] In 1905, this argument was essentially repeated when he explained that his work was done and that 'more new and successful work' could only be pursued by 'specialists, who have the necessary *clinical* and pathological material'.[53] Nevertheless, during the rest of his career Ehrlich tried to maintain this clinical contact and sometimes even tried to reintroduce the research context he had known under Frerichs. Serum standardization contributed to the promotion of work in clinical pathology, and although Ehrlich was drawn into theory-oriented laboratory work, he was able, at least initially, to combine human experimentation, animal experimentation and work in his laboratory while collaborating with active clinicians. Thus, the institutions and practices that Ehrlich promoted with his work on serum standardization also helped to shape twentieth-century medicine. Although Ehrlich never identified himself as a 'clinical pathologist', the profile of his work fits with this title, which began to be used around 1900, particularly in Anglo-Saxon countries. In this context, chemical pathology, bacteriology and experimental pathology were not exclusively domains of experimental science; they were developed with the practical aim of curing the sick. The 'clinical pathologist' analyzed bodily fluids and tissue samples from the living patient, integrating the patient into an expanding context of scientific measurements that in turn linked them to the new theories of scientific medicine. Clinical pathology thus exemplified a clear trend across the twentieth century that promoted physiological or functional medicine.[54] This new clinical pathology was closer to the functional ideal defended by Cohnheim and Ehrlich than that represented by Virchow, the traditional hero in histories of early-twentieth-century medicine. Indeed, Ehrlich personally contributed to the institutionalization of this field at the voluntary teaching hospitals in London. As a personal friend of the London bacteriologist William Bulloch, Ehrlich had a strong influence on Almroth Wright (1861–1947), who structured the discipline at St Mary's Hospital after 1902.[55] It was at Wright's institute in 1928 that Alexander Fleming first discovered the antibiotic power of Penicillin. But it was only after 1945 that clinical pathology was re-exported back to Germany.[56]

Ehrlich's *Wertbestimmung* and the collaboration between science, politics and industry

This new scientific medicine that Ehrlich's efforts to standardize diphtheria serum helped to establish was also characterized by networks of close relationships between physicians, businessmen and politicians.[57] Ehrlich and his wife were close personal friends of Althoff and his wife, and it was Althoff who enabled him to obtain an independent position with the Institute first in Steglitz (1896) and then in Frankfurt (1899). Althoff's efforts on behalf of Ehrlich were not simply born out of this friendship but reflected an interest in his promising work in serum standardization, an important issue for the pharmaceutical industry.[58] Indeed, Ehrlich had to negotiate with the pharmaceutical companies himself, not only to keep up the standard of the diphtheria sera, but also to establish and maintain his own authority in the field, which had to be accepted by both the state and the private sector.[59] Ehrlich's activities as a science administrator again show how he embodied a new type of researcher, pioneering a model that would be widely diffused in the twentieth century, comparable with the type of 'expert' described by Jonathan Harwood in his analysis of the genetics community in Germany.[60] Ehrlich was specialized, devoted to his scientific aims and saw his contribution to science as a sort of service to the state and society. Whereas von Behring sought to build up a 'serological empire' that would make his fortune, Ehrlich seemed most of all to want more time for research.[61] Whether new experts such as Ehrlich were apolitical is questionable, but he did succeed in making and maintaining connections between the universities, private companies and the state. The standardization of diphtheria serum was just the first of several domains in which he demonstrated his remarkable administrative or 'political' skills. The handling of these relationships also depended on his earlier collaboration with the chemical industry, whose dyes he had used as innovative tools in his laboratory and clinical research.[62] The collaboration with these companies and with the physicians who tested dyes in medical practice was maintained by Ehrlich even during the period when he was mostly engaged in immunological work. After 1910, he reactivated these relationships to conduct the chemotherapy programme leading to the discovery of Salvarsan, coordinating the activities of physicians, pharmaceutical companies and governmental representatives.[63]

Conclusion

Through a careful examination of his life and work, we can see that Ehrlich's serum standardization represents more than just a new technical procedure. As well as providing an approach for measuring drug efficacy in general, it also led Ehrlich into fundamental research to understand the toxin-antitoxin interaction. This theoretical investigation was at the basis of the receptor theory that has proved so influential through the twentieth and into the twenty-first centuries.[64] This serum research was also responsible for Ehrlich's institutional reinvention as a clinical pathologist. The immunological work that allowed him to make a living as a research physician without any patients encompassed clinical testing of products and medical diagnostics while remaining firmly anchored in the research laboratory. This kind of relationship between laboratory and practical medicine has been essential to the rise of a modern form of 'biomedicine' capable of transforming fundamental research into therapeutic applications.

Finally it is crucial to remember the importance of Ehrlich's place as a medical expert, facilitating the cooperation between medicine, industry and the state. This kind of networking played an important role in the introduction of the medical sciences into industrialized countries in the twentieth century, and one only needs to think of the chemist Fritz Haber to see that Ehrlich was not alone in combining these larger political and social perspectives in this period.[65] Thus, considering Ehrlich's career and the place of the standardization of serum therapy in it, we can better understand the processes that were shaping modern medicine in the decades around 1900.

Notes

1. Ehrlich's complete works can be found in: Fred Himmelweit (ed.), *The Collected Papers of Paul Ehrlich in Four Volumes Including a Complete Bibliography*, vol. I, *Histology, Biochemistry and Pathology*, London /New York (Pergamon), 1956; vol. II, *Immunology and Cancer Research*, London/New York/Paris (Pergamon), 1957; vol. III, *Chemotherapy*, London/Oxford/New York/Paris (Pergamon), 1960, with the assistance of Martha Marquardt, under the editorial direction of Sir Henry Dale. The best general account of Ehrlich's life is: Claude E. Dolman, 'Paul Ehrlich', in *Dictionary of Scientific Biography*, vol. 3, New York (Charles Scribner's Sons), 10th edn., 1981, pp. 295–305. Dolman also gives an overview of the literature on Ehrlich up to 1980. For several reasons, literature on Ehrlich is largely dominated by hagiographic accounts: See Ernst Bäumler, *Paul Ehrlich: Forscher für das Leben*, 3. edn., Frankfurt/M. (Wötzel) 1997; Ernst Witebsky, 'Ehrlich's Side-Chain Theory in the Light of Present Immunology', *Annals of the New York Academy of Sciences* 59 (1954), pp. 168–81; Bela Schick, 'Ehrlich and Problems of Immunity', in *Annals of the New York Academy of Sciences* 59 (1954), pp. 182–9; Hugo Bauer, 'Paul Ehrlich's Influence on Chemistry and Biochemistry', in *Annals of the New York Academy of Sciences* 59 (1954), pp. 150–67. For the history of research on Ehrlich, see also: Henry Hallett Dale, 'Introduction', in Martha Marquardt, *Paul Ehrlich*, London (Heinemann), 1949, pp. xiii–xx; Henry Hallett Dale, 'Introduction', in Himmelweit, vol. I, *Histology, Biochemistry and Pathology*, pp. 1–18; Bäumler, *Paul Ehrlich*, pp. 5–9. The serious recent historiography of Ehrlich includes papers mostly on specific aspects of his work. For recent literature on Ehrlich see also: Frederick Kasten, 'Paul Ehrlich: Pathfinder in Cell Biology. 1. Chronicle of His Life and Accomplishments in Immunology, Cancer Research, and Chemotherapy', *Biotechnic & Histochemistry*, 71 (1996), pp. 1–37.

2. Cf. Russell C. Maulitz, 'The Pathological Tradition', in William F. Bynum and Roy Porter (eds), *Companion Encyclopedia of the History of Medicine*, vol. 1, London / New York (Routledge), 1993, pp. 169–91.
3. Bäumler, *Paul Ehrlich*, pp. 24–8. Paul Ehrlich, 'Beiträge zur Kenntnis der Anilinfärbungen und ihrer Verwendung in der mikroskopischen Technik' (*Archiv für mikroskopische Anatomie*, 1877), in Himmelweit, vol. I, *Histology, Biochemistry and Pathology*, pp. 19–27.
4. Bäumler, *Paul Ehrlich*, pp. 28–30, 31–6; Ernst Jokl, 'Paul Ehrlich – Man and Scientist', *The Bulletin of the New York Academy of Medicine* 30 (1954), pp. 968–75, esp. 972; Margaret Goldsmith, 'Paul Ehrlich', in Hector Bolitho (ed.), *Twelve Jews*, London (Rich & Cowan Ltd.), 1934, pp. 65–81, here esp. 69.
5. For the history of staining, see: Robert Fox and Augustí Nieto-Galan (eds), *Natural Dyestuffs and Industrial Culture in Europe, 1750–1880*, Canton/MA (Science History Publications), 1999.
6. Concerning Ehrlich and chemistry, see also: Henry Hallett Dale, 'Introduction', in Himmelweit, vol. I, *Histology, Biochemistry and Pathology*, pp. 1–18, esp. p. 2; Goldsmith, 'Paul Ehrlich', pp. 69/70.
7. 'Beiträge zur Theorie und Praxis der histologischen Färbung'. Paul Ehrlich, *Beiträge zur Theorie und Praxis der histologischen Färbung*, Thesis, Leipzig, 1878, in Himmelweit, vol. I, *Histology, Biochemistry and Pathology*, pp. 29–64, English translation: pp. 65–98.
8. Franz Hermann Franken, *Friedrich Theodor Frerichs (1819–85). Leben und hepatologisches Werk*, Freiburg (Falk Foundation), 1994; M. Classen, F. H. Franken and D. Gericke, 'Friedrich Theodor Frerichs in Berlin', *Deutsche Medizinische Wochenschrift* 120 (1995), pp. 1334–7.
9. Anthony S. Travis, 'Science as Receptor of Technology: Paul Ehrlich and the Synthetic Dyestuffs Industry', *Science in Context* 3 (1989), pp. 383–408, here p. 393.
10. One example of this research is Ehrlich's study of glycogen in the healthy and in the diabetic human organism. See Dolman, 'Paul Ehrlich', p. 296; Paul Ehrlich, 'Über das Vorkommen von Glykogen im diabetischen und im normalen Organismus' (*Zeitschrift für klinische Medizin*, 1883), in Himmelweit, vol. I, *Histology, Biochemistry and Pathology*, pp. 103–12.
11. One example is his studies on bacteriological problems in the course of pleuritic exudations (effusions into the pleural cavity) in relation to the patient records of women in labour. The application of different staining methods on microbes allowed him to identify different infections. These results enabled Ehrlich to give diagnostic and prognostic advice. Paul Ehrlich, 'Beiträge zur Ätiologie und Histologie pleuritischer Exsudate' (*Charité-Annalen*, 1882), in Himmelweit, vol. I, *Histology, Biochemistry and Pathology*, pp. 290–310.
12. See as an overview: Paul Ehrlich and A. Lazarus, *Histology of the Blood*, Cambridge (Cambridge Univ. Press), 1900 rev. transl. of: P. Ehrlich and A. Lazarus, 'Die Anaemie', in H. Nothnagel, *Specielle Pathologie und Therapie*, Vienna (Hölder), 1898, in Himmelweit, vol. I, *Histology, Biochemistry and Pathology*, pp. 181–268.
13. Rolf Winau, *Medizin in Berlin*, Berlin/New York (Walter de Gruyter), 1987, pp. 198–200.
14. Dolman, 'Paul Ehrlich', p. 297. For Carl Gerhardt, see also: 'Professor Carl Gerhardt', *Lokal-Anzeiger* 20 (1902), No. 337, Berlin, July 22, 1902, p. 1, in *Acta betr. die Anstellung des Geheimen Medicinal Raths und Professors Dr. Gerhardt als dirigirender Arzt und Director der 2. medicinischen Universitäts-klinik*, 1885, Kgl. Charité-Direction, No. 437, Archive of the Humboldt-University, Berlin, p. 37.

15. Martha Marquardt, *Paul Ehrlich*, London (Heinemann), 1949, pp. 27/8; Bäumler, *Paul Ehrlich*, p. 68/9.
16. In what concerns Ehrlich's academic life, I am following the excellent account of Claude Dolman. See Dolman, 'Paul Ehrlich'.
17. Concerning Koch and his research programme, see: Christoph Gradmann, *Krankheit im Labor. Robert Koch und die medizinische Bakteriologie*, Göttingen (Wallstein), 2005.
18. Paul Ehrlich, 'Modification der von Koch angegebenen Methode der Färbung von Tuberkelbazillen' (*Deutsche Medizinische Wochenschrift*, 1882), in Himmelweit, vol. I, *Histology, Biochemistry and Pathology*, pp. 311–13; Paul Ehrlich, 'Referat über die gegen R. Koch's Entdeckung der Tuberkelbacillen neuerlichst hervorgetretenen Einwände' (*Deutsche Medizinische Wochenschrift*, 1882), in Himmelweit, vol. I, *Histology, Biochemistry and Pathology*, pp. 322–9.
19. Paul Ehrlich, 'Experimentelle Untersuchungen über Immunität I. Über Ricin' (*Deutsche Medizinische Wochenschrift*, 1891), in Himmelweit, vol. II, *Immunology and Cancer Research*, pp. 21–6; Paul Ehrlich, 'Experimentelle Untersuchungen über Immunität II. Über Abrin' (*Deutsche Medizinische Wochenschrift*, 1891), in Himmelweit, vol. II, *Immunology and Cancer Research*, pp. 27–30; Paul Ehrlich, 'Über Immunität durch Vererbung und Säugung' (*Zeitschrift für Hygiene und Infektionskrankheiten*, 1892), in Himmelweit, vol. II, *Immunology and Cancer Research*, pp. 31–44. See also: Adolf Lazarus, *Paul Ehrlich* (*Meister der Heilkunde*, vol. 2), Wien/Berlin/Leipzig/München (Ricola) 1922, pp. 34/5.
20. Goldsmith, 'Paul Ehrlich', p. 76.
21. Dolman, 'Paul Ehrlich', p. 297; Paul Ehrlich and Paul Guttmann, 'Die Wirksamkeit kleiner Tuberkulindosen gegen Lungenschwindsucht' (*Deutsche Medizinische Wochenschrift*, 1891), in Himmelweit, vol. II, *Immunology and Cancer Research*, pp. 7–12; Paul Ehrlich and A. Leppmann, 'Über schmerzstillende Wirkung des Methylenblau' (*Deutsche Medizinische Wochenschrift*, 1890), in Himmelweit, vol. I, *Histology, Biochemistry and Pathology*, pp. 555–8; Paul Ehrlich and Paul Guttmann, 'Über die Wirkung von Methylenblau bei Malaria' (*Berliner Klinische Wochenschrift*, 1891), in Himmelweit, vol. III, *Chemotherapy*, pp. 9–14.
22. Jonathan Liebenau, 'Paul Ehrlich as a commercial Scientist and Research Administrator', *Medical History* 34 (1990), pp. 65–78, esp. p. 66.
23. See Anne I. Hardy's contribution to this volume.
24. Bäumler, *Paul Ehrlich*, pp. 90–3. Dolman, 'Paul Ehrlich', p. 297. For Ehrlich and his early involvement in the development of serum therapy against diphtheria, see: Paul Ehrlich, Hermann Kossel and August Wassermann, 'Über Gewinnung und Verwendung des Diphtherieheilserums' (*Deutsche Medizinische Wochenschrift*, 1894), in Himmelweit, vol. II, *Immunology and Cancer Research*, pp. 56–60, esp. pp. 57–60; Paul Ehrlich and Hermann Kossel, 'Über die Anwendung des Diphtherieantitoxins' (*Zeitschrift für Hygiene und Infektionskrankheiten*, 1894), in Himmelweit, vol. II, *Immunology and Cancer Research*, pp. 61/2.; Paul Ehrlich and A. Wassermann, 'Über die Gewinnung der Diphtherie-Antitoxine aus Blutserum und Milch immunisirter Thiere' (*Zeitschrift für Hygiene und Infektionskrankheiten*, 1894), in Himmelweit, vol. II, *Immunology and Cancer Research*, pp. 72–9; Paul Ehrlich, 'Über Gewinnung, Werthbestimmung und Verwerthung des Diphtherieheilserums' (*Hygienische Rundschau*, 1894), in Himmelweit, vol. II, *Immunology and Cancer Research*, pp. 80–3.
25. See the descriptions of the clinical trials in: Ehrlich, Kossel and Wassermann, 'Über Gewinnung und Verwendung des Diphtherieheilserums', pp. 57–60; Ehrlich, Kossel, 'Über die Anwendung des Diphtherieantitoxins', pp. 61/2; Ehrlich,

Wassermann, 'Über die Gewinnung der Diphtherie-Antitoxine', pp. 72–9. See also Arthur Silverstein, *Paul Ehrlich's Receptor Immunology. The Magnificent Obsession*, San Diego/London etc. (Academic Press), 2002, pp. 44/45.

26. For Friedrich Althoff, see: Bernhard vom Brocke (ed.), *Wissenschaftsgeschichte und Wissenschaftspolitik im Industriezeitalter. Das 'System Althoff' in historischer Perspektive*, Hildesheim (Lax), 1991; Bernhard vom Brocke, 'Hochschul- und Wissenschaftspolitik in Preußen und im Deutschen Kaiserreich 1882–1907: das "System Althoff"', in Peter Baumgart (ed), *Bildungspolitik in Preußen zur Zeit des Kaiserreichs* (*Preußen in der Geschichte*, 1), Stuttgart (Klett-Cotta) 1980, pp. 9–118. See also: Wolfgang U. Eckart, 'Friedrich Althoff und die Medizin', in vom Brocke (ed.), *Wissenschaftsgeschichte und Wissenschaftspolitik*, pp. 375–404, on Ehrlich esp. pp. 398–401.

27. Goldsmith, 'Paul Ehrlich', p. 76. In the report on the work of Koch's Institute of 1892, Ehrlich is mentioned only once as a voluntary assistant. See: 'Ueber den Bericht des Koch'schen Instituts für Infectionskrankheiten', Leipzig (Thieme) 1892 (off-print from the *Deutsche Medizinische Wochenschrift*, Nos. 4–7 (1892)), in *Acta betr. die Einrichtung und die Verwaltung des (staatlichen) Institutes für Infektionskrankheiten in Berlin, vom Januar 1892 bis Dezember 1898*, in GstA PK I HA Rep. 76 Kultusministerium, VIII B, No. 2893, pp. 63–75, esp. p. 71. Ehrlich's name does not appear at all in the records of Koch's Institute, kept by the Berlin Charité Hospital, for the years 1893 to 1895. See: *Acta betr. das Institut für Infectionskrankheiten, Kgl. Charité-Direction.*, No. 2205, 1893–5, Archive of the Humboldt-University of Berlin.

28. The Prussian Minister of Science and Education to Robert Koch, 9 February 1895; *Bericht über die Thätigkeit des Kgl. Instituts für Serumforschung und Serumprüfung zu Steglitz. Juni 1896-September 1899. Zur Einweihung des Königl. Instituts für experimentelle Therapie* Frankfurt/M., Jena, Fischer, 1899, offprint from: *Klinisches Jahrbuch* 7 (1899), in *Acta betr. das Institut für experimentelle Therapie zu Frankfurt a.M., vom Februar 1895 bis Dezember 1900*, GStA PK I HA Rep. 76 Kultusministerium, Vc Sekt.1, Tit.XI, Teil II, No. 18, vol. 1, pp. 1, 189–203, here 203.

29. Goldsmith, 'Paul Ehrlich', p. 77/8.

30. Paul Ehrlich, 'Die Wertbemessung des Diphtherieheilserums und deren theoretische Grundlagen' (*Klinisches Jahrbuch*, 1897), in Himmelweit, vol. II, *Immunology and Cancer Research*, pp. 86–106, English transl. pp. 107–25. The quotations in the following footnotes refer to the German original. For the impact of this paper on Britain, see also: H. G. Plimmer, 'A Critical Summary of Ehrlich's Recent Work on Toxins and Antitoxins', *Journal of Pathology and Bacteriology* 5 (1897), pp. 489–98.

31. Ehrlich, 'Die Wertbemessung des Diphtherieheilserums', pp. 86/87.

32. Ibid., pp. 87–9.

33. Ibid., pp. 89/90.

34. Ibid., pp. 89–93.

35. Ibid., pp. 93, 96. See also: Lazarus, *Paul Ehrlich*, p. 36.

36. Ehrlich, 'Die Wertbemessung des Diphtherieheilserums', pp. 86–106, English transl. pp. 107–25. For a contemporary description of Ehrlich's side-chain theory, see: Ludwig Aschoff, *Ehrlich's Seitenkettentheorie und ihre Anwendung auf die künstlichen Immunisierungsprozese*, Jena (Gustav Fischer), 1902, esp. pp. 1–25. See also: Bruno Heymann, 'Zur Geschichte der Seitenkettentheorie Paul Ehrlichs', in: *Klinische Wochenschrift* 7 (1928), pp. 1257–60; Arthur Silverstein, *A History of Immunology*, San Diego/New York etc. (Academic Press inc.), 1989, esp. pp. 64–6, 94–9.

37. Paul Ehrlich, *Das Sauerstoff-Bedürfniss des Organismus. Eine farbenanalytische Studie*, Habilitation-thesis, Berlin, Hirschwald, 1885, in Himmelweit, vol. I, *Histology, Biochemistry and Pathology*, pp. 364–432, English transl.: pp. 433–96.

38. Paul Ehrlich, 'Studien in der Cocainreihe' (*Deutsche Medizinische Wochenschrift*, 1890), in Himmelweit, vol. I, *Histology, Biochemistry and Pathology*, pp. 559–66.
39. Ehrlich, 'Die Wertbemessung des Diphtherieheilserums', pp. 93–106 (immunological theory), p. 94 (key-lock mechanism); Heymann, *Zur Geschichte der Seitenkettentheorie*, p. 1258. For the history of the 'key-lock' metaphor in molecular biology, see: Friedrich Cramer, 'Emil Fischers Schlüssel-Schloß-Hypothese der Enzymwirkung – 100 Jahre danach', in Hans-Jörg Rheinberger, Michael Hagner and Bettina Wahrig-Schmidt (eds), *Räume des Wissens. Repräsentation, Codierung, Spur*, Berlin (Akademie-Verlag), 1997, pp. 191–212. The idea of cellular regeneration stemmed from Ehrlich's cousin Carl Weigert and was developed in contact with the latter, see: Rheinberger, Hagner and Wahrig-Schmidt (eds), *Räume des Wissens. Repräsentation, Codierung, Spur*. For the toxin-antitoxin reaction and Ehrlich's work, see also: Pauline M. H. Mazumdar, 'The Antigen-Antibody Reaction and the Physics and Chemistry of Life', *Bulletin of the History of Medicine* 48 (1974), pp. 1–21.
40. Ehrlich, 'Die Wertbemessung des Diphtherieheilserums', pp. 104–6.
41. Silverstein, *Paul Ehrlich's Receptor Immunology*, pp. 51/52.
42. For this period of Ehrlich's work, see also: Dolman, 'Paul Ehrlich', p. 298.
43. Paul Ehrlich and Julius Morgenroth, 'Über Haemolysine. Dritte Mittheilung' (*Berliner Klinische Wochenschrift*, 1900), in Himmelweit, vol. II, *Immunology and Cancer Research*, pp. 196–204, English transl. pp. 205–12. On Ehrlich's receptor concept, see also: Cay-Rüdiger Prüll, 'Part of a Scientific Masterplan? – Paul Ehrlich (1854–1915) and the Origins of his Receptor Concept', *Medical History* 47 (2003), pp. 332–56.
44. John Parascandola, 'The Development of Receptor Theory', in M. J. Parnham and J. Bruinvels (eds), *Discoveries in Pharmacology*, vol. 3, *Pharmacological Methods, Receptors & Chemotherapy*, Amsterdam/New York/Oxford (Elsevier Science Publishers), 1986, pp. 129–56, esp. pp. 134–41.
45. For Langley's receptor concept see: Andreas-Holger Maehle, '"Receptive Substances": John Newport Langley (1852–1925) and his Path to a Receptor Theory of Drug Action', *Medical History* 48 (2004), pp. 153–74.
46. See Dolman, 'Paul Ehrlich', pp. 295–305; Prüll, 'Part of a Scientific Masterplan', pp. 334/335.
47. For a general overview of the history of the receptor concept, see: John Parascandola and Ronald Jasensky, 'Origins of the Receptor Theory of Drug Action', *Bulletin of the History of Medicine* 48 (1974), pp. 199–220; Parascandola, 'The Development of Receptor Theory'; Andreas-Holger Maehle, 'Historical Foundations of the Receptor Concept in Pharmacology', *Gesnerus* 61 (2004), pp. 57–76; Cay-Rüdiger Prüll, Andreas-Holger Maehle and Robert Francis Halliwell, 'Drugs and Cells – Pioneering the Concept of Receptors', *Pharmacy in History* 45 (2003), pp. 18–30; Andreas-Holger Maehle, Cay-Rüdiger Prüll, Robert Francis Halliwell, 'The Emergence of the Drug-Receptor Theory', *Nature Reviews. Drug Discoveries* 1 (2002), pp. 637–41. See furthermore: Andreas-Holger Maehle, 'The Quantification and Differentiation of the Drug Receptor Theory, c. 1910–1960', *Annals of Science* 62 (2005), pp. 479–500; Cay-Rüdiger Prüll, 'Caught between the Old and the New – Walther Straub (1874), the Question of Drug Receptors and the Rise of Modern Pharmacology', *Bulletin of the History of Medicine* 80 (2006), pp. 465–89.
48. Since the 1920s and 1930s many physicians have argued for a new physiological medicine, like the specialist for internal diseases Louis R. Grothe (1886–1960), who differentiated between the 'morphological idea' and the 'functional idea'.

See Eva-Maria Klasen, *Die Diskussion über eine Krise der Medizin in Deutschland zwischen 1925 und 1935*, MD Mainz 1984, pp. 17–27. See the example of twentieth-century surgery for the shift of morphological to physiological medicine: Ulrich Tröhler 'Surgery (modern)', in: William F. Bynum and Roy Porter (eds), *Companion Encyclopedia of the History of Medicine*, 2 vols, vol. 2, London/New York (Routledge), 1993, pp. 984–1028.

49. 'eine neue bedeutungsvolle Richtung der biologischen Forschung eröffnet'. See: Paul Ehrlich, 'Über den Receptorenapparat der rothen Blutkörperchen', in *Gesammelte Arbeiten zur Immunitätsforschung*, Berlin (Hirschwald), 1904, in: Himmelweit, vol. II, *Immunology and Cancer Research*, p. 316–23, esp. p. 320. See also: Dolman, 'Paul Ehrlich', pp. 298/99.
50. Cay-Rüdiger Prüll, 'German Pathology and the Defence of Autopsy since 1850', in: Cay-Rüdiger Prüll (ed.), *Traditions of Pathology in Western Europe - Theories, Institutions and Their Cultural Setting (Neuere Medizin- und Wissenschaftsgeschichte. Quellen und Studien)*, Pfaffenweiler (Centaurus), 2003, pp. 139–61. Concerning Cohnheim, see: Cay-Rüdiger Prüll, 'La Patologia', *Storia della Scienza*. Estratto dal Volume VII. L'Ottocento, Capitolo LXXXI, Istituto della Enciclopedia Italiana. Fondata da Giovanni Treccani, 2003, pp. 875–86; Russell C. Maulitz, *A Treatise on Membranes: Concepts of Tissue Structure, Function and Dysfunction from Xavier Bichat to Julius Cohnheim*, Ann Arbor (Univ. Microfilms International), 1974; Russell C. Maulitz, 'Rudolf Virchow, Julius Cohnheim and the Program of Pathology', in *Bulletin of the History of Medicine* 52 (1978), pp. 162–82.
51. Paul Ehrlich, 'Zur Geschichte der Granula', repr. from *Farbenanalytische Untersuchungen zur Histologie und Klinik des Blutes*, Berlin (Hirschwald), 1891, in Himmelweit, vol. I, *Histology, Biochemistry and Pathology*, pp. 166–8, esp. 166.
52. 'Da ich selbst nicht in der Lage bin, derartige Untersuchungen an einem grösseren Krankenmaterial durchzuführen, habe ich es für meine Pflicht gehalten, die Gesichtspunkte klarzulegen und so die Basis für die Bearbeitung eines Gebietes zu schaffen, dessen Bedeutung für die Pathologie und Therapie vielleicht erst nach Jahren voll gewürdigt werden wird'. See: Paul Ehrlich, 'Die Schutzstoffe des Blutes' (*Verhandlungen der 73. Versammlung der Gesellschaft der Naturforscher und Ärzte*, 1901), in Himmelweit, vol. II, *Immunology and Cancer Research*, pp. 298–315, esp. p. 315.
53. Ehrlich to Althoff, January 1, 1905, in Nachlass Althoff B No. 33, note 60, pp. 150–6, here p. 153.
54. Cay-Rüdiger Prüll, 'Medizin am Toten oder am Lebenden ? – Pathologie in Berlin und in London 1900 bis 1945' (*Veröffentlichungen der Gesellschaft für Universitäts- und Wissenschaftsgeschichte*, 5) Basel (Schwabe), 2003, pp. 141–4, 202/203.
55. For more on Wright, see Michael Worboys' contribution to the present volume.
56. Ibid.; For Ehrlich and clinical pathology, see also: Cay-Rüdiger Prüll, 'Paul Ehrlich als klinischer Pathologe', *Praxis. Schweizerische Rundschau für Medizin* 93 (2004), pp. 1706–14.
57. See the papers of Anne I. Hardy, Axel Hüntelmann and Jonathan Simon in this volume. For the impact of Ehrlich's collaboration with representatives of politics and industry on his standardization procedure, see also: Anne I. Hardy, 'Paul Ehrlich und die Serumproduzenten: Zur Kontrolle des Diphtherieserums in Labor und Fabrik', *Medizinhistorisches Journal* 41 (2006), pp. 51–84.
58. Liebenau, 'Paul Ehrlich as a Commercial Scientist', pp. 65–78. For more on the issue of serum standardization and the pharmaceutical industry, see Axel Hüntelmann's contribution to the present volume.

59. Bäumler, *Paul Ehrlich*, pp. 95–7.
60. Jonathan Harwood, 'The Rise of the Party-Political Professor? Changing Self-Understandings among German Academics, 1890–1933', in: Doris Kaufmann (ed.), *Geschichte der Kaiser-Wilhelm-Gesellschaft im Nationalsozialismus. Bestandsaufnahme und Perspektiven der Forschung*, vol. 1, Göttingen (Wallstein), 2000, pp. 21–45.
61. The disagreement between Ehrlich and Behring concerning the organization and direction of diphtheria serum research and standardization cannot be described here, above all since it is not part of the topic of this paper. For more details, see: Silverstein, *Paul Ehrlich's Receptor Immunology*, pp. 49–51.
62. Travis, 'Science as Receptor of Technology', pp. 383–408.
63. For Ehrlich's contacts with Politics and Industry in general, see: Timothy Lenoir, 'A Magic Bullet: Research for Profit and the Growth of Knowledge in Germany around 1900', *Minerva* 26 (1988), pp. 66–88. Concerning Salvarsan, see: Lutz Sauerteig, 'Salvarsan und der 'ärztliche Polizeistaat'. Syphilistherapie im Streit zwischen Ärzten, pharmazeutischer Industrie, Gesundheitsverwaltung und Naturheilverbänden (1910–1927)', in: Martin Dinges (ed.), *Medizinkritische Bewegungen im Deutschen Reich (ca. 1870–ca. 1933)*, Stuttgart (Steiner), 1996, pp. 161–200; Lutz Sauerteig, 'Mit Zauberkugeln gegen die Syphilis: Paul Ehrlich (1854–1915) und das "Salvarsan"', *Praxis* 88 (1999), pp. 1841–9.
64. Joseph D. Robinson, *Mechanisms of Synaptic Transmission. Bridging the Gaps (1890–1990)*, Oxford (Oxford University Press) 2001.
65. Margit Szöllösi-Janze, *Fritz Haber 1868–1934. Eine Biographie*, Munich (Beck), 1998; Margit Szöllösi-Janze, 'Lebens-Geschichte – Wissenschaftsgeschichte. Vom Nutzen der Biographie für Geschichtswissenschaft und Wissenschaftsgeschichte', *Berichte zur Wissenschaftsgeschichte* 23 (2000), pp. 17–35.

2
Evaluation as a Practical Technique of Administration: The Regulation and Standardization of Diphtheria Serum

Axel C. Hüntelmann

In April 1896, a one-and-a-half-year-old boy called Ernst Langerhans died shortly after receiving a prophylactic dose of diphtheria serum. The day after his death, his father, a well-known Berlin pathologist, claimed in public that his son had been poisoned by Behring's serum. In the weeks that followed, the 'Langerhans Case' was a topic of controversy in the public press and in medical journals. It was in these circumstances that the diphtheria serum had to prove its harmlessness. The boy's dead body was examined, as was the serum he had received, which underwent both chemical and bacteriological analysis.[1] In this context, Paul Ehrlich, director of the newly founded Royal Prussian Institute for Serological Research and Serological Survey (henceforth ISRSS), prepared a report concerning the serum, which formed part of batch No. 216. Ehrlich introduced his report by describing the duties and responsibilities of his institute, explaining the principles of evaluation.

The main difficulty of the method is to keep the test-toxin, which serves as a unit for measuring the serum, unchanged, because the serum is a complicated mixture that can easily lose its strength. Only a constant control of the test-toxin, new series of tests with the test serum and regularly repeated measurements of the toxicity can prevent errors. If these precautions are taken, the method is safe. Paying attention to the uniformity of the conditions (animals of the same weight, same kind of intoxication etc.), ensures that the method works exactly like a chemical titration.[2]

In what follows, I will describe evaluation as a technique of administration in the context of the production, standardization and regulation of diphtheria serum around 1895/1896, shortly after the serum first became available in German pharmacies. Diphtheria serum was a remedy of biological origin, taken directly from the blood of a horse. Here, I will focus on the process that turned the blood serum into a state-certified pharmaceutical, inscribed in the *Pharmacopeia Germanica*. This process is described as a transformation accompanied by certain 'rites of passage'.[3] These rites – the evaluation of the serum and the related administrative techniques – were required to effect this transformation. Following the institutionalization of evaluation, the ISRSS

31

became an obligatory point of passage that every batch of serum offered on the pharmaceutical market had to pass to become a state-approved remedy.[4] Besides state regulation, evaluation also had an impact on standardization, guaranteeing a standardized pharmaceutical product as well as regulating serum production.

The chapter is divided into two parts. First, I will describe the production and evaluation of serum as well as the regulations concerning evaluation as they were specified in the official instructions from around 1896. In the second part, I will analyze the administrative techniques of evaluation under the following categories: actors, techniques of regulation, order of space and time.

The production, standardization, evaluation and regulation of diphtheria-serum

To master the complex process of evaluation, Paul Ehrlich describes a method characterized by control of the process, surveillance, monitoring, inspection, regularity, repetition, uniformity and coding – an approach ideally suited to bureaucratic administration. Standardization does not imply a natural regularity, rather an artificial, instituted order that has to be codified in regulations and fixed in norms and standards, to become regular.[5] The object of the different administrative techniques was the monitoring and control of the production process and the minimization of risk and the assessment [and stabilization] of uncertain situations.[6]

The 'production' of diphtheria serum was an innovation in two respects. The launching of the serum onto the pharmaceutical market represented a new type of drug and a new principle for therapy. Previously, the control and legislation of drugs was aligned with the pharmacies and the production of drugs was performed only on a small scale. The pharmacopoeia regulated only the purity of the ingredients. This changed in the 1890s when new medicaments had to be tested and their inoffensiveness and potency had to be proven before they could be brought onto the market. Once introduced onto the market, the serum was produced in large quantities.[7] Second, the serum was a biological agent – an extract of a living organism – that could be of variable quality depending on the organism that produced it. With this variability emerged a new problem: to find a universal standard, a serum unit had to be defined and implemented because the physicians in private practice and in the hospitals needed a criterion for the usage of the serum. To achieve therapeutic success, a precise dosage was needed.[8] Furthermore, a value was necessary to compare the different results of examinations and experiments, not only in the scientific context but also in the economic context, enabling the comparison of the potency of different sera on the pharmaceutical market.

The problem was solved by the transformation of an undefined, imponderable, variable effect into an objective, quantifiable and comparable value.

Evaluation was seen as an objective criterion for determining the serum's 'value', indicating its physiological or therapeutic impact in terms of antitoxin or immunization units. The number that defined the potency of a serum was generated by a process that formed part of an institutional order based on regulation, standardization and specialization. Indeed from April 1895 on, the value of diphtheria serum was certified by a state-run control station that guaranteed the curative effect of the serum.[9]

In December 1894, an article in the *Münchener Medizinische Wochenschrift* described the new bacteriological department at the Hoechst Dye Company (Farbwerke Hoechst)[10] (Figure 2.1). The production of serum started with the production of the toxin that would be injected into the serum-producing horses. In the front building, the largest room was the breeding chamber for the bacteria cultures (A). Instead of one large incubator there were a few small incubators, allowing different conditions, taking into account the individuality and sensitivity of the diphtheria bacteria. These small incubators were used to breed the diphtheria cultures that produced the toxin. After the cultures were sufficiently rich in toxin, the bacilli were killed using a disinfectant and the toxin extracted. In the other part of the front building there were rooms for sterilization, cooking, cleaning, cooling and a stable for the laboratory animals (B). The laboratories and the workshops were located in the rear of this front building (C). In these rooms, the technicians completed the registration and entered the records in the minute book. A passage (D) led from this building to the stables, the feeding rooms (F) and the operating room (E), in which the horses were injected with diphtheria toxin and bled for serum.

Figure 2.1   Bacteriological department 1894
Source: Emil Behring Archive in Marburg.

The horses were injected with toxin that activated an immunization process, which became apparent when the horse exhibited the typical symptoms of an illness, such as elevated temperature or fever (Figure 2.2). After the injection of toxin, the horse 'produced' antibodies in more or less regular 'waves' that depended on the strength of the toxin and the health of the individual horse. The concentration of antitoxins would level off before rising again. This process recurred until the toxin was neutralized by antitoxins with an additional surplus of antibodies.[11] This immunization process was prolonged with successive injections of increasing quantities of toxin. After a certain concentration of antibodies had been 'produced', the concentration would start to decline. The ideal moment for the bloodletting was just this turning point, when the serum would have its maximum value in terms of immunization units. In the beginning, this immunization process lasted nearly three months, and following intensive bloodletting, the horse was allowed a period for recuperation before the whole process started again. With the steady iteration of the immunization process and the growing experience of the stable and laboratory staff, the productivity of the horses was increased and the intervals between intoxication and bloodletting reduced.[12] Subsequent to the bloodletting, the blood was collected, filtered and conserved using 0.5 per cent carbolic acid.

In the article about the bacteriological department at the Hoechst Dye Works, the author underlined the modernity of the production plant, based on the division of labour. Only reliable and qualified personnel were

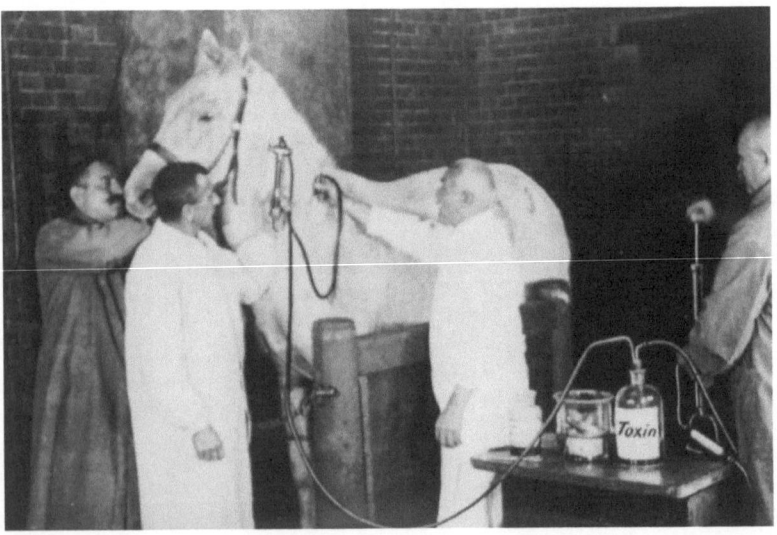

Figure 2.2   The injection of (diphtheria) toxin into a horse, from circa 1900
Source: Emil Behring Archive in Marburg.

employed. Even in the stables they had achieved a 'very meticulous cleanliness and purity', ensured by a regiment of stablemen under the surveillance of a 'military educated stable master'.[13]

After April 1895, when a national set of security measures was implemented, the state control of the process took place at two different sites. First, a medical official, appointed by the state, worked at the production plant itself. This medical official first surveyed the bloodletting, then supervised the collection of serum in a large receptacle, before ensuring the preparation of a representative sample. Both the filled receptacle and the sample were labelled with an operation number, closed and sealed by the medical official. While the receptacle was kept in a cool and secure place at the production plant, the sample was sent to the ISRSS, with all relevant dates being recorded in several ledgers.[14]

At the ISRSS, two medical officials tested the sample simultaneously and independently. The assay had to prove the harmlessness of the serum and determine its curative value. The serum was tested to ensure its sterility, to check the preservative that had been added, and to verify that the product was clear, without any particles or sediment. The test procedure to ascertain the official value in immunization (antitoxin) units was very tricky. Around 1896, the procedure was as follows: the submitted serum was mixed with a fixed standard toxin, which was the lethal dose, or the minimum amount capable of killing a guinea pig of approx. 250 grams within four days. The mixture of standard toxin and serum was then injected into a guinea pig.[15] The curative value was determined in terms of the number of immunization units based on the mixture at which the guinea pig showed only a very mild reaction at the site of injection, which faded away within four days.[16] The results of the test procedure were certified and sent to the producer, while the samples were stocked for several months. The tests were automatically repeated at certain intervals. If the serum stopped being sterile during this time, or if its potency was determined to be ten percent less then the originally determined value, the serum would be withdrawn from circulation.[17]

After receiving the certificate from the ISRSS, the medical official at the production plant unsealed the receptacle containing that batch and the serum was poured into phials, which were corked, all under his supervision. Every phial received a label bearing the operation number and the date of filling. Every step was carefully noted in a register, which made it possible to trace back the serum and identify the host-animal and the day of bloodletting. In the end, every phial was packaged in paper and then in a wooden box to protect the serum, before being closed with a seal bearing the official notification 'state-controlled'.[18]

If the serum had a lower value than necessary, or was not sterile for some reason, the ISRSS would then give instructions concerning what should happen to this serum. Either it could be improved or else it needed to be destroyed. In either case, the receptacle containing the serum at the production plant was officially returned by the on-site medical official to the producer, and the producer had to report the actions taken in accordance with the instructions from the ISRSS.[19]

Interpretation – transformation – analysis

In this section I will focus on the process between the bloodletting of the horse and the distribution of a certified and state-approved drug, especially the administrative parts. This process was initiated, carried out and completed by a fixed set of actions – 'rites of passage' – that ensured the transformation.[20] After the blood had been collected in a receptacle and brought through the building's passage into the laboratory, it was placed in the cooling room where the red blood cells coagulated and the serum was extracted and filtered from the blood. Nothing else happened to the serum itself, but at the end of the process one had a remedy against diphtheria, protected by imperial laws and defined in the B register of the *Pharmacopoeia Germanica*, which meant that it could only be obtained with a prescription in a pharmacy. Between the bloodletting and the distribution of serum, there was an interval of more than one week. During this time, the serum underwent a transformation from a highly symbolic and mystical material[21] to a commercial and pharmaceutical consumer good. What happened during this time? In what sense did a transformation take place?

The epistemological process of evaluation, originally developed by Behring and Ehrlich, was promoted by the state, scientifically consecrated and bureaucratically fixed.[22] Some were uncomfortable with this relationship between science and state; as one critic of the new serotherapy complained, science had blundered into an area of government medicine where state-approved therapy had become a dogma.[23] This dogma implied norms that had to be rigidly followed, and only at the end of the process would a successful transformation have taken place. If the serum broke the rules – independently of its effect on the patient – the serum was denied the status of a pharmaceutical. The regularities and symbolic rituals were sanctioned by bureaucratic actions, but the rigid rituals contrasted with the tentativeness of the people involved in the process. The standardization of the test procedure was meant to minimize the risks and insecurity at four levels: first at the level of the source (the host-animal, the horse); second at the level of the bio-indicator (the guinea pig); third at the level of the translator (the medical official); and finally at the level of the recipient (the patient).[24]

Actors

Numerous species of animals could be used for producing diphtheria serum: goats (Paul Ehrlich),[25] sheep (Emil Behring),[26] horses or even dogs (Erich Wernicke),[27] provided that the animal was susceptible to diphtheria. The serum producers preferred horses for the large-scale production of serum because of their size and the quantity of blood that could be taken. The horse also had to pass through a multilevel process of standardization prior to being defined as a serum host. In the beginning there was an arbitrary horse, bought at the market. The horse then stayed for several weeks under observation in a stable far removed from the production plant. The horse received an individual number or name and was measured and weighed.[28] A (district) veterinarian examined the horse several times under the supervision of the medical official to confirm its good health, before granting it a new identity as a potential serum host.[29] The new recruit then moved from the periphery to the centre and the host was now included in the production plant.[30] All data and specifications about the serum host were written down in a 'Kontrollbuch – Formular B'. Every horse/serum host had its own journal, detailing its number and name, age, sex, height, weight, colour and special characteristics. In the journal's first column the operation number was noted that referred to the main database, the so-called 'Hauptkontrollbuch'. Subsequent columns contained the serum host's date of entry into the quarantine stable, the date of transfer into the production plant and the ongoing findings of veterinary examinations, including the pulse, temperature, general form and feeding information. The days of immunization with diphtheria toxin, the days when a blood sample was taken (as well as the results of its analysis) and the days of bloodletting were noted in different columns. The last column space was left for general remarks. Every horse had its own temperature chart with information about the horse's health status.[31]

In combination with this journal, the horse became more than a horse; with the assignment of characteristics and information about the immunization process, it turned into a serum host. Thus, the horse and journal necessarily became a single unit, with the former transformed into a numerical code, which allowed calculations and a forecast into the future about the expected highest quantity of immunization units. From this perspective, the entries in the journal had to be continuous; if a stableman forgot just a single entry, the procedure would lose its credibility and veracity.

The personnel involved in the process of production and evaluation also had to be standardized – all the more so as the process was specialized and differentiated. Thus, the stablemen at the Hoechst Dye Works were instructed by a 'military educated stablemaster',[32] accountable to the director of the sero-bacteriological department, who in turn was supervised by the medical official. The different levels of control guaranteed that all administrative duties were taken care off, with the medical official reporting to the district president at regular intervals.[33] At the top of the pyramid, the district president or the Ministry of Culture could at any time ask for information concerning the proper filling out of the journals or revise the operation of the production plant.[34]

Even within the ISRSS the medical officials controlled each other, with independent verification of the serum's value. Despite official guidelines and instructions, the results could differ because of an individual's mode of operating or the difference in perception between the different workers, and so, to compensate, the medical officials worked simultaneously and independently from one another. Only if the results were in agreement did the value become official; in case of disagreement, the final authority was the director of the ISRSS.[35]

In his role interpreting the results of the evaluation, the technically highly qualified medical official held a key position with institutional supremacy over the value of the serum. He also ensured the continuity of the evaluation process, testing sera from different producers over an extended period of time. Finally, the medical official was the model for an absolutely upstanding, loyal and reliable person, and the tests he performed were synonymous with precision, accuracy, incorruptibility and truthfulness.[36] This was important as the medical official in the production plant was in a dangerous situation. Paid by and answerable to the state, there was always the risk that he could fall under the influence of the producer to whose plant he was assigned.

Despite all the control mechanisms and detailed instructions, there could still be misinterpretation and differences between the value announced by the producer and the official test result. The guinea pig, for example, which served as the indicator of potency, was a living organism that was difficult to regulate and interpret.[37] Guinea pigs differed in their constitution (age, weight, sex) and how they reacted to a serum sample, reactions that were open to interpretation. How should one determine a swelling at the point of injection? Also, in case of a loss of weight, there were many possible explanations such as a change in temperature or seasonal variation. As Ehrlich mentioned in a letter, the absolute determination of toxin or serum was neither easy and nor always exact, especially in winter.[38] To handle these problems of regulations and transparency,[39] every guinea pig – like the horses producing the serum – was accompanied by an individual record book, providing similar information as that described above for the host. Later on, standardized guinea pigs – the same age, same weight and same constitution – were bred specifically for the experiments and test procedures, and were all kept in identical living conditions including food, light and hygiene.[40] Nevertheless there was always room for doubt concerning the results of the evaluation.

Even if the serum was absolutely standardized, there was a final actor, the patient, who provided another source of incertitude. The value of the serum depended on the patient's reaction, which ultimately determined whether or not it was effective. The value of immunization units, fixed in the laboratory by the reaction of a certain test animal after an injection of a toxin/antitoxin mixture, was the result of a model that was intended to predict the effect of the serum on the patient. But the injection of a toxin-antitoxin mixture did not reflect the 'reality' or the 'natural course' of infection.[41] As a curative remedy, diphtheria serum was injected several days after the infection had occurred, and as a preventive treatment, it was injected before the bacteria invaded the human body. While the toxin/antitoxin mixture was a model for the infection and the remedy, the guinea pig was a model for the patient.[42]

But in practice, despite all of the security measures, there was no guarantee that the serum injected into a patient would have any effect as treatment or prophylactic. To avoid failures, the recipient's environment was also standardized. The pharmacist had to store the serum in a dark and cold place so the serum would not lose its potency. A detailed instruction leaflet was included to help the physician administer the serum correctly.[43] In some ways, the disease was also standardized; the diagnosis of 'diphtheria' was increasingly determined definitively only in a laboratory. Anyone presenting the clinical symptoms associated with the disease was now only suspected of having diphtheria, and it was only the microbiological examination in the laboratory that determined the diagnosis.[44] But all these precautions gave no guarantee that the serum would in fact cure, because the producer had no influence over the patient's behaviour and reaction. If the patient showed no reaction despite the use of a serum that was certified as potent, it had to be the patient's fault; it was the patient who was not standardized, meaning that the disease was accompanied by complications or a multiple infection. Furthermore, the patient's living conditions could accelerate or delay the healing process.

In the article about the new bacteriological department at the Hoechst Dye Works, the author presented the motives of the producer, explaining why all the safety and control measures were necessary.[45] In the case of failure, the company had to prove the inoffensiveness of the serum – the harmlessness of every single phial – and that security measures had been implemented.[46] The 'Kontrollbücher' were cited as giving the possibility of reducing one's losses or averting any further damage. The control of the production process also made it possible to advance with serum production while being able to trace back every single step. The need for this kind of absolute control made it obvious that the company was not as confident about the serum as they pretended to be; they tried to calculate every incalculable risk.

Institution and techniques of regulation

The complete production process was recorded without interruption. The spatial arrangements and the process of production have been described above, and in this section I will characterize the techniques of administration and regulation. We can note several crucial aspects of the technique of regulation: first the 'journalisation' and the logging of information in the 'Kontrollbücher', and second the ISRSS and the medical official.

Their interaction at all levels of production and control transformed the blood serum into a pharmaceutical and made the production of a complex living organism as close as possible to a chemical titration.[47] Originally incalculable contexts were standardized, calculated and 'journalised'. The production process became transparent and all the personnel interchangeable, including the medical official. Only the 'Kontrollbuch' was unique, obligatory and could not be substituted.

I will now consider the process of production and evaluation with the focus on standardization and administration. Commentators emphasized the modernity of the incubators at the new production plant of the Hoechst Dye Works.[48] Having several independent incubators allowed the producers to take into account the 'individuality' of the bacteria cultures producing the toxin. The regulation of temperature, time and other conditions were intended to guarantee the production of a standardized toxin. In the ongoing 'production process' the serum was standardized, via the journal, for the serum host. After the horse had been repeatedly injected with diphtheria toxin at regular intervals, test bleedings were made to examine the amount of antitoxin units. All dates of the toxin injections and the findings of the test bleedings were recorded in the logbook for the horse so that the point in time of the (expected) highest value of immunization units could be predicted. Beyond this, the procedure of evaluation was standardized and had to follow an explicit set of instructions.[49]

All was defined in detail for the creation of the test toxin, with specifications concerning its handling and other technical considerations. The most suitable culture and nutrient solution were specified, as were the methods for killing the bacteria and filtering the toxin and the dilution of the toxin-antitoxin mixture. Special pipettes were constructed for the dilution of small quantities along with special devices for sterile bloodletting.[50] Members of the ISRSS communicated regularly with the serum producers, sharing both its ongoing practices and the established knowledge about the evaluation of serum.[51] The technicians working for the various producers were instructed and trained at the ISRSS on how to evaluate the serum in accordance with the official guidelines.[52] Finally, if the results of the evaluation of any serum diverged, the serological institute sent its own test toxin to the serum producer.[53] The production and evaluation process was standardized continuously by means of the working procedures. Problems of measurement and precision emerging from the processes of control and the procedure of evaluation were handled using a system of ongoing correction and calibration,[54] with every step noted in different journals[55] to provide complete traceability.

Every step was precisely defined and independent from the person carrying it out, although this did not mean that just anyone could control the process. The process demanded a growing degree of specialization including a thorough training in bacteriological and other laboratory techniques. Ideally, the executing laboratory staff in the production plant belonged to a single bacteriological thought collective, with a bacteriological thought style that made it easier to communicate with the members of the ISRSS and to analyze any problems that might arise. Whereas for an outsider the procedure was mysterious (a ritual), the transformation was easily interpreted by an inner circle – the serum producer and the medical officials of the ISRSS. For the inner circle, the procedure of evaluation was decipherable at any time in the procedure, using just the codes entered in the journals (Figures 2.3 and 2.4).

The main logbook – the 'Hauptkontrollbuch' – collected together all the information about the production process (Figure 2.4). The operation number recorded in the first column served as the link between the different journals, with other columns documenting the bloodletting: the identification number of the serum host, the date of bloodletting and the quantity of serum. The table also included details on the test animal, the amount of immunization units and the sterility of the serum, the quantity of preservative added and the date of the test procedure at the production plant. Further columns showed the date and the results of the official audit by the ISRSS. If the serum was rejected, there was a column for remarks. The day of the official sealing was noted in the last columns with a place left for general remarks. The medical officials in the state-run institute had their own journals in which they entered the test results of the evaluation of the serum.[56]

The journals had a number of functions; for the producer they could prove that the rules were being followed and the serum was harmless and effective in case of litigation,[57] and they allowed the optimization of serum production by indicating the level of immunization of the horses. They also provided the means for correcting or improving any deficiency in the production process. On the side of the state administration, these journals could be used to monitor all the aspects of production. Indeed, the controller became visible in the main logbook, representing the material of his daily routine. Thus, he was not just a passive observer of a productive process, but was also producing something: lists, tables, journals. The control of the controller became material and manifest in the *Kontrollbuch*, a countermovement to the material arrangements described by the regulations,

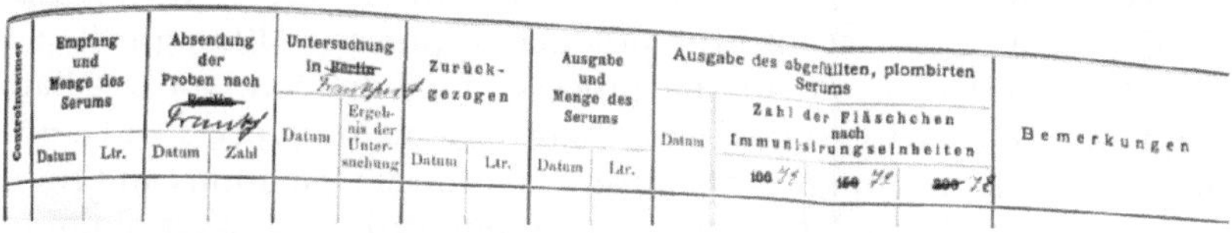

Controlnummer	Empfang und Menge des Serums		Absendung der Proben nach ~~Berlin~~ Frankfurt		Untersuchung in ~~Berlin~~ Frankfurt		Zurückgezogen		Ausgabe und Menge des Serums		Ausgabe des abgefüllten, plombirten Serums				Bemerkungen
												Zahl der Fläschchen nach Immunisirungseinheiten			
	Datum	Ltr.	Datum	Zahl	Datum	Ergebnis der Untersuchung	Datum	Ltr.	Datum	Ltr.	Datum	100	150	200	

Figure 2.3 Draft of the logbook for the medical official
Source: APEI, Dept. Va, No. 1.

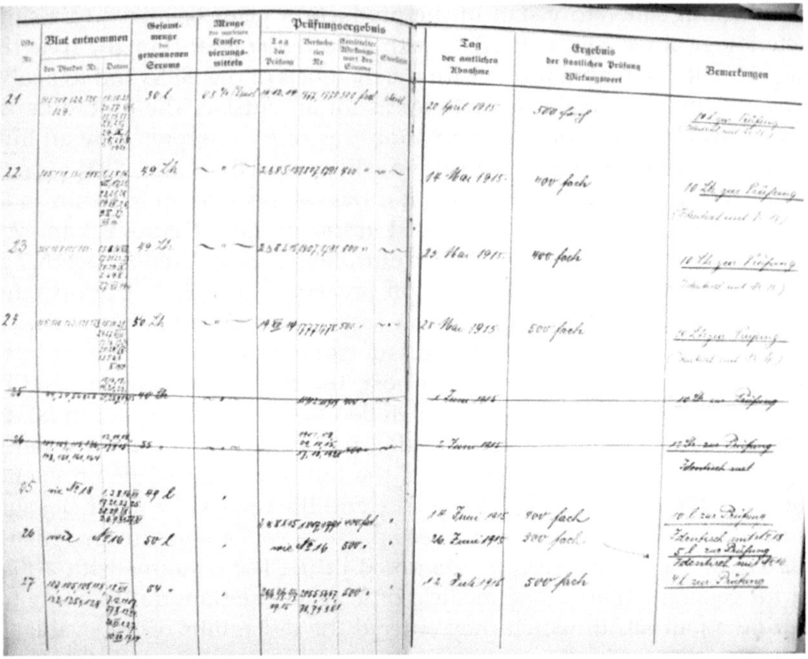

Figure 2.4 Main production logbook for diphtheria serum at the Behringwerke 1915
Source: Archive Behringwerke in Marburg.

whose standardizations and techniques of administration represent a form of dematerialization that rendered laboratory techniques invisible.

The space of control

The 'Kontrollbuch' also performed an important spatial function, with the transformation taking place in a special space; the material places, rooms, and institutions in which the process of evaluation took place. As we have already explained, the journal connected the company's findings and the official test results of the ISRSS. The problem of the German system, de-localized production with centralized control, was solved by the introduction of a medical official into the production process. Although the state was not directly involved in the process of production, the mediating medical official gave the state a view of the production process and a direct influence over it, ensuring compliance with official regulations and norms. In case of any problems, the constant communication between serum producers and the ISRSS served to bridge the distance between these different locations.

The logbook could also be read as if the entire process of evaluation had been carried out in one place. The distance between the medical official and the ISRSS was abolished, giving one virtual space of control.[58] While the distance between the on-site medical official and the ISRSS was minimized, a virtual distance was constructed between the medical official in the production plant and the producer. This distance became apparent in the symbolic transfer of the serum: the medical official took the serum after it had been extracted from the blood, closed the storage vessel and kept it apart. After the medical official had received the certificate from the ISRSS, he returned the serum to the producer. The medical official and ISRSS constituted one unit leading to a virtual frontier between the producer and the medical official at the production plant.

Chronological order – institutions are founded on experiences

> Past experience is encapsulated in an institution's rules so that it acts as a guide to what to expect from the future. The more fully the institutions encode expectations, the more they put uncertainty under control, with the further effect that behaviour tends to conform to the institutional matrix: if this degree of coordination is achieved, disorder and confusion disappear.[59]

The institution of evaluation was built on past experience. On the production side the leading company, the Hoechst Dye Works, gained much experience of the handling of the new serum.[60] Starting at the end of October 1894, representatives of the Prussian and Imperial ministerial bureaucracy met fortnightly with scientists and producers to discuss the production, distribution and regulation of the new drug.[61] The state bureaucracy and public health administration had already had some experience with the regulation of food production in the 1880s, another complex and sensitive industrial consumer good that could represent a public health risk. In implementing the resulting food act, the Imperial Health Office had defined 'normal' food and developed appropriate skills and techniques to measure it, training experts and founding institutions for control.[62]

Similar problems emerged with the industrial production of high precision instruments and electronic and optical devices. With a growing science-based industry, new technical requirements like high precision devices as well as the generation of national standards and norms became necessary. Founded in the 1880s, the Physikalisch-Technische Reichsanstalt was given the task of compiling the scientific basis of the domains of optics, electricity and high-precision mechanics, before approving and certifying technical instruments, tools and commodities. They ensured that these goods corresponded to the official norms and issued instructions to ensure the uniformity, equality and normality of devices, measuring units and materials as well as generating standardized physico-technical norms.[63]

At the end of the nineteenth century, the chemical and pharmaceutical industries were also prospering and were confronted with comparable standardization problems. To fabricate a standardized drug, the output depended on exact definitions of pure, 'normal' chemical ingredients. In Germany, therapeutic substances, their definition as pharmaceuticals and their grade of purity were regulated by the *Pharmacopeia Germanica*, which was regularly revised, being updated in 1883 and 1890.[64] While pharmaceuticals, sera, food and physico-technical devices are quite different products, they had several things in common from the perspective of risk, regulation and standardization. In all cases, new problems emerged in the context of large-scale, industrial production, when qualitatively variable products had to be transformed into standardized mass products conforming to national standards. This transformation was accompanied by official instructions concerning a (valid) certification procedure and the introduction of an administrative system including reporting and journals. In all cases, the state was involved and became a central actor in the generation of national standards aimed at the minimization of public (health) risks. Thus, the Imperial authorities already had considerable experience of how to set up a system of indirect control over a free market system, how to implement an information network and how to process different data.

There was also the negative experience of the tuberculin affair. A few years earlier, around 1890, Robert Koch had launched a new drug against tuberculosis. After a few months it became clear that the new pharmaceutical was more or less ineffective, but in the meantime several people had died after injections with tuberculin. After an enthusiastic reception, the new pharmaceutical came to be rejected.[65] To avoid a similar disaster in the case of the new serological therapy, the state installed several control procedures and it was in this context that the ISRSS was founded.[66] But despite several test series, there remained an incalculable risk due to the side or long-term effects of serum therapy. To monitor and evaluate this risk, the authorities decided to gather statistics on the use of the diphtheria serum.[67]

Evaluation as a technique of administration

After the bloodletting of the serum host in the stables of the bacteriological department of the serum producer, a sample was sent to the ISRSS, while the serum itself was stored in a dark, locked storeroom. After a week, during which nothing had happened to the stocked serum, the medical official unlocked the store room and, if the serum's value had been confirmed by the ISRSS, the serum was no longer just a serum, it had been transformed into a pharmaceutical product, defined in the *Pharmacopeia Germanica* with a precisely determined value that allowed a prediction concerning its therapeutic efficacy. Within this ontological shift the administrative techniques and the fixed set of actions – the 'rites de passage' – played a major role, constituting a scientific ritual of initiation. This transformation was made possible by the records kept in the various journals, with the process being made visible by the continuous records, which unified the different spaces of production and control.

To transform a method based on a living agent into a method that worked as precisely as a chemical titration, it was necessary constantly to control the process of production and evaluation. As a result, the main characteristics of evaluation are regulation, regularity, standardization, the decoding and transmission of varying biological data into calculable, comparable and uniform information. Institutions served to produce the required uniformity. The standardization of diphtheria serum was an example of the construction of uniformity in institutions. As an obligatory point of passage, the ISRSS was there to guarantee that all the commercialized blood serum had passed through this transformation process.

Unlike the standardization in experimental and epistemological systems, the standardization described in this paper is of a different type. In experimental settings, the arrangement, surroundings, devices, indicators, measuring units and the different actors have to be absolutely uniform except for one factor that varies to produce different results. This introversive standardization in experiments is intended to produce difference.[68] But in the case of large scale production and regulation of diphtheria serum the individuality of the serum host, the test animal and the toxin has to be taken into account and factored out in order to produce a standardized serum. In this case the output has to be absolutely standardized and uniform. The aim of this external standardization is to minimize difference.

This process of evaluation is a theoretical ideal; in fact, the process was full of resistance and contradictions. The actors lived their own lives, but the contradictions diversified and improved the process of administration as well as the process of evaluation (Häußermann calls this the organization of resistance)[69] – which was why an ISRSS was founded as an obligatory point of passage. The result was a dynamic process, a system of self-regulation that sought to exclude the contradictions.[70] The contradictions are not only necessary; they are the motor for systematic development. The impulse of the process was given by an indefinable and impervious mixture of unspecified desires, scientific ambitions, public needs, economic aims and the need to minimize risk. Once started, the administrative system developed its own dynamic self-regulating bureaucratic mechanism (bürokratischer Amtsmechanismus), with a life of its own.

The techniques of administration and evaluation described above transformed the blood of a horse into a standardized, mass-produced and officially certified drug. This transformation was promoted by the state. The legislation produced official regulations and instituted a specific technique and then enforced their relevance and significance. This transformation process could also be described as a technology of trust. In a rationality-based, secular, modern society these rituals help to handle incertitude and risk; in practice they helped to convince the public that the diphtheria serum injected into Ernst Langerhans was harmless.

Notes

This article has been written in the context of two projects financed by the German Research Foundation – DFG HE 2220 and DFG GR 2116. I would like to thank Christoph Gradmann, Volker Hess and Jonathan Simon for helpful suggestions.

1. Cf. Axel C. Hüntelmann, 'Das Diphtherie-Serum und der Fall Langerhans', in: *MedGG* 24 (2006), pp. 71–104.
2. 'Die Hauptschwierigkeit der Methode beruht darauf, die als Masstab für das Serum dienende Testgiftdosis unverändert zu halten, da es sich hier um *complicirte* Gemische handelt, welche leicht einer Abschwächung ausgesetzt sind. Nur ein stetes *Ueberwachen* des Testgiftes, erneute *Versuchsreihen* mit dem Testserum und *regelmässig wiederholte* Toxicitätsbestimmungen schützen vor Täuschungen. Werden aber diese Cautelen *geübt*, so ist die sonstige Ausführung der Methode eine ganz sichere. *Trägt man Sorge*, die *Versuchsbedingungen gleichmässig* zu gestalten (Wahl von gleichgrossen Thieren, gleiche Intoxicationsart etc.), so arbeitet die Methode *nahezu so genau, wie eine chemische Titration.*' Paul Ehrlich, 'Die staatliche Controle des Diphtherieserums', in: *Berliner Klinische Wochenschrift* 33 (1896), pp. 441–3.
3. For further information about the concept of rites of passage, originally invented by the ethnologist Arnold van Gennep, *Übergangsriten (Les rites de passage)*, Frankfurt am Main: Campus, 2005 (OE 1909); see Victor Turner, *Das Ritual. Struktur und Anitstruktur*, Frankfurt am Main: Campus, 2000 (English transl. *The Ritual Process. Structure and Anti-Structure*, Chicago: Aldine, 1969).
4. For the obligatory passage point see Bruno Latour, *Reassambling the Social: An Introduction to Actor-Network-Theory*, Oxford: Oxford University Press, 2005; the ISRSS as an obligatory passage point sketched in Axel C. Hüntelmann, 'Ways of Evaluation. Therapeutic Agents between Standardisation and Institutionalisation', in: Jean-Paul Gaudillière/Volker Hess (eds), *Ways of Regulating Therapeutic Agents between Plants, Shops and Consulting Rooms (Drug Trajectories V)*, Berlin: Max Planck Institute for the History of Science, 2008, pp. 65–75.
5. Cf. Peter Lundgreen, *Standardization – Testing – Regulation. Studies in the History of the Science-Based Regulatory State (Germany and the USA, 19th and 20th Centuries)*, Bielefeld: B. Kleine 1986; M. Norton Wise, 'Introduction', in: M. Norton Wise (ed.), *The Values of Precision*, Princeton: Princeton University Press, 1995, pp. 3–13.
6. Diphtheria serum was applied as a treatment to cure a patient, but was also used when a patient was suspected of having diphtheria, or in the patient's entourage in case of an epidemic. In the context of biopolitics the preventive impact of diphtheria serum was important to avoid the outbreak from the very first. This preventive measurement was again made safer by a preventive security procedure – the installation of serum regulation that should guarantee the potency, purity and harmlessness of the serum to strengthen confidence in the new serum therapy.
7. Cf. Erika Hickel, *Arzneimittel-Standardisierung im 19. Jahrhundert in den Pharmakopöen Deutschlands, Frankreichs, Großbritanniens und der Vereinigten Staaten von Amerika*, Darmstadt: Wissenschaftliche Verlagsgesellschaft, 1973; Wolfgang Wimmer, 'Wir

haben fast immer war Neues'. Gesundheitswesen und Innovation der Pharma-Industrie in Deutschland, 1880–1935, Berlin: Duncker & Humblot, 1994.

8. The development of diphtheria serum at the end of the nineteenth century is described in Carola Throm, *Das Diphtherieserum. Ein neues Therapieprinzip, seine Entwicklung und Markteinführung*, Stuttgart: Wissenschaftliche Verlagsgesellschaft, 1995.

9. The control station for diphtheria serum was first housed in the Institute for Infectious Diseases, headed by Robert Koch, before it became independent in 1896 as the ISRSS. After the institute moved to Frankfurt on Main in 1899, it was renamed in Institute for Experimental Therapy. For a short history, see Wilhelm Kolle, 'Das Staatsinstitut für experimentelle Therapie und das Chemo-therapeutische Forschungsinstitut "Georg Speyer-Haus" in Frankfurt a. M. Ihre Geschichte, Organisation und ihre Arbeitsgebiete', in: *Arbeiten aus dem Staatsinstitut für experimentelle Therapie und dem Georg Speyer-Hause zu Frankfurt a. M.* 16 (1926), pp. 1–67.

10. Arnold Eiermann, 'Die Einrichtung zur Darstellung des Diphtherie-Heilserums in den Höchster Farbwerken', in: *Münchener Medicinische Wochenschrift* 41 (1894), pp. 1038–40.

11. Cf. Paul Ehrlich/Ludwig Brieger, 'Beiträge zur Kenntniss der Milch immunisierter Thiere', in: *Zeitschrift für Hygiene und Infectionskrankheiten* 13 (1893), pp. 336–46; Throm, *Diphtherieserum*, 1995, pp. 52–3.

12. This is only a short, simplified overview; for a full description, see Throm, *Diphtherieserum*, 1995.

13. Cf. Eiermann, 'Einrichtung', 1894.

14. For more detail, see Throm, *Diphtherieserum*, 1995.

15. To assure the poisonous action, the standard toxin was injected reclusively into a guinea pig that should die within four days, cf. Robert Koch/Emil Behring/Paul Ehrlich/Stabsarzt Dr. Weisser, *Vorschläge, die Prüfung des Diphtherieserums und seinen Vertrieb betreffend*, November 1894, Bundesarchiv Berlin (Federal Archive henceforth BA Berlin), R 86/1646; Richard Otto, *Die staatliche Prüfung der Heilsera (Arbeiten aus dem Königlichen Institut für experimentelle Therapie zu Frankfurt a. M, Vol. 2)*, Jena: Gustav Fischer, 1906, pp. 39–40.

16. Cf. Otto, *Prüfung*, 1906, pp. 39–40; Ehrlich, 'Controle', 1896. The different drafts for the test procedure in the Archive of the Paul Ehrlich Institute (henceforth APEI), Dept. Va, No. 1, Vol. 1. Already in 1897 the method of evaluation changed due to problems resulting from variances and the sinking poisonous of the standard toxin. Since mid 1897, the scale basis was a dried and vacuum sealed standard serum, cf. Axel C. Hüntelmann, 'The Dynamics of "Wertbestimmung"', in: *Science in Context* (21) 2008, pp. 229–52.

17. Cf. Otto, *Prüfung*, 1906. The confiscation of serum from the market in BA Berlin, R 86/1646, passim; APEI, Dept. IV, No. 1, Vol. 1, passim; Prussian Secret Central Archive Berlin (Geheimes Staatsarchiv – Preussischer Kulturbesitz, Berlin – hereafter GStA PK), HA 1, Rep. 76 VIII B, No. 3747–51, passim; and GStA PK, 1. HA, Rep. 76 Vc, Sekt 1, Tit. XI, Part II, No. 18, Vol. 1 f., passim.

18. Cf. Throm, *Diphtherieserum*, 1995; for the production side Eiermann, 'Einrichtung', 1894.

19. Cf. Otto, *Prüfung*, 1906.

20. Cf. van Gennep, *Übergangsriten*, 2005 (OE 1909); Turner, *Ritual*, 2000 (English transl. 1969).

21. Cf. Christina von Braun/Christoph Wulf (eds.), *Mythen des Blutes*, Frankfurt am Main: Campus, 2007.

48   *Evaluation as a Practical Technique of Administration*

22. Emil Behring, Paul Ehrlich, Robert Koch and surgeon major Weisser, member of the Imperial Health Office, had worked out the guidelines for the state control and the evaluation of diphtheria serum and thus codified the method of Behring and Ehrlich in October 1894. Moreover Behring, Ehrlich and Koch were the advising experts on the proceedings in the Imperial Health Office in early November 1894 concerning the state control of diphtheria serum, cf. the guidelines and the minutes of the meeting in BA Berlin, R 86/1646. For further information see Heinz Zeiss and Richard Bieling, *Emil von Behring. Gestalt und Werk*, Berlin: Bruno Schultz, 1941.
23. 'Alle diese Vorgänge, die sich seit vielen Monaten vor unseren Augen abspielen, wären nicht möglich und erklärlich, wenn wir nicht schon seit Jahren in ein Stadium ministerieller Staatsmedizin hineingerathen wären, wo jede wissen-schaftliche Bestrebung, falls sie nur die Billigung und Zustimmung der zuständi-gen Behörde findet, sofort als staatlich konzessionirtes Dogma verkündet wird'. Article about the death of Ernst Langerhans in the Berliner Neueste Nachrichten, 10 April 1896, BA Berlin, R 86/1182.
24. Actors as part of a network in Latour, *Reassambling*, 2005.
25. Cf. Paul Ehrlich, Hermann Kossel, August Wassermann, 'Ueber Gewinnung und Verwendung des Diptherieheilserums', in: *Deutsche Medicinische Wochenschrift* 20 (1894), pp. 353–5.
26. Cf. Emil Behring and Erich Wernicke, 'Ueber Immunisierung und Heilung von Versuchsthieren bei der Diptherie', in: *Zeitschrift für Hygiene und Infectionskrankheiten* 12 (1892), pp. 10–44.
27. Cf. Ulrike Klöppel, 'Stabilizing the Diphtheria Serum in Germany and France. Regulating its Production', in: Jean-Paul Gaudillière and Volker Hess (eds), *Ways of Regulating: Therapeutic Agents between Plants, Shops and Consulting Rooms (Drug Trajectories V)*, Berlin: Max Planck Institute for the History of Science, 2008, pp. 77–93; on Wernicke see Erika Schulte, *Der Anteil Erich Wernickes an der Entwicklung des Diptherieantitoxins. Eine medizinhistorische Untersuchung zur Entwicklung der Serumtherapie am Beispiel des Diphtherieantitoxins unter Berücksichtigung der Bioergographie des Geheimen Medizinalrates Professor Dr. Erich Wernicke*, Med. Diss., Free University Berlin 2000.
28. Cf. Throm, *Diphtherieserum*, pp. 87–98.
29. The process is described for the company Ruete & Enoch in Hamburg in 1895 in BA Berlin, R 86/1646; and APEI, Vd, No. 4, Vol. 1.
30. The transfer was also noted in the journal for the serum host, see APEI, Dept. Va, No. 1.
31. The draft for the serum host's journal in BA Berlin, R 86/1646; and APEI, Dept. Va, No. 1. The fever curve for horse 'Fritz' in the Emil Behring Archive, Marburg.
32. Cf. Eiermann, 'Einrichtung', 1894.
33. Some reports in the Emil Behring Archive, Marburg, Corr. Farbwerke Hoechst; and GStA PK, HA 1, Rep. 76 VIII B, Nr. 3747 f.; GStA PK, 1. HA, Rep. 76 Vc, Sekt 1, Tit. XI, Part II, No. 18, Vol. 1 f.
34. The public health administration had an interest in the strict surveillance of the production process not only for public health reasons and the minimization of public health risks but also to guarantee that for every quantity of serum pro-duced the test fees were paid.

Unfortunately, cultures of control have mainly been examined in terms of their technical aspects (control of machines or technical systems), cf. Miriam R. Levin (ed.). *Cultures of Control*, Amsterdam: Harwood academic publishers, 2000. Despite Michael Power's focus on accountability in the original sense of the word and his discussion of contemporary financial/economic aspects, his research offers many stimulating ideas concerning the role of audits as rituals of verification, Oxford: Oxford University Press, 1999; Michael Power, *Organized Uncertainty: Designing a World of Risk Management*, Oxford: Oxford University Press, 2007.

35. Cf. Otto, *Prüfung*, 1906.
36. Concerning the attitude and behaviour of German civil servants in Otto Hintze, *Beamtentum und Bürokratie*, Göttingen: Vandenhoeck & Ruprecht, 1981 (OE 1911); Tibor Süle, *Preußische Bürokratietradition. Zur Entwicklung von Verwaltung und Beamtenschaft in Deutschland 1871–1918*, Göttingen: Vandenhoeck & Ruprecht, 1988. Bureaucracy as such was originally installed as an independent, objective power, standing for truthfulness in contrast to aristocratic corruptibility, see Bernhard Wunder, *Geschichte der Bürokratie in Deutschland*, Frankfurt: Suhrkamp, 1986; on the importance and uniformity of bureaucracy, see Max Weber, *Wirtschaft und Gesellschaft. Grundriss der verstehenden Soziologie*, 5th edn. (OE 1922), Tübingen: J. C. B. Mohr, 1980.
37. The standardization of laboratory animals in the context of cancer and genetic research is described in Cheryl A. Logan, 'Before There were Standards: The Role of Test Animals in the Production of Empirical Generality in Physiology', in: *Journal of the History of Biology* 35 (2002), pp. 329–63; Karen A. Rader, *Making Mice. Standardizing Animals for American Biomedical Research, 1900–1955*, Princeton: Princeton University Press, 2004.
38. Paul Ehrlich to Angelo Knorr, 21 February 1897, APEI, Dept. Va, No. 1, Vol. 1.
39. In any case but especially for deviant results, evidence had to be provided to make the process transparent and the products traceable in the whole test procedure.
40. The Imperial Health Office was already in the 1880s known as the most important breeder for laboratory animals, Stenographic Report of the Prussian State House, Session from 16 April 1883, quote from the Prussian Minister for Cultural Affairs von Goßler.
    In the first years the ISRSS had difficulties in acquiring sufficient guinea pigs for the test procedures. As a consequence any available guinea pig was used – even the population of the rural neighbourhood had been asked to breed guinea pigs at the expense of the institute, but the use of inhomogenous laboratory animals caused several other difficulties, Throm, *Diphtherieserum*, 1995. With the ongoing cancer research and the research in chemotherapy in the 1920s the Institute for Experimental Therapy planned its own breeding department, see Alexander von Schwerin, *Experimentalisierung des Menschen. Der Genetiker Hans Nachtsheim und die vergleichende Erbpathologie 1920–1945*, Göttingen: Wallstein, 2004, pp. 145–8; for the US-context Rader, *Making Mice*, 2004.
41. This was criticized by the Viennese Professor Rudolf Kraus around 1908, BA Berlin, R. 86/2711.
42. For the different meanings and problems of model systems cf. 'Introduction', in: Angela N. H. Creager, Elizabeth Lunbeck and M. Norton Wise (eds), *Science without Laws. Model Systems, Cases, Exemplary Narratives*, Durham: Duke University Press, 2007, pp. 1–20.
43. Instruction leaflets for diphtheria serum (Schering and Farbwerke Hoechst) in BA Berlin, R 86/1646 and 1182.
44. Cf. for example the report concerning the experiences with diphtheria serum in Bremen by the Director of the bacteriological institute Dr Kurth, 9 December 1894 and enclosure 3: Anweisung über die Einsendung von Krankheitsstoffen zur bakteriologischen Untersuchung, BA Berlin, R 86/1182. Also on the instruction sheet

for diphtheria serum produced by the Pasteur Institute in Paris (1895) the advice is given to make a bacteriological diagnosis. From June 1895 the physicians in Berlin could send blood samples and throat swabs of persons who had been clinically diagnosed with diphtheria or were suspected of having diphtheria for bacteriological diagnosis to the Institute for Infectious Diseases, the notification about the establishment of an examination centre within the Institute for Infectious Diseases in Nationalzeitung No. 353, 5.6.1895, cf. BA Berlin, R 86/1180. The delivery of tubes for free bacteriological diagnosis with an instruction sheet on how to proceed (as simple as taking a temperature) by the Berliner Waarenhaus. To relieve the diagnostic procedure, the 'Institute for Medical Diagnostic' in Berlin provided – as was common practice in many other cities – tubes for free for the blood samples or throat swab that only had to be sent back to the Institute, the notification in Reichsanzeiger 19.12.1900, cf. BA Berlin, R 86/1180. As a matter of course, while the tubes were given away for free, the physicians had to pay for the diagnosis.

45. Eiermann, 'Einrichtung', 1894.
46. As was done after the death of Ernst Langerhans, cf. Ehrlich, 'Controle', 1896; Hüntelmann, 'Diphtherie-Serum', 2007.
47. As announced in Ehrlich, 'Controle', 1896. After the method of evaluation had been improved around 1897/1898 and vacuum dried standard serum was taken as the basis for the scale – cf. Paul Ehrlich, 'Die Wertbemessung des Diphtherieheilserums und deren theoretische Grundlagen', in: *Klinisches Jahrbuch* 6 (1898), pp. 299–326 – Richard Otto compared the standard serum stocked in the ISRSS with the prototype metre held in Paris, Otto, *Prüfung*, 1906, p. 28.
48. Eiermann, 'Einrichtung', 1894.
49. The test procedure constantly repeated the experimental path that had led to the successful invention of diphtheria serum that Behring and Ehrlich had followed on their own – with the difference that the test procedure represented the successful outcome ignoring the research work and failures of several years. The instructions concerning the evaluation of serum had to be followed exactly. With the steady iteration of the successful development their names were commemorated with every test procedure and via the technical arrangement inscribed in the collective memory of the scientific community.
50. The pipettes and other devices had been developed in cooperation with the instrument maker Lautenschläger and Co., cf. Otto, *Prüfung*, 1906, figures 1, 4 and 5; for the preparation of the test toxin see also Throm, *Diphtherieserum*, 1995.
51. Davis Baird, *Thing Knowledge: A Philosophy of Scientific Instruments*, Berkeley: University of California Press, 2004 describes measuring instruments as encapsulated knowledge. In the context of evaluation I would enlarge the explanation of encapsulated knowledge on measuring procedures.
52. The owner of the serum producing company Ruete and Enoch, Carl Enoch, was instructed after problems concerning the procedure of evaluation emerged, see for example the correspondence in August 1896 between Carl Enoch and Wilhelm Dönitz, member of the ISRSS, APEI, Dept. Vd, No. 4, Vol. 1. Later on, this instruction was institutionalized in the guidelines as a voluntary training course in the laboratory of the ISRSS, see APEI, Dept. III, No. 6, Vol. 1.
53. This emanates from the correspondence between the ISRSS and the company Merck in autumn 1896/winter 1896, APEI, Dept. Vd, No. 3, Vol. 1.
54. For the importance of working knowledge Baird, *Thing Knowledge*, 2004; examples in Hüntelmann, 'Dynamics', 2008.
55. Cf. the connection between information management and (working as well as encapsulated) knowledge (in a contemporary context) Max H. Boisot, *Knowlege*

Assets. Securing Competitive Advantage in the Information Economy, Oxford: Oxford University Press, 1998.

56. The drafts for the different journals in APEI, Dept. Va, No. 1, Vol. 1; BA Berlin, R 86/1646; and completed journals in the production plant for the period after 1914 in the Archive of the Behringwerke, Marburg.

57. Hüntelmann, 'Diphtherie-Serum', 2006.

58. Current discussions about virtual space mainly concern the Internet or geography. The virtual space of control represents not only two connected geographical spaces but also a medial space of information, in this respect we have used Martina Löw, *Raumsoziologie*, Frankfurt am Main: Suhrkamp, 2001, Chap. 5; Markus Schroer, Räume, Orte, Grenzen. *Auf dem Weg zu einer Soziologie des Raums*, Frankfurt am Main: Suhrkamp, 2006, pp. 252–76; Barney Warf and Santa Arias (eds), *The Spatial Turn: Interdisciplinary Perspectives*, New York: Routledge, 2008; Jörg Döring and Tristan Thielmann (eds), *Spatial Turn: Das Raumparadigma in den Kultur- und Sozialwissenschaften*, Bielefeld: transcript, 2008.

59. Mary Douglas, *How Institutions Think*, Syracuse: Syracuse University Press, 1986, p. 49.

60. Eiermann, 'Einrichtung', 1894; Throm, *Diphtherieserum*, 1995; the files of Hoechst in the APEI, Dept. Vd, No. 1, Vol. 1. As a result of their early experiences in the development and large-scale production of diphtheria serum the Dye Industry Hoechst was also involved in the considerations and consultations about the state regulation of diphtheria serum between November 1894 and February 1895.

61. Cf. the files with the protocols and memos in BA Berlin, R 86/1646; GStA PK, HA 1, Rep. 76 VIII B, No. 3747–9. The process is described in detail in Axel C. Hüntelmann, *Hygiene im Namen des Staates: Das Reichsgesundheitsamt 1876–1933*, Göttingen: Wallstein, 2008, Chap. 3.

62. Cf. for the implementation of the food act and the difficulties that arose with the adjustment of a system of control Jutta Grüne, *Anfänge staatlicher Lebensmittelüberwachung in Deutschland. Der 'Vater der Lebensmittelchemie' Joseph König (1843–1930)*, Stuttgart: Franz Steiner, 1996.

63. Cf. David Cahan, *An Institute for an Empire. The Physikalisch-Technische Reichsanstalt 1871–1918*, Cambridge: Cambridge University Press, 1989.

64. Cf. Hickel, *Arzneimittel-Standardisierung*, 1973; Wimmer, *Neues*, 1994.

65. Christoph Gradmann, *Krankheit im Labor: Robert Koch und die medizinische Bakteriologie*, Göttingen: Wallstein, 2005.

66. According to Barbara Elkeles, *Der moralische Diskurs über das medizinische Menschenexperiment im 19. Jahrhundert*, Stuttgart: Gustav Fischer, 1996, the Tuberculin affair influenced the development of diphtheria serum in that over nearly four years countless series of animal and clinical tests were made and discussed by scientists before the remedy was sold in the pharmacies.

67. Cf. the results of a statistical survey concerning the practical experiences with diphtheria serum, 'Veröffentlichungen aus dem Kaiserlichen Gesundheitsamte' published and summarized as 'Ergebnisse der Sammelforschung über das Diphtherieheilserum für die Zeit vom April 1895 bis März 1896', in: *Arbeiten aus dem Kaiserlichen Gesundheitsamte* 13 (1897), pp. 254–92.

68. Hans-Jörg Rheinberger, *Experiment – Differenz – Schrift. Zur Geschichte epistemischer Dinge*, Marburg: Basilisken-Presse, 1992.

69. Hartmut Häußermann, *Die Politik der Bürokratie. Einführung in die Soziologie der staatlichen Verwaltung*, Frankfurt am Main: Campus, 1977.

70. These dynamics for the period after 1896 are described in Hüntelmann, 'Dynamics', 2008.

3
From Diphtheria to Tetanus: The Development of Evaluation Methods for Sera in Imperial Germany

Anne I. Hardy

The rapid establishment of a system of control for anti-diphtheria serum in Imperial Germany in 1895 is often seen as a result of the Prussian preference for administrative solutions. Yet, diphtheria was the exception until, in 1903, there was a second official regulation covering sera against tetanus and erysipelas. The anti-tetanus serum had been controlled on a voluntary basis from 1896 onwards, anti-erysipelas serum and tuberculin starting in 1899. Although the evaluation of anti-tetanus serum was 'definitive and obligatory'[1] – in contrast to sera that were controlled on a 'facultative and provisory' basis – it was not integrated into the pharmacopoeia until 1910, after years of negotiation. At first glance, this might seem surprising, as the possibility of curing animals suffering from diphtheria or tetanus by serotherapy was demonstrated by Behring and Kitasato in the same paper published in 1890. Judging by their publications, Behring, Ehrlich and other immunologists worked predominantly on tetanus until 1894, when they began to concentrate on anti-diphtheria serum.

I will argue that the main reasons for this shift were financial resources, the importance of diphtheria for public health, and the inconclusive results of clinical trials. Behring, Ehrlich and their co-workers needed the support of the pharmaceutical industry if they wanted to produce enough serum for clinical evaluation, and anti-diphtheria serum promised to be profitable since nearly every third infant death was due to diphtheria.[2] By contrast, tetanus infections occurred relatively seldom, and their most common victims were gardeners, grooms and wounded soldiers. The more marginal epidemiology of the disease explains why scientists, industrialists and state authorities did not collaborate nearly as intensively over the tetanus serum as they had for the diphtheria antitoxin. In principle, the legislation for the diphtheria serum could have been extended to tetanus and tuberculin, but the state authorities had little interest in going through the same time-consuming procedures once again. The main economic importance of the tetanus-antitoxin lay in veterinary medicine where it was used for the treatment of horses, but even here, the cure seemed less effective than with the serum against diphtheria. For these reasons the production of anti-tetanus serum was less attractive for the industry, and, in 1896, the Hoechster Farbwerke in Frankfurt was the first and only company to produce anti-tetanus serum in Imperial Germany. Two years earlier the chemical factory Merck in Darmstadt had begun to sell serum produced by Tizzoni and Cattani in Italy.

The fact that the tetanus-antitoxin played a minor role until WWI, when it was widely used for the (passive) immunization of wounded soldiers, may be the reason why it has generally been neglected by historians. But look-ing at the case of tetanus sheds new light on the development of evaluation methods for sera and vaccines in Imperial Germany. Additionally, a study of the early research on tetanus reveals that Behring as well as Ehrlich had done intensive work in the field before they began joint research on the anti-diphtheria serum in 1893/4. In a first section I want to discuss the influ-ence of this early work on the efforts to obtain an anti-diphtheria serum of sufficient potency for clinical application. I will then describe the problems that arose in fixing standards and establishing an evaluation method.

This process was more complicated than in the case of the anti-diphtheria serum since from 1895 on, Behring and Ehrlich followed different career paths. Behring was appointed professor in Marburg, although his research continued to be supported by Hoechst, for whom he produced the anti-tetanus serum. Ehrlich pursued his scientific career in the 'Control Station' of Robert Koch's 'Institute for Infectious Diseases' and later in his own 'Institute for Serum Research and Evaluation', enjoying the status of a civil servant answerable to the Prussian Health Ministry. This divergence in career paths created tensions and rivalry that hindered a quick settlement of the state control of anti-tetanus serum, but the state authorities had little interest in entering this arena (section three) until they began to consider introducing government control for a variety of other sera applied in human and veterinary medicine. In the fourth section I develop my thesis that the anti-tetanus serum played a significant role in the establishment of Paul Ehrlich's institute as a national institution for the evaluation of sera and vac-cines in general. Finally, section five takes a critical look at the significance of evaluation for clinical practice.

Early research on the anti-diphtheria and anti-tetanus serum

On 4 December 1890, Behring and Kitasato reported in the *Deutsche Medizinische Wochenschrift* on a specific therapy for tetanus and diphtheria. Both diseases had been treated by the transfusion of blood serum taken from animals previously immunized with the corresponding toxins. They suggested that the effect was due to antitoxic substances capable of neutral-izing the toxins produced by the diphtheria or the tetanus germs.[3] Behring was confident that serotherapy could be applied to a variety of infectious diseases, but according to his own account he concentrated on tetanus first because Kitasato had already done a lot of work in this field.[4] Moreover, tetanus was ideal for study in the laboratory: the symptoms were easy to recognize, the tetanus toxin was more accessible and potent than the diphtheria toxin, experiments on the white mice used for tetanus research were cheaper than diphtheria experiments on guinea pigs, and finally Behring was more successful in immunizing animals against tetanus than against diphtheria.[5] This is why between 1890 and 1893 most research was done on tetanus.

Another important aspect was funding. Behring quickly recognized that to extend his therapy to humans he needed large quantities of serum, but Koch's *Institut für Infektionskrankheiten* could not afford to keep large animals capable of producing serum in sufficient quantities. In August 1891, Behring's assistant Erich Wernicke bought several sheep on his own account and started keeping them at his home near Berlin.[6] With these animals, Behring and Wernicke made their first attempts to immunize larger animals against diphtheria, but they soon had to face the fact that their private means were insufficient for the larger scale experiments they wanted to undertake. The first opportunity to work with a larger group of horses was offered to Behring a few months later, when the veterinary school of the Prussian state, directed by Professor Schütz, took an interest in serotherapy. The Ministry for Agriculture had given 10,000 Marks for research on tetanus because it frequently affected domesticated animals.[7] It was agreed that Behring should provide the tetanus toxins and Schütz the horses. After a long and laborious process and more than 100 bleedings, Behring succeeded in obtaining serum whose immunizing power he estimated to be $1:10^6$.[8]

This initial work with horses represented a partial success for Behring, which he needed badly after having suffered an embarrassing setback with the public demonstration of his anti-diphtheria serum in December 1891. In order to persuade a larger circle of practitioners, he had infected guinea pigs with diphtheria cultures under the eyes of Professor von Bergmann and his staff in the surgical department of the university hospital in Berlin. Half of the guinea pigs were also given anti-diphtheria serum, but in the following days not one animal showed any symptoms of diphtheria. Although Koch came to Behring's aid by explaining that the cultures had probably suffered from the low winter temperatures during the transport from his institute to the hospital, the clinicians had lost confidence in serotherapy.[9] It was more than a year later that Behring made another such attempt – this time with tetanus-infected mice which he presented in Emil du Bois-Reymond's Institute for Physiology.[10] This time, he succeeded in demonstrating the therapeutic effect of the serum. In the same month, Behring reported on extensive experiments on the evaluation of the curative qualities of the serum he had undertaken with his co-worker Angelo Knorr. As it was very difficult to obtain reproducible results in animal experiments,[11] they chose one particular serum as a standard and attributed it the value of 1. Conscious of the fact that results from different laboratories were only comparable with reference to a fixed standard, Behring suggested that other laboratories send him their sera for evaluation.[12] At the same time, he tried to define a standard for the anti-diphtheria serum as well.[13]

Behring was keen to establish his claims to priority in the area of sera, since he was not the only one working in the field. Paul Ehrlich, who had joined Koch's research team at the *Institut für Infektionskrankheiten* in 1891, had carried out a very significant study on the reaction of mice with respect to the vegetable poisons Abrin and Ricin.[14] Ehrlich worked with these toxins rather than those produced by bacteria because they were easier to dose with precision, and his aim was to describe the immune reaction of the body 'in a more mathematical way'.[15] In the beginning, Behring happily acknowledged this work because it supported his own hypothesis of specific immune reactions, but Ehrlich soon turned into a serious competitor. Shifting his attention to tetanus, Ehrlich had begun to study the transmission of antibodies to maternal milk – first in mice[16] and then goats.[17] Because milk was more accessible than blood, Ehrlich could determine the antitoxin concentration frequently – every second day or even daily. In detailed graphical representations, he demonstrated, together with Ludwig Brieger, that the number of antitoxins in maternal milk dropped shortly after the injection of tetanus-toxin only to increase dramatically a few days later. Having passed a maximum, the level decreased again until it reached a plateau at a level superior to the one prior to the injection of the toxin. Ehrlich's explanation of the initial phenomenon was that the existing antitoxins were used up neutralizing the injected toxins. The later increase was due to an immune reaction that overcompensated for the initial loss. Based on these detailed observations, Ehrlich could deduce how to augment the immunity of his goats most effectively by injecting doses of increasing toxicity. The ideal time for re-injecting was just before the concentration of antitoxins in the body had reached its maximum.[18]

In this way, Ehrlich had solved a problem for the tetanus-antitoxin that Behring was still struggling with for the diptheria-antitoxin. Furthermore, Behring was now under pressure from Hoechst who wanted to see a return on its investments, particularly as his first clinical trials had not been convincing because the serum was not potent enough. Not willing to accept that Ehrlich had gone further in obtaining high quality sera by using his elaborate immunization method, Behring accused him of having stolen his intellectual property. In return, Ehrlich, who had always given credit to Behring for the discovery of serotherapy, insisted on his priority in developing an effective immunization method;[19] 'the date of our publication proves that we began our studies on the subject long before Behring's publication appeared'.[20] As Ehrlich claimed that he could produce very potent anti-diphtheria serum by the same immunization method, Behring – possibly under pressure from Hoechst – consented to clinical trials using Ehrlich's serum.[21] In October 1893, both agreed to work together on the production of anti-diphtheria serum,[22] and five months later, in March 1894, Ehrlich also signed a contract with Hoechst, awarding him 75 per cent of the profits from the serum he produced for the company.[23] Shortly afterwards, Ehrlich, Hermann Kossel and August Wassermann reported on 220 cases of diphtheria treated with their serum, with 76 per cent of the patients being cured.[24] This was the first clinical trial considered to be successful and marked the turning point in the research on the diphtheria serum. From this point on, things moved very quickly. In 1894, Schering as well as Höechst put their first commercial anti-diphtheria sera on the market and in 1895 the quality control by the 'Control Station' of Koch's institute was made mandatory.

In September 1895, when Behring was invited to the annual meeting of the Society of German Natural Scientists and Physicians (GDNÄ)[25] in Lübeck, he mainly spoke about diphtheria. With respect to tetanus, he mentioned that his co-worker, Angelo Knorr, had made considerable progress in producing high-potency sera, so 'that the treatment of tetanus promises to enter a new stage. But, as on the one hand the disease occurs only seldom and on the other hand the remedy is expensive to produce, I doubt that the industry will be prepared to undertake considerable work on it'.[26]

The problem of standards and evaluation methods

In October 1896, Paul Ehrlich, now director of the Institute for Serum Research and Serum Evaluation, received a letter from the Hoechster Farbwerke containing instructions for the use of its new tetanus antitoxin which was to be put on the market within the next week. To his surprise, Ehrlich read that the new serum was controlled by his institute according to an evaluation method developed by Behring and Knorr.[27] Ehrlich immediately sent a cable saying that Hoechst must first make a formal request to the Prussian Ministry for Medical Affairs.[28] In December, he was officially authorized to control the tetanus serum produced by Hoechst.[29]

The main reason that Hoechst sought the seal of the Institute for Serum Research and Serum Evaluation was that it increased public confidence in the quality of the product.[30] By this means they hoped to gain an advantage over the Italian serum that had been imported and sold by Merck since 1894. To the satisfaction of Hoechst and Behring a comparative evaluation of the competing anti-tetanus sera by Ehrlich clearly favoured the German product.[31]

As for the creation of standards and evaluation methods, the situation was complicated. After being given a chair in Marburg in 1894, Behring had continued his research for Hoechst and taken over the production of the anti-tetanus serum. Angelo Knorr, Behring's assistant, developed an evaluation method for anti-tetanus serum in 1895,[32] based on the same principle as the evaluation method for anti-diphtheria serum – mixing toxin and antitoxin before injecting the mixture into the animal. In his work, Knorr defined antitoxin and toxin standards which were subsequently used for evaluation in Ehrlich's institute. Nevertheless it was Behring who provided Hoechst and Ehrlich with the test toxin and test antitoxin needed for evaluation.[33] Problems arose in 1899, when a controversy over the 'correct' standard developed between Ehrlich and Behring. The controversy began when the supplies of test toxin and serum in Ehrlich's institute were depleted in April 1898. Following a request, Ehrlich was sent a new test antitoxin by Behring's assistant, Ransom.[34] He carefully compared it with his old test antitoxin before using it as a new reference for his evaluations.[35] It is interesting to note that by this time Ehrlich had already given up the use of a test toxin as a reference. His experience of the diphtheria-toxin's instability, complemented by intensive research on the patterns of decay, had led him to conclude that they were not suitable as standards.[36] But the evaluations in Marburg still referred to a tetanus toxin (labelled No. 2), which Ransom had also sent to Ehrlich and the Hoechst Company the year before. Trouble began when the toxin content of the test-toxin kept in Marburg started to diminish without Behring or Ransom noticing.

In October and November 1899, there was an extended correspondence between Ehrlich and Behring about the marked differences in their evaluations. Behring had sent an anti-tetanus serum to Ehrlich which he considered ten-fold, but Ehrlich determined it to be only five-fold.[37] For Behring and Hoechst this difference was of considerable economic importance; the more potent the serum the more profit could be made from its sale. Similar differences in evaluation had already occurred the year before, and in the period from October to December 1898, Hoechst had even refrained from voluntary control.[38] But Ehrlich had complained about the inferior quality of the serum in a letter to the Prussian Minister for Educational and Medical Affairs and in January 1899 the control was re-instated.[39]

What was the origin of these differences in evaluation? As Ehrlich had supposed and Behring later confirmed, the test toxin in Marburg had diminished in strength by 50 per cent.[40] As a response, Behring produced a new test-toxin (labelled No. 5) which he declared to be very stable. Ehrlich proposed to evaluate this new toxin with reference to the antitoxin he preserved in his institute, following the reverse procedure used for the evaluation of anti-diphtheria serum. Ehrlich had developed an elaborate conservation method to keep the standard antitoxin stable,[41] but Behring was not willing to admit that his own standard had changed, and suspected that the error arose because Ransom had evaluated the antitoxin sent to Ehrlich using the decaying test toxin No. 2.[42] It was only later that Ehrlich was able to prove that Ransom's evaluation had been done before the toxin had started decaying.[43] In any case, Behring thought it necessary to re-establish the standard by creating a new test antitoxin which he labelled No. 60. This and the test toxin No. 5 were sent to Ehrlich to serve as the new basis for evaluation.[44] Ehrlich avoided a discussion about who was to fix the standard for anti-tetanus serum, but insisted that if a new standard was to be introduced its relation to Knorr's 'old' unit had to be determined.[45] Behring for his part declared that the unit defined on the basis of his new substances was exactly the same as Knorr's. This unit, also called the Marburg antitoxin unit, was to be used from then on in the scientific literature. As far as the evaluation results were concerned, Behring was prepared to retain Ehrlich's unit (the Frankfurt antitoxin unit).[46] For some time, the potency of the serum on the labels of their phials was expressed in both units, a solution that caused widespread confusion.[47]

Throughout the controversy, Ehrlich remained convinced that his standard was congruent with the one defined by Knorr but, because he was in the middle of moving his institute from Berlin to Frankfurt, he could not prove it by making any new measurements. After he was able to resume his scientific work, Ehrlich and his co-worker, Dönitz, carried out extensive studies to compare the two antitoxins in a series of parallel animal experiments (one series for each antitoxin). Within a series, the animals were all given the same doses of toxin but the amount of antitoxin being mixed with it was gradually increased. This arrangement had the advantage that the results were independent of the absolute value of the test toxin. The behaviour of the animals ranged from 'dying from tetanus within a few days', to 'showing tetanic symptoms', all the way to 'staying entirely healthy'. As the animals in both series reacted in exactly the same way, Ehrlich was convinced that the antitoxins being compared were of equal potency. He ended his letter with the statement:

> I must therefore declare most formally – also in the name of Privy Councillor Dönitz – that the serum we use has an effective value of 50 Knorr antitoxin units.
> That the original preparation which you now possess differs from this, is weaker, cannot be doubted – but this can only be due to a diminishment.
> I simply emphasize once again that in all our examinations we have strived most strenuously to comply with your requests. However, it is necessary that we keep to the numbers we have in our books.[48]

It appears that Behring paid little attention to these arguments. In January 1900, he described his evaluation method based on the new test substances in the *Deutsche Medizinische Wochenschrift*, and announced that the test antitoxin No. 60 and test toxin No. 5 were now sold by Hoechst.[49] Thus, Behring tried to force his new definition of the standard upon Ehrlich.

The administrative side of the problem

Shortly after the Hoechster Farbwerke had applied for the optional evaluation of their anti-tetanus serum in 1896, Ehrlich wrote to the Prussian minister for

Medical Affairs to recommend the introduction of such voluntary controls. In his opinion, the anti-tetanus serotherapy had become mature enough for the treatment of humans, and appropriate state control could guarantee that only potent and safe products were administered.[50] Before reaching a decision, the Ministry asked for more information on the immunizing and curative powers of the serum.[51] In a second letter, Ehrlich explained that French experiments by Edmond Nocard on horses and those of Professor Netter on humans in a Paris hospital had proved that the serum was quite effective as a form of (passive) immunization. As for the curative success, he admitted that it was difficult to judge due to the specific character of the tetanus toxins. He explained that once the toxins had begun to damage the nervous system their action was difficult to neutralize by the injection of antitoxins. Moreover, it seemed as if more potent sera were needed for the treatment of tetanus than for diphtheria. Ehrlich's most important argument for state control was that according to his evaluation the anti-tetanus serum provided by Tizzoni (and sold by Merck) was about 12 times weaker than the serum produced by Hoechst. He ended his letter with a patriotic appeal:

> It would be a pity if Germany, the country in which Behring's serotherapy was discovered, lagged behind other countries in its practical application.[52]

The Prussian Minister consented to the voluntary control of anti-tetanus serum in December 1896. The next step towards legislation for anti-tetanus serum came in May 1898, when the Prussian Ministry for Medical Affairs asked Ehrlich if it was advisable to place it on the list of drugs which were to be sold only in pharmacies. Before reaching a decision, the minister wanted Ehrlich to compare the different sera on the German market,[53] covering the serum from Hoechst, Merck and a German Pasteur Institute in Stuttgart. In August 1898, Ehrlich reported that the quality of these sera was very diverse, estimating Behring's serum at Hoechst as 100-fold, the Pasteur serum as only 12-fold, and Tizzoni's sold by Merck as no more than 7-fold. This meant that for a curative dose the patients had to be given as much as 700 cubic centimetres of the Tizzoni serum. The difference in the prices was even more striking; for a curative dose of 500 antitoxin units (Behring's recommendation) Hoechst charged 30 Marks, Pasteur in Stuttgart 280 Marks, and Merck as much as 370 Marks for the Tizzoni serum. Ehrlich offered the following commentary:

> If the sera from Hoechst and from foreign countries are considered equivalent in the treatment of tetanus, there is not only a lack of a basis for correct statistics, but any assessment is misleading.[54]

Ehrlich therefore requested that the sale of anti-tetanus serum be restricted to pharmacies and that the product be controlled by the state. The Prussian

Minister passed this letter on to the Imperial Health Office and for almost a year nothing happened. In July 1899, it became clear why the Imperial Health Office had waited such a long time to answer. First, the question arose as to whether all sera on the German market, and not just the anti-tetanus serum, should be sold in pharmacies and controlled by the state. This question led on to the more practical one of whether it was possible to develop evaluation methods for all the existing sera. Before undertaking a long and complicated administrative process, Köhler, the director of the Imperial Health Office, wanted to make sure that it was worth making the effort. He therefore asked Ehrlich how much serum was sold and what quantities were needed for the treatment of human disease.[55]

In his answer, Ehrlich repeated the arguments he had given the Prussian Minister a year earlier concerning the considerable differences in the potency and prices of the different sera. He then discussed in detail the sera for which an appropriate assay could be found, envisaging no problem for any of the antitoxic sera (against diphtheria, tetanus, snake poison and 'Wurstgift' or food poisoning, literally 'poison in a sausage').

The precondition for the production of these sera is that one can procure the relevant poisons in sufficient quantity and toxicity. In these cases, it is always possible, with the help of the poison, to determine the value of the sera obtained.[56]

Ehrlich saw no problem with evaluating the sera against typhus and cholera, for they were caused by well-known microorganisms; but he was sceptical about the case of streptococcal infections which involved more than one pathogenic strain. Thus, the treatment of this disease would require the production of a 'polyvalent' serum by immunizing animals against the different strains. Only then could the value of such a serum be determined by animal experiments. Finally for the diseases with no known cause, such as scarlet fever, measles or syphilis, Ehrlich recommended imposing control procedures on the producers rather than on the serum they produced. For the anti-tetanus serum, he made it clear that even though the demand was much smaller than for the anti-diphtheria serum this was not sufficient reason to leave it uncontrolled. In his opinion, the state had to guarantee an effective treatment even if the disease occurred relatively seldom.[57] As Ehrlich was later informed by the Imperial Health Office this letter 'started the ball rolling',[58] leading the Imperial Ministry of the Interior to convene a commission to debate the various questions raised by Ehrlich, who was in turn asked to submit his suggestions.

First, Ehrlich argued that the serum-producing factories had to be controlled as was the case for the production of the anti-diphtheria serum, before going on to discuss whether the authorities should differentiate between sera for use in humans and sera for use in veterinary medicine. In 1896, the Berlin Veterinary School had reported good curative results in the treatment of horses with tetanus, but by 1897 the veterinarians complained that the

potency of the serum had considerably declined.[59] Ehrlich thought it necessary to check the serum, evaluating it using the method developed by Knorr,
which he had refined, and which involved performing two parallel series of
experiments with eight white mice in each. He pointed out that this method
was more reliable than the evaluation method employed by Behring, and
insisted that the antitoxin had to be used as a standard and that it was to be
conserved in evacuated glass tubes like the diphtheria antitoxin standard.
As a concession to Behring's work, Ehrlich proposed using the Marburg test
antitoxin No. 60 as the antitoxin standard.[60] These suggestions were sent to
the Ministry in February 1900.

Which sera and vaccines should be controlled and by whom?

The fact that Ehrlich underlined his competence in the technical and scientific aspects of the evaluation procedure in his letter to the Ministry of
Medical Matters is understandable, as Behring had recently hinted at his
own desire to take control of the anti-tetanus serum. Early in January 1900,
when Ehrlich had asked Hoechst about their wishes concerning the draft
law, he was told to direct his questions to Behring, as Hoechst was only
responsible for serum sales. In his reply to Ehrlich, Behring complained that
his advice had not been solicited in the past when important modifications
of the state control for anti-diphtheria serum had been decided, and insisted
that the matter be discussed in personal negotiations between experts:

> I want especially to point out that following my experience I have serious
> reservations about simply transferring the protocols for the control of the
> anti-diphtheria serum to the anti-tetanus serum. A discussion of these
> reservations and a sufficient explanation is only possible in personal
> negotiations, in which we should consider, among other things, to whom
> the protocols for the control of anti-tetanus serum should be assigned.[61]

Clearly Behring thought that he was the right person for this job, although
from the point of view of the administrative procedure, granting the control
of anti-tetanus serum to Behring would have meant that not all the different
sera would be treated in the same way. The alternative solution was to put
Ehrlich's institute in charge of the state control and to establish a general procedure for the evaluation of sera. This solution seems to be what the experts
had in mind when, in November 1900, they negotiated which sera were to be
submitted to state control as a precondition for their being sold in pharmacies. Köhler, the President of the Imperial Health Office, opened the meeting
by informing the delegates that the sera for use in human medicine had been
included in the draft of a new imperial decree (section B), which meant that
in future they were only to be sold in pharmacies. 'Therefore, the question

is whether these sera should all or only partly be submitted to state control and how this control should be exercised'.[62] Sera for veterinary medicine and vaccines were also to be included in the negotiations. Hermann Kossel, who had joined the Imperial Health Office in 1899, discussed the pros and cons of the introduction of state control for these various biological therapeutic and prophylactic products. In support of control, he argued that it could prevent the sale of useless or harmful products, and against, he emphasized the fact that evaluation methods did not exist for all sera, and that prescriptions could hinder the exploitation of newly discovered remedies.

It was Richard Pfeiffer, Professor of hygiene in Königsberg, who made the proposition that would later be adopted, demanding that the inventor of any such product be obliged to specify its properties. In the case of drugs that could not be evaluated using animal experiments, the producer was required to prove their harmlessness. With respect to the public authority that was to be charged with the control of the products, Pfeiffer thought it suitable to assign the task to Ehrlich's institute in Frankfurt.[63] Behring was not happy with Pfeiffer's intervention, as he had expected that this question would be resolved by a representative of the Prussian Ministry for Medical Affairs.[64] In contradiction with the historical facts, Behring declared that in 1896 it had been impossible to find an institution which considered itself competent to control his anti-tetanus serum.[65] Nevertheless it seems clear that Behring was the only one present who thought that someone other than Ehrlich would be assigned the task, since the subject was dropped in the course of the discussion.

In the end, the recommendation was that in a first wave the sera against tetanus, erysipelas, swine flu, swine fever, avian cholera and foot-and-mouth disease should be submitted to mandatory control, to be followed later by the sera against cholera, typhoid fever and the plague. The therapeutic value of this second list of products had still to be proved, but it was nevertheless considered useful to determine their potency in case they were needed to combat epidemics. Schütz from the Veterinary School in Berlin was asked to draw up a protocol for the evaluation of swine flu, swine fever and avian cholera, while Ehrlich was put in charge of the treatments for the other infectious diseases.[66] It took another three years for the law requiring the control of anti-tetanus serum to be passed, and the requirement to be sold in pharmacies only came into effect seven years later.

Indeed, in January 1910, it was Ehrlich himself who reminded the Ministry of Medical Affairs that the conditions for the sale of anti-tetanus sera in pharmacies, as well as the prices, had not yet been fixed.[67] These questions were settled a month later during a meeting at the Ministry in Berlin attended by representatives of the Hoechster Farbwerke and the Behringwerke (founded by Behring in 1903). Merck, who continued to sell imported Italian serum, was not invited to participate. Since it was difficult to produce sera of high potency, it was agreed that the anti-tetanus serum should be at least four-fold and that it should be free from carbolic acid.[68] Finally, in May 1910, a Prussian Ministry official, Martin Kirchner, informed the district presidents of the German states that not only did anti-tetanus serum have to be submitted to state control, but it was also to be sold exclusively in pharmacies upon presentation of a prescription from a physician or veterinarian.[69] Paradoxically, Merck continued selling the Tizzoni serum which could not be submitted to German state control because it was produced in Italy, although now it was sold exclusively by Merck's pharmacy in Darmstadt. Merck stopped importing Italian serum only in 1915, probably because it was then needed by the Italian army. Starting in 1912, Merck became a distributor for Behring's anti-tetanus serum as well,[70] probably because Merck's worldwide distribution network could not be matched by a small enterprise like the Behringwerke in Marburg.

The significance of evaluation for clinical practice

Ehrlich's main arguments for the introduction of state control had been that the sera on the market had to be comparable if one wanted to correctly judge the effect of the treatment. But from the clinician's point of view the situation looked quite different. In August 1897, Max Engelmann, from the medical clinic in Leipzig, published a review of all the cases between 1891 and 1895 in which the anti-tetanus serum had been used. He reported very positive results from the Italian serum,[71] but the serum produced by Behring had problems. Unlike the Tizzoni serum, its potency varied and it was generally less potent. This assessment dated from before Hoechst put the serum on the market and had it tested by Ehrlich. Only after Behring had succeeded in developing a more potent (100-fold) dry anti-tetanus serum, which was commercialized in 1896,[72] was the German serum considered equivalent to the Italian product.[73] This result seems rather puzzling as Ehrlich had shown in his own comparison that the Tizzoni serum was weaker. This paradox may be in part due to the fact that – in contrast to the treatment of diphtheria – clinicians did not stick to a fixed dose of the serum. Tizzoni recommended injecting half the contents of one of its phials and then distributing the rest over four doses to be given in the following days. But in severe cases one was to start with the whole phial and continue with several high doses until the patient felt better or died.[74] Since adverse reactions to tetanus-antitoxin were very rare, the administration of such high quantities of serum was possible.

The veterinarians limited themselves to Behring's antitoxin and simply followed the accompanying instructions to inject the contents of the phial in one curative dose. This regimen also reflected the economics of veterinary medicine, since 35 Marks per dose was considered rather a high price to pay for treating a horse.[75] Professor Dieckerhoff from the medical clinic of the Veterinary School in Berlin, who treated four horses in November and December 1896, initially formed a favourable opinion of the new treatment.[76]

One year later, however, his assistant Dr E. Brass complained that the quality of the serum had very much declined and that he had been unable to save a single horse. He attributed the failure to a new, weaker serum that Hoechst was producing for veterinary use. Since it was difficult to obtain high potency anti-tetanus serum, the company thought it advisable to produce a weaker but cheaper serum that could be administered in larger quantities. The Prussian Army experienced similar problems when it introduced serum therapy for its horses in 1897. Unable to decrease tetanus mortality among the horses, 'Oberroßarzt' König put an end to the experimental initiative.[77]

Interestingly a second review of the use of anti-tetanus serum in human medicine, conducted by F. Köhler from the medical clinic in Jena in November 1898, reached less positive conclusions than the first. Comparing his own data with those of Engelmann, Köhler found that fewer tetanus patients had been cured by serotherapy during the preceding two years.[78] One may suspect that this was because Köhler also took into account patients treated with weaker serum produced by Pasteur and different British laboratories. This had been one of the theories to explain why serotherapy for diphtheria had originally been less successful in England than on the continent.[79] Limiting the analysis to cases treated with the Tizzoni and Behring sera, the mortality of tetanus did not appear to change significantly due to the serotherapy. Furthermore, Köhler had been unable to correlate the amount of antitoxin administered with the progression of the disease. Even treatment at an early stage, which was essential for the successful treatment of diphtheria, did not give any better results.[80]

In 1908, after having debated the issue of which sera should be introduced into military medicine, the Kaiser-Wilhelms-Akademie proposed the interaction between the tetanus-toxins and the nervous system as a possible explanation for the lack of clinical response: 'The action of the antitoxins probably fails because the toxin moves through the nerves into the central nervous system where it is virtually isolated from the antitoxins that circulate in the whole body'.[81] The experts saw little value in using anti-tetanus serum for curative purposes, but, given the fact that no other treatment was available, it was agreed that prophylactic injections should be given to soldiers with soiled wounds. A similar conclusion was reached by the experts discussing the state control of the anti-tetanus serum in February 1910. While they accepted its protective power, the curative value remained doubtful.[82] In conclusion, the evaluation of the anti-tetanus serum seems not to have been a very urgent matter from the clinical point of view because there was no evident correlation between its potency and its curative value.

Conclusion

The standardization and introduction of state control for the anti-diphtheria serum has usually been taken as the only model for the evaluation of subsequent biological pharmaceuticals.[83] But the comparison with the case of the anti-tetanus serum reveals that the diphtheria model is valid only in terms of the general prescription for evaluation. Looking more closely, we can see that there were many differences between the cases of the anti-tetanus and the anti-diphtheria sera. Since tetanus was of minor importance for public health before WWI, neither the pharmaceutical industry nor the state authorities had much interest in creating laws and procedures for its systematic control, and so it was left up to Behring and Ehrlich to make informal agreements, leading to conflicts following their divergent careers and increasingly divergent interests. While Behring, the pharmaceutical entrepreneur, wanted serum production to be as profitable as possible, Ehrlich was more concerned with its quality.

As I have suggested, the discussions about the evaluation of anti-tetanus serum also served as a model in the sense that they initiated a process to find general regulations for the evaluation of sera and vaccines, not only for human but also for veterinary medicine. Most of these subsequent biological pharmaceuticals were even less effective than the anti-tetanus serum, so that a provisional and voluntary system of control was established for the period during which they were being tested. For Ehrlich, the tetanus case was especially important because his institute, which had initially only been charged with controlling the anti-diphtheria serum, was promoted to the status of a national institution for drug evaluation. Nonetheless, the most important aspect of Ehrlich's work for the practical concerns of the clinician lay in the evaluation of anti-diphtheria serum, which, for a significant period of time, was the only antitoxin that could legitimately be assigned a significant role in public health.

Notes

1. R. Otto, 'Die staatliche Prüfung der Heilsera', *Arbeiten aus dem Königlichen Institut für Experimentelle Therapie zu Frankfurt a. M.* Heft II (1906), pp. 337–8.
2. *Real-Encyclopädie der gesammten Heilkunde*, ed. Albert Eulenburg, 3 edn., vol. 6 (Wien/Leipzig: Urban & Schwarzenberg, 1895).
3. Emil Behring and Shibasaburo Kitasato, 'Ueber das Zustandekommen der Diphtherie-Immunität und der Tetanus-Immunität bei Thieren', *Deutsche Medicinische Wochenschrift* 16 (1890).
4. Emil Behring, 'Ueber Immunisirung und Heilung von Versuchsthieren beim Tetanus', *Zeitschrift für Hygiene und Infectionskrankheiten* 12 (1892).
5. Emil Behring, *Geschichte der Diphtherie. Mit besonderer Berücksichtigung der Immunitätslehre* (Leipzig: Thieme, 1893), pp. 164–5.
6. Erika Schulte, *Der Anteil Erich Wernickes an der Entwicklung des Diphtherieantitoxins* (Berlin, 2001).
7. Schütz, 'Versuche zur Immunisirung von Pferden und Schafen gegen Tetanus', *Zeitschrift für Hygiene und Infectionskrankheiten* 12 (1892), p. 58.
8. Emil Behring, 'Ueber die Verschiedenheit der Blutserumtherapie von anderen Heilmethoden und über die Verwendung des Tetanusheilserums zur Behandlung

des Wundstarrstrampfes beim Menschen', in *Die Blutserumtherapie*, ed. Emil Behring (Leipzig: Thieme, 1892). Here p. 48.

9. Schulte, *Der Anteil Erich Wernickes an der Entwicklung des Diphtherieantitoxins.*

10. 'Verhandlungen der Berliner Physiologischen Gesellschaft', *Archiv für Anatomie und Physiologie. Physiologische Abtheilung* (13. 1. 1893). Here p. 199.

11. The effect of the serum depended on the evaluation technique used; whether the white mice were infected or intoxicated, whether the serum injection was preceded or followed by the infection. The time that elapsed between the two injections also played a role, as well as the doses of toxin and antitoxin.

12. 'Wir erklären uns zum Schluss bereit, im Institut für Infectionskrankheiten solche vergleichende Prüfungen auch mit Tetanusheilserum aus anderen Laboratorien mit wirklichem Heilwerth anzustellen, eventuell Proben von unserem Normalheilserum an andere Centralstellen für solche vergleichende Untersuchungen abzugeben'. See 'Verhandlungen der Berliner Physiologischen Gesellschaft', pp. 201–2.

13. Emil Behring, O. Boer and Albrecht Kossel, 'Zur Behandlung diphtheriekranker Menschen mit Diphtherieheilserum', *Deutsche Medicinische Wochenschrift* 19 (1893).

14. Paul Ehrlich, 'Experimentelle Untersuchungen über Immunität. I. Ueber Ricin, II. Ueber Abrin', *Deutsche Medicinische Wochenschrift* 17 (1891).

15. Ibid., p. 977.

16. Paul Ehrlich, 'Ueber Immunität durch Vererbung und Säugung', *Zeitschrift für Hygiene* 12 (1892).

17. Paul Ehrlich and Ludwig Brieger, 'Beiträge zur Kenntniss der Milch immunisirter Thiere', *Zeitschrift für Hygiene und Infectionskrankheiten* 13 (1893); Paul Ehrlich and Ludwig Brieger, 'Ueber die Uebertragung von Immunität durch Milch', *Deutsche Medicinische Wochenschrift* 18 (1892).

18. Ehrlich and Brieger, 'Beiträge zur Kenntniss der Milch immunisirter Thiere'.

19. 'Behring (Blutserumtherapie I) reclamirt das Princip der Immunitätssteigerung durch immer größer werdende Injectionen vollgiftiger Culturflüssigkeiten mir gegenüber als sein geistiges Eigenthum. Ich muss diesen Anspruch als einen nicht berechtigten zurückweisen. [...] In meiner Arbeit über Ricin und Abrin habe ich [...] als *erster* eine genaue *zahlenmässige* Untersuchung über Immunität und Immunitätssteigerung vorgenommen. [...] Ich hätte also weit eher die Berechtigung, Herrn Behring gegenüber Reclamation zu erheben'. Ehrlich and Brieger, 'Beiträge zur Kenntniss der Milch immunisirter Thiere', Note 1, pp. 337–8.

20. 'Das Datum unserer Veröffentlichung beweist, dass wir uns schon lange vor Behring's Veröffentlichung mit diesem Gegenstande beschäftigten.' Ehrlich and Brieger, 'Beiträge zur Kenntniss der Milch immunisirter Thiere', Note 1, p. 343.

21. Paul Ehrlich, 'Vortrag vor der Deutschen Gesellschaft für öffentliche Gesundheitspflege in Berlin', *Hygienische Rundschau* IV (1894).

22. Ernst Bäumler, *Paul Ehrlich. Forscher für das Leben* (Frankfurt am Main: Societäts-Verlag, 1979), p. 106.

23. Ibid., p. 107.

24. Paul Ehrlich, Hermann Kossel and August Wassermann, 'Ueber Gewinnung und Verwendung des Diphtherieheilserums', *Deutsche Medicinische Wochenschrift* 20 (1894).

25. The German name is 'Gesellschaft Deutscher Naturforscher und Ärzte'.

26. 'dass die Behandlung des Wundstarrkrampfes beim Menschen auch in ein neues Stadium zu treten verspricht. Ob freilich die Industrie bei der Seltenheit dieser

Krankheit einerseits, bei der Kostspieligkeit der Herstellung des Mittels andererseits, die Arbeit im grossen wird ausführen wollen, das ist mir noch zweifelhaft'. Emil Behring, 'Leistungen und Ziele der Serumtherapie', *Deutsche Medicinische Wochenschrift* 21 (1895), p. 633.

27. Instructions for Tetanus-Antitoxin by Hoechst: 15 Oktober 1896, Paul Ehrlich Institut (PEI), Abt. VI No. 1, sheet 2. Knorr developed the evaluation method in his 'Habilitationsschrift'. Angelo Knorr, 'Experimentelle Untersuchungen über die Grenzen der Heilungsmöglichkeiten des Tetanus durch das Tetanusheilserum' (Marburg, 1895).

28. Ehrlich, telegram to the Höchster Farbwerke, Berlin Steglitz, 9. 19. 1895. PEI, Abt. VI No. 1, sheet 3.

29. The Prussian Ministry for Medical Affairs (PMMA) to Ehrlich, Berlin, 2. 12. 1896. PEI, Abt. VI No. 1, sheet 18.

30. In France, the name of the Pasteur Institute stood for the quality of a serum. Since there were no competing producers, the situation was simpler. See Jonathan Simon, 'Quality Control and the Politics of Serum Production in France' in the present volume.

31. Ehrlich to the PMMA, 26. 10. 1896. PEI Abt. VI No. 1 Bd. I, sheet 17.

32. Knorr, 'Experimentelle Untersuchungen über die Grenzen der Heilungsmöglichkeiten des Tetanus durch das Tetanusheilserum'.

33. Ransom to Ehrlich, Marburg, den 7. 4. 98. PEI Abt. VI No. 5 Bd. 1, sheet 11, 12, and Ehrlich's report to the PMMA for April 1898, 16. 5. 98, PEI Abt IV No. 1 Bd 1, sheet 57.

34. Ransom to Ehrlich, 7. 4. 1898, PEI Abt. VI No. 5, sheets 11 and 12.

35. Ehrlich explained this in detail more than a year later in a letter to Behring, when he was trying to resolve the differences between them. Ehrlich to Behring, Frankfurt, 29. 10. 1899. PEI Abt. VI No. 5, sheets 27 and 28.

36. Paul Ehrlich, 'Die Werthbemessung des Diphtherieheilserums und deren theoretische Grundlagen,' *Klinisches Jahrbuch* 6 (1897).

37. Behring to Ehrlich, Marburg, 17. 10. 1899. PEI Abt. VI No. 5, sheet 26.

38. Ehrlich to the PMMA. Berlin, 27. 10. 1898. PEI Abt. VI No. 1 Bd. 1, sheet 29. In November, Libbertz announced that he would submit anti-tetanus to another control: Libbertz to Ehrlich, 28. 11. 98. PEI Abt. VI No. 1 Bd. 1, sheet 35. But Ehrlich does not mention having controlled anti-tetanus serum from Höchst in his November report; see PEI Abt. IV No. 1 Bd. 1 sheet 69.

39. Ehrlich's reports to the PMMA, PEI Abt. IV No. 1 Bd. 1 sheet 70, 71.

40. For this and the following, see the correspondence between Ehrlich and Behring from October and November 1899. PEI Abt. VI No. 5 Bd. 1, sheets 26–36.

41. The serum was reduced to a powder which was conserved in evacuated glass tubes. Ehrlich, 'Wertbemessung'.

42. Behring to Ehrlich, Marburg, 30. 10. 1899. PEI Abt. VI No. 5 Bd. 1, sheet 30.

43. Ehrlich´s reply to Behring´s letter from 15. 11. 1899 (handwritten copy without a date). PEI Abt. VI No. 5 Bd. 1, sheets 33–6.

44. Behring to Ehrlich, Marburg, 30. 1. 1899. PEI Abt. VI No. 5 Bd. 1, sheet 30.

45. Ehrlich to Behring, Frankfurt, 1.11.1899. PEI Abt. VI No. 5 Bd. 1, sheet 31.

46. Behring to Ehrlich, Marburg, 1.11.1899. PEI Abt. VI No. 5 Bd. 1, sheet 32.

47. Höchster Farbwerke to PMMA, Höchst, 23.3.1900. Geheimes Staatsarchiv Preußischer Kulturbesitz (GStAPK) Rep. 76 VIII B Nr. 4125 Bd. I. Höchster Farbwerke to Ehrlich, 2.6.1900. PEI Abt. VI No 5 Bd. 1, sheet 38, 39. The President of the Imperial Health Office to Ehrlich, Berlin, 9.8.1901. PEI Abt. VI No 6, sheet 30.

48. 'Ich muss daher auf das feierlichste erklären – auch im Namen von Herrn Geh. Rath Dönitz, dass das von uns verwandte Serum den effektiven Werth von 50 I.E. Knorr besitzt.
    Dass das jetzt in Deinem Besitz befindliche Originalpräparat Knorr davon abweicht, schwächer ist, ist nicht zu bezweifeln – kann aber doch nur auf einer Abschwächung beruhen. Ich möchte nicht umhin, nochmals zu betonen, dass wir bei den ganzen Untersuchungen mit dem ernstesten Bestreben verfahren sind, Deinen Wünschen zu entsprechen. Aber wir müssen uns nothwendigerweise an die Zahlen halten, die wir in unseren Büchern haben'. Ehrlich to Behring, undated handwritten copy. PEI Abt. VI No. 5 Bd. 1, sheet 34.
49. Emil Behring, 'Die Werthbestimmung des Tetanusantitoxins und seine Verwendung in der menschenärztlichen und thierärztlichen Praxis', *Deutsche Medicinische Wochenschrift* 26 (1900).
50. Ehrlich to the PMMA, Steglitz 10. 10. 1896. PEI Abt. VI No. 1, Bd. 1, sheets 4 and 5. This had also been the main reason for the introduction of the control of anti-diphtheria serum. See Axel Hüntelmann's contribution to this volume.
51. PMMA to Ehrlich. Berlin, 26. 10. 96. PEI Abt. VI No. 1, Bd. 1, sheet 7.
52. 'Es wäre zu bedauern wenn Deutschland, das Vaterland der Behringschen Entdeckung in der praktischen Durchführung hinter andern Ländern zurückbliebe'. Ehrlich to PMMA, 26. 10. 1896. PEI Abt. VI No. 1 Bd. 1, sheet 17.
53. PMMA to Ehrlich. Berlin, 21. 5. 1898. PEI Abt. VI No. 1, Bd. 1, sheet 24.
54. 'Wenn bei der Behandlung von Tetanus Höchster u. ausländische Praeperate als gleichwerthig angenommen werden, so fehlt nicht nur jede Basis für eine richtige Statistik, sondern das Urtheil wird geradezu gefälscht'. Dönitz to PMMA, Steglitz, 29. 8. 98. PEI Abt. VI No. 1 Bd. 1, sheet 28.
55. Wusehold to Ehrlich, Berlin, 28. 7. 1898. PEI VI No. 5 Bd. 1, sheets 16 and 17.
56. 'Die Präparation dieser Sera hat zur Voraussetzung die Möglichkeit sich die betreffenden Gifte in genügenden Mengen und in ausreichendem Stärkegrad zu beschaffen. Es ist also in diesen Fällen stets die Möglichkeit gegeben, den Wert der erzielten Sera mit Hilfe des Giftes zu bestimmen'. Ehrlich to Wusehold, Frankfurt, 8. 8. 1899. PEI Abt. VI No. 5 Bd. 1, sheet 18–24, here, sheet 19.
57. Ehrlich to Wusehold, Frankfurt, 8. 8. 1899. PEI Abt. VI No. 5 Bd. 1, sheets 18–24.
58. Wusehold to Ehrlich, Berlin, 16. 11. 1899. PEI Abt. VI No. 5 Bd. 1, sheets 37 and 38.
59. E. Brass, 'Ueber das Tetanus-Antitoxin nach Beobachtungen in der medicinischen Klinik der thierärztlichen Hochschule zu Berlin', *Berliner Thierärztliche Wochenschrift* (1897), W. Dieckerhoff and B. Peter, 'Zur Serumbehandlung des Starrkrampfes beim Pferde mit Tetanus-Antitoxin (Behring)', *Berliner Thierärztliche Wochenschrift* (1896).
60. Ehrlich to PMMA, 22. 2. 1900. Rep. 76 VIII B Nr. 4125 Bd. I. Draft in PEI Abt. VI No. 1 Bd 1, sheet 47–71.
61. 'Ganz besonders möchte ich noch hervorheben, dass nach meinen Erfahrungen einer einfachen Übertragung der Bestimmungen über die Diphtherieserum-Controle auf das Tetanusserum die schwersten Bedenken entgegen stehen. Eine Erörterung dieser Bedenken und ihre ausreichende Begründung ist aber nur in mündlicher Diskussion möglich, in welcher u. a. auch darauf Rücksicht zu nehmen sein wird, wem die Bestimmungen über die Tetanussserum-Controle übertragen werden sollen'. Behring to Ehrlich, Marburg, 18. 1. 1900. PEI Abt. VI No. 1 Bd. 1, sheets 40 and 41.

62. 'Es liege deshalb die Frage vor, ob diese Sera ganz oder theilweise unter staat-
    liche Kontrole zu stellen seien und auf welche Weise diese Kontrole zu üben
    sei'. Protocol 24. 11. 1900, p. 2. Bundesarchiv (BA), Akte R 86 Nr. 2713. The
    Chancellor of the German Reich had already asked in November 1899 that the
    sale of anti-tetanus serum be restricted to pharmacies, as the Prussian Minister for
    Medical Affairs informed Ehrlich in a letter, Berlin, 21. 11. 1899. PEI Abt. VI No. 1
    Bd. 1, sheet 38.
63. Protocol, 24. 11. 1900, p. 4. BA R 86 Nr. 2713.
64. He probably meant Friedrich Althoff, who was not present at the meeting.
65. Protocol, 24. 11. 1900, p. 5. BA R 86 Nr. 2713.
66. Ibid., p. 7.
67. Ehrlich to PMMA, 4. 1. 1910. GStAPK Rep. 76 VIII B Nr. 4125.
68. Besprechung im Kultusministerium Berlin über die staatliche Prüfung des
    Tetanusheilserums und seine Abgabe in den Apotheken, 26. 2. 1910. PEI, Abt.
    VI No. 1, sheet 152–6. Ehrlich, who was not present at the meeting, later recom-
    mended sticking to conservation with carbolic acid.
69. Kirchner to the district presidents, Berlin, 10. 5. 1910. PEI Abt. VI No. 1, sheet
    159–61.
70. See Merck's price lists from 1902–17. Merck Archiv. Bestand W 35, No. 2–5.
71. Only 8 out of 36 patients died, and of these three suffered from septic complica-
    tions. The other five cases, Engelmann concluded, had received insufficient doses
    of the serum or were treated too late. Max Engelmann, 'Zur Serumtherapie des
    Tetanus', *Münchener Medicinische Wochenschrift* 32 (1897), p. 919.
72. Behring and Knorr announced their product and its voluntary control in the
    Deutsche Medicinische Wochenschrift in October 1896. Emil Behring and Angelo
    Knorr, 'Tetanusantitoxin für die Anwendung in der Praxis', *Deutsche Medicinische
    Wochenschrift* 22 (1896).
73. Engelmann, 'Zur Serumtherapie des Tetanus', p. 942.
74. Instructions in Merck's annual report from 1894, p. 96.
75. 'Ich bin deshalb der Ansicht, dass die von der Fabrik zu Höchst am Main verab-
    folgte Normaldosis nicht vergrößert zu werden braucht, was bei dem theueren
    Preise des Mittels für die Privatpraxis der Thierärzte gewiss beachtenswerth ist'.
    W. Dieckerhoff, 'Zur Behandlung des Starrkrampfes bei Pferden mit Tetanus-
    Antitoxin', *Berliner Thierärztliche Wochenschrift* (1897), p. 302.
76. Dieckerhoff and Peter, 'Zur Serumbehandlung des Starrkrampfes beim Pferde mit
    Tetanus-Antitoxin (Behring)'. *Berliner Thierärztliche Wochenschrift* (1896).
77. 'nach Ansicht des Oberrossarztes König [dürften] nunmehr die Akten geschlossen
    sein'. Staatlicher Veterinär-Sanitäts-Bericht über die preußische Armee für das
    Rapportjahr 1898 (Seite 87) II. Theil Nr. 12: Starrkrampf. Copy in PEI Abt. VI
    No. 1 Bd. 1, sheets 70 and 71.
78. Köhler calculated the mortality with respect to the incubation time, since a short
    period of incubation (1–10 days) usually meant a worse prognosis than a longer
    one. According to Engelmann's data, 66.6 per cent of the patients with a short
    incubation time were cured by serum therapy and 93.6 per cent of the patients
    with a long incubation time. According to Köhler's data only 43 per cent and
    63.6 per cent were cured, respectively. F. Köhler, 'Zum gegenwärtigen Stand der
    Serumtherapie des Tetanus', *Münchener Medicinische Wochenschrift* 45 (1898).
79. The Lancet Special Commission, 'Report on the Relative Strengths of Diphtheria
    Antitoxic Serums', *The Lancet* II (18. 7. 1896).
80. Köhler, 'Zum gegenwärtigen Stand der Serumtherapie des Tetanus', p. 1473.

81. 'Die Antitoxinwirkung versagt wahrscheinlich deshalb, weil das Toxin gewisser-maßen geschützt vor dem im ganzen Körper kreisenden Antitoxin, in den Nerven nach dem Zentralnervensystem wandert'. Kaiser Wilhelms-Akademie für das militärärztliche Bildungswesen (ed.), *Ueber die Anwendung von Heil- und Schutzseris im Heere*, vol. 37, Veröffentlichungen aus dem Gebiete des Militär-Sanitätswesens (Berlin: Hirschwald, 1908), p. 19.
82. Besprechung im Kultusministerium Berlin über die staatliche Prüfung des Tetanusheiserums und seine Abgabe in den Apotheken, 26. 2. 1910. PEI, Abt. VI No. 1, sheets 152–6.
83. Carola Throm, *Das Diphtherieserum. Ein neues Therapieprinzip, seine Entwicklung und Marktführung* (Stuttgart: Wissenschaftliche Verlagsgesellschaft, 1995).

4
The Construction of a Culture of Standardization at the *Institut Pasteur* (1885–1900)

Gabriel Gachelin

The initial wave of conception and production of animal and human vaccines supervised by Louis Pasteur (1822–95), although considered a great achievement in France, suffered from several theoretical, methodological and technical weaknesses that threatened the reproducibility of the results as well as part of the validity of Pasteur's approach. While these problems were mostly pointed out by German scientists, there were also French critics of Pasteur's work.

While the existence of problems was initially vigorously denied, these problems were progressively resolved, often years after the first announcement of a success. This strategy could be considered a kind of *a posteriori* reconstruction of data that, had they been known at the time of their announcement, could have led to the initial positive conclusions. This progressive construction of acceptable practices, formed part of a confused medical methodology during the period that Mirko Grmek has termed the 'heroic age' of the *Institut Pasteur* (hereafter, the Pasteur Institute).[1] By contrast, an entirely different scientific attitude, now involving quality control, quantification at every step of the production process, closer contacts with medical structures and a more positive attitude towards German science, was put in place at the beginning of the 1890s. The development of serotherapy as a specific treatment against diphtheria is the earliest example of this change: it involved the introduction of quantification and norms largely adapted from German procedures, albeit present in an approximate form in Pasteur's earlier protocols. This approach ultimately led to the building up of a Pasteurian bio-industry that could compete with that of Germany.

The present paper deals primarily with the manner in which a 'culture of quantification and standardization' was introduced at the Pasteur Institute or, more precisely, was imposed by Emile Roux (1853–1933) and his co-workers as part of an approach to tackling infectious diseases. This was based on the direct coupling of research in microbiology with the production of drugs of biological or chemical origin, and to medical care in the hospital on the site of the Pasteur Institute.

The weaknesses of Pasteur's approach to vaccination

The distinctive place occupied by Pasteur in the scientific landscape of France's Third Republic has been amply described elsewhere,[2] as have been the conditions and uncertainties of Pasteur's approach to vaccination, following the unveiling of Pasteur's laboratory notes in 1971.[3] The present paper focuses on the evolution of laboratory practices at the Pasteur Institute during the 10 years that followed its creation in 1887, as French researchers tried to narrow the gap between Pasteur's statements and Koch's methods. The Koch–Pasteur dispute has been extensively discussed elsewhere,[4] and it can be assumed that Koch's indictment of Pasteur's approach to vaccination should primarily be considered as the enunciation of rules of good laboratory practice and methods in microbiology. This positive methodological discussion has been obscured in France by the violence of the exchange between the protagonists, as well as by hagiography and nationalism, although its value was clearly perceived by foreign scientists as well as by certain scientists close to Pasteur.[5]

A short discussion of the anthrax and rabies cases will help to define the scientific problems at issue. The etiological agent of anthrax[6] had been characterized by Robert Koch (1843–1910) in 1877, leading to a priority dispute with Pasteur.[7] Several 'vaccines' against the disease had been tested in France before 1880, with no clear evidence of effective protection. The results of the vaccination of sheep against anthrax using oxygen-attenuated anthrax bacilli were first communicated on February 28, 1881.[8] The experiment was publicly repeated on sheep in May 1881 at Pouilly-le-Fort, and the results were highly publicized in France as a success-story. However, experiments carried out in 1881–2 in Germany met with no success and German scientists – particularly Koch and Friedrich Loeffler (1852–1915) – remained more than sceptical about the efficacy of Pasteur's vaccine. Apart from emotional and nationalistic comments in France and in Germany, several scientific criticisms were enunciated. It was pointed out that the observed protection might have been non-specific, and other criticisms were voiced concerning the irreproducibility of the results of vaccination as the initial oxygen-based attenuation process did not reproducibly inactivate the bacilli. Moreover, the protocol for the attenuation of the anthrax bacilli used in the Pouilly-le-Fort experiment was only published in 1883. A chemically induced attenuation procedure previously rejected by Pasteur but perfected by Chamberland and Roux had been used in 1881 without being disclosed, and the substitution was only rendered public 50 years later.[9] Finally, the statistical analysis of the results depended simply on the comparison of large numbers, with only a rough estimate of the mortality in the absence of vaccination. Only several years[10] of work using Chamberland's attenuated vaccine would later enable researchers at the Pasteur Institute[11] to collect a large enough number of cases to conclude at the International Congress of

Hygiene in Vienna in 1887 that cattle were protected against anthrax by vaccination to a level of 80–90 per cent.[12] Even today complete protection against anthrax is more difficult to obtain than was claimed in the 1880s. Doubts concerning Pasteur's vaccine were due to three factors; first, uncertainties about the quality of the vaccine and the protocol for its preparation; second, the premature release of 'positive' data, and, finally, insufficient statistical analysis of the results. Pasteur's co-workers were aware of at least the first two of these problems.

Similar methodological problems emerged in the rabies case.[13] A critical discussion of the conditions of preparation of the successive anti-rabies vaccines and the conditions of their first use on humans in 1885 has been published elsewhere.[14] Despite the opposition of Roux to a premature extension of the experiment to humans and a controversy in France over the effectiveness and danger of the anti-rabies vaccination, the outcome of the first injections for two badly bitten children were widely interpreted in 1885 as signalling the defeat of the disease. The wave of emotion that followed these first treatments led to the creation of the Pasteur Institute in Paris in 1887 and later on, the creation of numerous other institutions around the world that dubbed themselves '*Institut Pasteur*',[15] all primarily devoted to vaccination against rabies. However, the treatment did not always prove effective against the disease.

The occasional absence of effectiveness was attributed to the vaccine being injected too late after infection or to the localization of bites on the body. Starting in 1885, the techniques of preparation of the 'virus-vaccine' were modified several times and the protocols used to inject the drug were subjected to major changes ('intensive' vs. 'normal' treatment). Finally, no statistical analysis could be carried out in the absence of negative controls and of data concerning the propagation of rabies in humans. The increasing number of 'cured cases' was considered as the proof of the vaccine's effectiveness, as shown by the monthly description of treated cases published in the *Annales de l'Institut Pasteur*. The weakness of the demonstration evidently needed to be addressed, and the task was undertaken by N. Gamaleya, vice-head of the Bacteriological Institute of Odessa – the frequency of rabies being much higher in Russia than in France.[16] In the end, the notion of success in humans was a kind of extrapolation of data obtained from infecting genuinely vaccinated animals with a virulent form of the virus, rather than a rigorous demonstration of the development of a protection in presumably infected humans.[17] As in the case of anthrax, one could observe a combination of uncertainty concerning the vaccine and its administration, the premature announcement of success with a treatment on humans and the absence of any proper statistical analysis of the treated cases, all problems that were tentatively answered by the empirical adjustment of procedures and reagents, without, however, delivering any definitive proof of the therapeutic effectiveness of the procedure in man.

Of course, microbiology was still under construction and the shift to immunology proper had not yet occurred.[18] The description of the host-microbes relationships and the methods of study, including the use of animals, were far from the high standard of rationality that prevailed in physiology.[19] Statistical analysis of results was not widespread. The lack of precision described above seems characteristic of a discipline emerging from a kind of chaos of observations and results. Whatever its causes, the restrictive or protective attitude of Chamberland, Roux and Duclaux strongly suggests they were aware of the problems from the beginning, though they never publicly interfered with Pasteur's approach to vaccination or with his emotional way of communicating with the public.[20] The adoption of a more rigorous policy after the creation of the Pasteur Institute and Pasteur's retirement from activity, suggests that the weaknesses in the previous regime had been identified as serious enough to threaten the scientific, medical and economic development of the Institute, in particular when compared with the approach of its main competitor, Koch's Institute of Hygiene in Berlin.

A new sociological environment for French microbiology

The building of the Pasteur Institute was financed by a public subscription launched by the French Academy of Medicine in 1886 to create a rabies research and treatment centre. The Pasteur Institute – inaugurated in 1888 – had a different project and was more of an institute for general microbiology, with at most one-sixth of its space devoted to rabies consultation and vaccination.[21] From the very beginning of the Pasteur Institute, Roux defined plans with the architects and supervised the construction, designing the institute so that the various aspects of microbiology could be dealt with in one building (the present *batiment historique*). Starting in 1887, Pasteur's declining health forced him gradually to withdraw from his official positions and pushed Roux towards the intellectual leadership of the Pasteur Institute.[22]

The plans and the organization of the institute reflect the major changes that had occurred in Parisian microbiology. Before the years 1887–8, research was carried out in Pasteur's own laboratories and facilities at the *Ecole Normale Supérieure* under his close personal supervision, or at the veterinary school of Maisons-Alfort where Nocard had established an unofficial branch of Pasteur's laboratory. Pasteur imposed a rigid hierarchical structure, but after 1888 laboratory space was allocated to heads of laboratories, whose publications indicate they had become scientifically independent. These heads of laboratories were mostly young: Roux was 35, Chamberland 36, Metchnikoff 42, Grancher 44; Yersin (the youngest) 24 and Duclaux, the oldest, was 47. While Duclaux was elected Director of the institute after Pasteur's death in 1895, Roux and Metchnikoff were clearly the scientific leaders. The new team leaders were all civilians (in contrast to Koch's staff) and had an open attitude towards European, particularly German, science. Roux, who had left the army due to his insubordination, brought the medical thinking that Pasteur lacked; he conducted research with a rigorous approach to science, including technological developments (photography, temperature-controlled incubators etc.). Elie Metchnikoff (1845–1916), a zoologist of Ukrainian origin, was based in Russia until 1888, during which time he had enjoyed long stays in several European laboratories. During the period 1864–6 Metchnikoff was in Germany where he visited Kohn in Heligoland and Leuckart in Giessen, before spending 1867 in Naples with Kovalevsky, and visiting several marine laboratories in France, Spain and Italy from 1869 to 1873. Metchnikoff brought a wealth of knowledge about diverse biological systems, including the invertebrates' defence system against microbes. Alexandre Yersin (1863–1943) had first studied medicine in Marburg and had attended Koch's lessons in Berlin in 1889. Emile Duclaux (1840–1904), a chemist specialized in fermentation, taught biological chemistry at the Sorbonne and represented a link with the university. Charles Chamberland (1851–1908), a physician and biologist, and a close associate of Pasteur's work on anthrax, was familiar enough with German science to be sent to Germany in 1887 to discuss results with Koch and Loeffler. A number of non-permanent members of the Pasteur Institute worked in the laboratories or attended the *cours de microbie technique* given by Roux starting in 1889, among whom were numerous foreigners including Germans. In a few years, the Pasteur Institute had become an interdisciplinary and international research centre.

In general, the 1880s saw a reduction in the tension between France and Germany. Missions to study the functioning of German universities, initiated by chemists in 1866 (before the Franco-Prussian war) were resumed in 1877 and their reports, for example on chemistry or embryology were published and widely discussed in the context of the reform of French universities.[23] It was argued that France had been defeated by German science and technology making German universities models for strengthening French university research and teaching, with the ultimate goal of avenging the defeat of 1870. Reports were enthusiastic about the quality of the teaching, the organization of faculties and the relations between professors and students.

At the Pasteur Institute, the attitude of Pasteur – engaged in conservative political circles and strongly anti-German – contrasted with that of most of his colleagues, who had different political beliefs, and neither their official statements nor their personal letters reveal an anti-German attitude. Roux and Behring were friends and Roux and Metchnikoff were the godfathers of one of Behring's sons. Letters exchanged between Roux and Metchnikoff and between Roux, Behring and Metchnikoff, indicate that Roux and Metchnikoff highly esteemed Koch's scientific knowledge and competence. Metchnikoff and Yersin even submitted data or preparations to Koch to confirm some of their most important results, although the two of them were amused by the reluctance of the scientists around Koch to express a personal opinion.[24] Politically, most of the Pasteur Institute's researchers had socialist or radical sympathies, with Roux a well-known Dreyfusard. Duclaux, a pacifist, was very sensitive to questions of human rights and defended Dreyfus as early as 1898, as well as contributing to the creation of '*La ligue des droits de l'homme*'. Chamberland was elected as a Deputy to the French parliament in 1885 and, like Grancher, belonged to the *républicain radical* group, defending laws on public hygiene. Metchnikoff was a committed socialist and pacifist, and later a theosophist, which explains the serious problems he had with the tsarist administration in Odessa.[25] Thus, the majority of the scientific staff of the Pasteur Institute belonged to the republican meritocracy of the Third Republic impregnated with social-democrat thinking and more open to foreign influences.

The translation of foreign scientific publications was by no means new at the end of the 1880s, but the effort made by the Pasteur Institute to follow science outside France is shown by the excellent coverage of foreign publications in the *Annales de l'Institut Pasteur* launched in 1887 by Emile Duclaux (1840–1904).[26] Although the rationale underlying the creation of the journal clearly concerned auto-promotion in the context of a rivalry with Koch's journal (*Zeitschrift für Hygiene und Infektionskrankheiten*),[27] the style of the papers soon became less polemical. Thus, the *Annales* reported summaries of important work carried out elsewhere, particularly in Germany, making microbiological information published in Germany available to French microbiologists.[28] Indeed, in their papers concerning diphtheria, Roux and his colleagues commented positively on the successive and decisive contributions of their German colleagues; for example, a paper written by Roux and Louis Martin (1864–1946) in 1894 that dealt with the preparation and testing of anti-diphtheria serum, was preceded by an extensive bibliography of papers published abroad and analysis of the articles by Emil von Behring (1854–1917), Shibasaburo Kitasato (1852–1931) and co-workers on the production of sera able to neutralize bacterial toxins. The authors acknowledged German contributions by plainly writing: *Nous pouvons déclarer que nos résultats confirment, dans ce qu'ils ont d'essentiel, ceux de M. Behring et de ses collaborateurs'*.[29]

In other words, the years 1885–90 saw Pasteur's microbiologists come out of the 'heroic age'. Members of the staff of the Pasteur Institute had gained scientific and practical autonomy in their work. Moreover, both the work and the deliberations of the Board of Administrators show that they were now conscious of the economic context and the competition in which the medical or veterinary use of biological derivatives was now embedded.[30]

Competition over antitoxins and the adaptation of German standardization procedures

Rabies was more a political target than a medical problem for Pasteur.[31] In contrast to rabies, other human infectious diseases like tuberculosis, diphtheria, tetanus, and syphilis, were major public health problems in Western

Europe and thereby attracted research in France as well as in Germany. Despite early promising results, work on tetanus did not progress well and was soon overtaken by diphtheria research.[32] The work carried out on diphtheria at the Pasteur Institute bears the hallmark of Roux's personal approach and should be considered in a double context. Firstly, the context of competition with German scientists, as has been underlined by P. Weindling.[33] Analysis of publications shows that very similar or complementary results were being obtained and published by German and French scientists at nearly the same time. The French work seemed to be influenced by a desire to avoid the criticisms raised against the anti-anthrax and anti-rabies vaccines: it had to be impeccable, particularly with respect to the German approach to the same problem(s). Secondly, Roux's scientific strategy was underwritten by a rationalized medical project, proposed in 1894 as the conclusion of the successful trial on diphtheria and which resulted in 1900 in the construction of a dedicated hospital, the *Hôpital de l'Institut Pasteur* located on the Paris site, in the vicinity of the research laboratories.[34] To a certain extent, the first 20 years of existence of the Pasteur Institute were shaped by Roux's medical project, which later extended far beyond diphtheria and soon included tropical diseases such as yellow fever, malaria and sleeping sickness as well as the testing of treatments for these diseases.[35]

The causal agent of diphtheria is the Klebs-Loeffler bacillus (*Corynebacterium diphtheriae*), identified in 1883 in the throat lesions of patients. In 1892–4, in the absence of treatment, diphtheria was responsible for an annual death toll of 6–7 per 10,000 persons in Europe.[36] The suffocating form of diphtheria in children (named the 'croup') inspired much fear among parents, making it not only a deadly disease but also a socially significant target for researchers.[37] Two main scientific breakthroughs were made between 1885 and 1890 concerning the mechanisms of action of bacteria and of vaccines, respectively.[38] It was shown that the microbial agents of diphtheria and tetanus acted through the soluble toxins they released into the infected organisms. Later, Behring and Kitasato demonstrated that protective immunity could be induced against the toxins by injecting them into appropriate recipient animals, a notion soon extended by Roux and his co-workers to other soluble molecules. Behring and Erich Wernicke worked on the idea that protection was associated with the serum of the immunized animals and transferable to other animals. Serotherapy was the development of this initial and decisive discovery.

The first period of research on diphtheria at the Pasteur Institute (1888–9) corresponded to the verification by Roux[39] that the Klebs-Loeffler bacillus was the genuine cause of diphtheria and the demonstration that the symptoms of the disease could be reproduced by injecting animals with the poison secreted by bacteria grown in vitro. In 1889, Roux and Alexandre Yersin (1853–1943) described some properties of the toxin,[40] and in 1892, Wernicke published a paper describing the cure of three cases of diphtheria in children by injection of the sera of dogs that had survived toxin and bacteria injections. This paper was discussed at the institute soon after its publication,[41] and the preparation of hyper-immune sera in horses was initiated at the end of 1892 by Edmond Nocard (1850–1903) at the veterinary school of Maisons-Alfort.[42] From the autumn of 1894, large scale serum production was organized mostly in the annexe of the Pasteur Institute in Marnes-la-Coquette near Paris.[43]

The first paper written by Roux and Martin in 1894, dealt with the preparation and testing of the horse sera. The quality and protective features of serum produced in various animals had already been tested by German scientists, the dog serum initially used by Wernicke being a mere by-product of earlier experimental infection experiments. The choice of horses as serum producers suggested by German scientists was promptly verified in Paris. Horses were large and hardy enough to provide the quantities of sera needed, but more importantly they were susceptible to diphtheria allowing their immunization against the toxin. In contrast with earlier articles on anthrax or rabies vaccines, which were rather imprecise about the technical procedures used, the protocols concerning the isolation and inactivation of diphtheria toxin were extensively described making them reproducible elsewhere; the methods used to control the steps leading to the selection and tests of serum by using *in vivo* assays on guinea pigs and rabbits were fully described. The toxicity of the toxin was defined *in vivo* as the volume needed to kill a 500g guinea pig in 48 hours, and it could be chemically inactivated using iodine trichloride. This method, which could provide 'diluted' toxin for the immunization process was largely inspired by the disinfectants used by German army physicians during the Franco-Prussian war and systematically studied afterwards.[44]

The protocol of injection into horses of increasing doses of the inactivated toxin, and then of pure toxin was described in great detail: the health and behaviour of each injected horse was monitored in laboratory files[45] and some exemplary cases were reported in the paper, including the immunization of a cow. More importantly, the anti-toxic properties of the sera of each individual horse were tested *in vivo* on a panel of guinea pigs challenged with a defined quantity of toxin, along with the appropriate negative and positive controls, which allowed the determination of protective units. The definition of the units used by Roux and Martin in 1894 (one ml of horse serum capable of neutralizing 20 ml of a toxin solution, 0.1ml of the latter killing a 500g guinea pig within 48 hours) was preceded by a one-page description of the three successive types of units used by Behring and Paul Ehrlich (1854–1915). The need for international units was emphasized: Roux speculated that standardization would be of more general use if comparative studies could be carried out using Behring's toxin and sera.[46]

The amount of serum to be administered to children was defined on a weight- (guinea pig) to-weight (child) basis. The entire production and testing process thus obeyed a defined and published protocol. The units may have been different in France and Germany, but they were linked by a conversion factor that made their comparison easy. Standardization of anti-diphtheria serum in France was definitively associated with the fact that German scientists had already worked out their own standards. Indeed, Roux wrote: 'to tell the truth, we do not attach much importance to all these complicated definitions (...) However it was necessary to speak of these units because they are continually employed in German publications.'

Standardization of biological reagents became an integral part of Pasteurian culture as early as the beginning of 1894. The Public Health Authority (*Direction de l'hygiène*) of the Ministry of the Interior, in a note dated 10 February 1895, defined the conditions of delivery (20ml vials) of the reagent by the Pasteur Institute and pharmacies, with no mention of other institutions allowed to produce the serum. The 20ml dose was the dose injected by Martin into incoming patients prior to bacteriological diagnosis. It corresponded to the ability to neutralize 200ml of toxin supernatant in the *in vivo* guinea pig test.[47] The use of anti-diphtheria serum was officially regulated in France as of 25 April 1895.[48,49]

A more rational statistical analysis of the data

The second important step in the use of a therapeutic agent is the proper statistical analysis of the results obtained on humans. We mentioned in an earlier section that this was one of the weaknesses of Pasteur's initial testing of vaccines. Indeed, statistical analysis of results from therapeutic trials was not commonly used in medicine,[50] but the progressive assessment of the efficacy of serotherapy led in 1898 in Denmark to what Ian Chalmers has dubbed the first random trial in human medicine.[51] Roux took an important step towards the definition of homogeneous patient groups in 1890 when he introduced systematic bacteriological diagnosis for those suspected of having diphtheria. Indeed, physicians were faced with difficulties in diagnosing true diphtheria because of the heterogeneity of the clinical signs shown by patients. The introduction of bacteriological testing had two main consequences; first, these tests allowed the monitoring of the presence of the bacterium during the evolution of the disease, including during convalescence. Second, the patient groups were better defined, with those suffering from proven diphtheria being further subdivided into two groups, those with diphtheria bacilli alone and those with diphtheria bacilli associated with other pathogens, an association found to offer a poor prognosis.[52] This meant that cases could be organized into cohorts defined on a bacteriological basis for testing the efficacy of the serum.[53] Bacteriological diagnosis was carried out in a laboratory set up in the *Hôpital des Enfants-malades* by Louis Martin (1864–1946), a physician selected by Roux from among Grancher's students.

An article by Roux and Martin from 1894 describes large-scale clinical trials of the anti-diphtheria serum.[54] There is no need to present the results communicated by Roux at the International Congress of Hygiene of Budapest in

September 1894 once again, except to note that priority had been established by a preliminary communication given in Lille in spring 1894. At the same Budapest meeting, Hans Aronson from Berlin confirmed that similar results had been obtained in Germany on diphtheria patients. The discussion that followed oral presentations showed a general consensus concerning the efficacy of the therapeutic method and pointed to the significant numbers of cases that had been reported in the comparison of treated versus untreated patients.[55]

In their study Roux and Martin compared two cohorts of patients identically defined in terms of bacterial infection: members of the treated group received serum in Grancher's own ward at the *Hôpital des Enfants-malades* in Paris, and the members of the second group, here used as a negative control, were hospitalized at the *Hôpital Trousseau* in the East of Paris and did not receive any serum. However, neither the patient population, nor the organization of the hospital, the hygienic rules applied in the wards, the frequency of tracheotomy etc., were necessarily the same at *Trousseau* as at the *Enfants-malades*. Although intuitively suggestive of a decrease in the mortality in treated patients, the results could have been statistically less significant than they appeared.

Starting in the autumn of 1894, anti-diphtheria serum was widely distributed in Europe by French and German institutes. A two- to five-fold decrease in mortality rates observed through most countries in Europe during the following two years was attributed to the use of the serum,[56] although the effects of new isolation rules in hospitals may have contributed just as much. The adverse effects of serum-therapy (serum sickness) were known but barely questioned in France.[57] Accidents, however, soon limited the usage of the antiserum, by casting doubt on the benefit of using it. The positive effects of the sera on the outcome of the diseases in diphtheria patients were only proven later, in 1898, by a Danish physician of the Blegsdamhospitalet in Copenhagen, Johannes Fibiger (1867–1928), who wanted to determine if anti-diphtheria serum was worth using considering its serious side effects. He introduced a genuine randomized trial (same hospital, same diagnosis, one day incoming patients were treated, the next day incoming patients were not etc.), using sera of Danish, German and French origin and thus demonstrated the existence of a significant benefit for serum-treated patients, thereby confirming earlier conclusions.[58] Subsequently, statistical analyses of test results were of particular significance in the evaluation of the efficacy of several sera and vaccines produced by the Pasteur Institute (against tetanus, typhoid, gangrene, typhus, staphylococci), and tested in the field during WWI by army physicians.

State-defined norms of quality, or the Pasteur Institute's control of the quality of French producers of sera

A fundamental question concerning the significance of the norms used in France stems from the quasi-monopoly of production and trade of the sera granted in France to the Pasteur Institute – a situation which contrasted with the state-defined standardization procedures used in Germany.[59] In other words, was the control of quality of the antitoxin ensured by some independent supervisory office or institution? The serum committee of the Academy of Medicine, set up for that purpose in May 1895 by the health administration, was largely controlled by influential members of the Pasteur Institute. Thus, the Pasteur Institute was both issuing norms and controlling the conformity to these norms. It is beyond doubt that the Pasteur Institute occupied a privileged position in France concerning infectious diseases, and an exceptional place in the French public health landscape.[60] However, the situation that prevailed before WWI was less an official monopoly granted to the Institute than a kind of institutionalized tutorship of the Pasteur Institute over other private and public health institutions. Also, with the exception of the original agreement granted by government agencies for drugs and reagents for therapeutic use against infectious diseases, the production and trade in the latter were left to the initiative of private structures and local administrations, provided that their activity was approved by the above-mentioned committee, and thus indirectly by the Pasteur Institute.

The Pasteur Institute had managed to place itself at critical places in the network of public health institutions organized as a response to the requests of different components of the French administration. As an example, the bacteriological diagnosis of diphtheria had to be carried out either at the Pasteur Institute or in trustworthy laboratories, such as that established in the *Hôpital des Enfants-malades*, which in practice meant laboratories headed by people trained at the Pasteur Institute. A similar observation could in turn be made about the *Cours de microbie technique* created by Roux as early as 1888, not only as an original, high-level training course, but also as a powerful tool for the organization of a network of people and institutions associated in some way with the Pasteur Institute.[61]

The production of anti-diphtheria serum was not coordinated at a national level or transferred to industry under the quality control of state agencies, even though a kind of control was granted to the Pasteur Institute. Production in Paris responded to local regional needs (Paris and the surrounding areas). Dried serum could later be transported to more distant places in and outside France, but local production of antiserum was preferred, although always in some sense under the supervision of the Pasteur Institute at Paris. In distant sites, the serum was produced under the supervision of people trained and approved by the Pasteur Institute on behalf of the committee; several *instituts sérothérapiques* engaged in the production of serum were created in France at the end of the nineteenth century (Marseille, Bordeaux, Rennes, Lyon etc.)[62] and abroad, such as in Geneva.[63] Some, because of their name, are not easily identified as such, like the *Institut Bouisson Bertrand* created in Montpellier on 6 February 1897 thanks to a private initiative. The *Institut sérothérapique* de Lille, renamed the

Institut Pasteur de Lille in 1898, was created by Albert Calmette (1863–1933) to produce anti-diphtheria serum at the request of the municipality of Lille. Calmette was a well-known 'Pasteurian' who had created the Pasteur Institute of Saigon in 1890.[64] In this particular case, the use of the name 'Pasteur' was granted to an institution placed under the administrative control of the city of Lille and the department of the Nord.

The well documented case of Nancy illustrates the complexity of the production and trade of anti-diphtheria serum in France.[65] Nancy, a city in which the University of Strasbourg had settled after the Franco-Prussian war of 1870, had developed a strong tradition of hygiene and social care at the municipal level, as well as teaching at the university. Following the 1894 meeting in Budapest, in a manner highly reminiscent of the national initiative of the newspaper *Le Figaro* which had led to the creation of the annexe of the Pasteur Institute in Marnes-la-Coquette, a regional subscription was launched by Pierre Parisot (1854–1938), a Professor of legal medicine at the University of Nancy, first to purchase serum from Paris, then to create a centre for its production and distribution. The serum could obviously have been purchased from nearby German factories, and the initiative of creating a production centre in Nancy, so close to Germany, may thus have been political as has been suggested by Jonathan Simon.[66] It also appears to participate in a broader strategy to create a number of similar centres throughout France. The *Institut de sérothérapie de l'Est* started producing anti-diphtheria serum in November 1894 by using toxin provided by Roux. Later, the toxin as well as the serum was produced locally, providing serum to all departments of Eastern France. The new building of the *Institut de sérothérapie de l'Est, Fondation Osiris*, was located in Nancy's general hospital and opened in June 1896 under the supervision of Eugène Macé (1856–1938), a microbiologist trained in 1894 by Roux at the Pasteur Institute.

The conditions of the opening and the administrative situation of the *Institut de sérothérapie de l'Est* were complex. It appears to have been a semi-public institution, subsidized by the municipality of Nancy and the department, and run by the board of directors of the '*Société privée de l'Institut sérothérapique de l'Est,*' located in a general hospital and it was soon included in the *Institut d'Hygiène* of the university, and later placed under the scientific and medical control of professors at the faculty of medicine. It is not clear at this time whether the very same biological criteria were used to define the protective units of the sera produced in Nancy as in Paris. For example, vials containing smaller volumes of serum (10ml, no titre indicated) were delivered to pharmacists. This example shows the difficulty of imposing strict standardization of serum across France and its colonies, as the Pasteur Institute could not apparently dictate the quality of the serum issuing from these regional centres of production.

Two central conclusions can be drawn from this survey of the introduction of norms at the Pasteur Institute, considered as one of the main consequences of the development of serotherapy. First, less than five years were required for all the steps of production and clinical tests to become standardized and included in a procedure developed by the Pasteur Institute, which was then transmitted to other French health institutions. A culture of standardization, representing a genuine break from earlier practices, was achieved as a require- ment for production of standardized and controlled drugs. The whole process had important long-term consequences for the functioning of the Pasteur Institute. Following the initiation of the production of anti-diphtheria serum at Marnes-la-Coquette in the fall of 1894, a dual scientific culture developed at the Pasteur Institute. Fundamental research in microbiology and related disciplines continued to be carried out primarily at the Pasteur Institute in Paris, while research aimed at producing and improving standardized phar- maceutical agents was primarily carried out at the Marnes-la-Coquette annexe which started as a horse-stable in 1894 and ended as an integrated production centre for a variety of sera and vaccines. This so-called applied research and the associated production of a variety of reagents, although remaining under the supervision of scientists working in Paris, was locally dominated by vet- erinarians, including Alexis Prévot (d. 1926) and Gaston Ramon (1886–1963). Despite the genuine technological and scientific importance of the results obtained in Marnes-la-Coquette (such as the discovery of anatoxins and the assay of specific antitoxin antibodies by flocculation tests, both discovered by Ramon soon after WWI), and despite the positive financial consequences of these activities for the Pasteur Institute, research carried out in Marnes was largely, though unofficially, considered as requiring inferior scientific aptitudes and skills. These two clear-cut cultural identities coexisted until the disappearance of the *Institut Pasteur Production* section in the early 1970s.

Our second remark concerns the manner in which standard procedures were put in place at the Pasteur Institute. Without considering the admin- istrative context which will be dealt with elsewhere in this volume,[67] stand- ardization of the whole process of production of anti-diphtheria serum needed the quantitative assessment of several stages including the charac- terization of the virulence of well defined microbial strains, of toxicity (of toxins and of immunization reagents), controlled attenuation of virulence, and the evaluation of the protective effects of the sera. The various objec- tions raised against the first wave of Pasteur's vaccines had been answered by a step-wise, ordered approach to the production of therapeutic agents. However, the final result did not emerge from the Pasteur Institute alone, but also required a precise knowledge of what was going on at the same moment in Germany.

It has often been argued that the French merely copied the Germans, but our analysis of the laboratory practices shows a reciprocal use of data, most probably made easier by a change in the attitude of the Pasteur Institute scien- tists after 1887. The opposition between French and German scientists during the 'heroic age' of the Pasteur Institute had indeed lost its intensity after 1890 and in some instances, such as the relationship between Roux, Metchnikoff, Behring and Pfeiffer, exchanges became rather friendly. Although some of the protagonists may have remained competitive, the frank discussions between German and French scientists significantly contributed to improving the concepts and practices of vaccination. The different steps in the production of effective biological reagents were clearly individualized, and standards and controls were introduced at the proper levels more or less simultaneously in the two countries, before merging through the use of Frankfurt's standards.

Although such a conclusion may appear provocative, especially as the way of conducting research in the context of nascent medical microbiology was different in the laboratories of Koch-Behring from that of Roux-Duclaux, we are inclined to admit that the development, testing and rapid diffusion of serum therapy against diphtheria in Europe were the consequences of recip-rocal transfers of knowledge from a local context to another and vice versa. The development of serum therapy in France was not the mere mimicking or transplantation of techniques and protocols established in Germany, but rather a succession of adaptive moves. A similar reciprocal transfer of knowl-edge between German and French laboratories has been studied in several other cases, notably immunology[68] and therapeutic chemistry[69] at the Pasteur Institute, and experimental psychology at the Sorbonne.[70] Thus, the history of serum therapy needs to be placed in the general context of the ambiguous and eventually positive relations between French and German researchers and universities which developed soon after the 1870 war and lasted until WWI.[71]

Notes

1. M. D. Grmek, 'L'âge héroïque: Les vaccines de Pasteur', in *L'aventure de la vaccina-tion*, dir. A.-M. Moulin, Fayard, Paris, 1996, pp. 143–59.
2. C. Salomon-Bayet, *Pasteur et la révolution pasteurienne*. Payot, Paris 1998; P. Debré, Louis Pasteur. Flammarion, Paris, 1994; R. Dubos, *Louis Pasteur, franc-tireur de la science*. La Découverte, Paris, 1995; Bruno Latour, *Pasteur: Guerre et paix des microbes*. La Découverte, Paris, 2001.
3. G. Geison, The *Private Science of Louis Pasteur*, Princeton University Press, Princeton, 1995.
4. T. D. Brock, *Robert Koch: A Life in Medicine and Bacteriology,* ASM press, Washington, 1998; K. C. Carter, 'The Koch-Pasteur Dispute on Establishing the Cause of Anthrax', *Bull Hist Med.*, 1988 62: 42–57; C. Gradmann, *Krankheit im labor. Robert Koch und die medisinische Bakteriologie*, Göttingen, 2005.
5. Anonymous, 'Dr Robert Koch latest estimate of Pasteur's methods and discover-ies and the present position of the general inoculation problem.' Editorial of the *Boston Medical and Surgical Journal*, 1883 Vol. CVIII, No. 3, 18 January. Pasteur's reply to Koch was published in the same journal, 1883 Vol CVIII, No. 9, 1 March.
6. Anthrax, or *charbon* in French, is an infectious disease caused by the sporulating bacterium, *Bacillus anthracis*. The disease exists in animals and humans. In the absence of vaccination, the death toll is about 10 per cent among cattle bred in infected areas. Spores persist for several years in the soil and are responsible for

sudden outbreaks of anthrax. This explains in part why discussions about the efficacy of the vaccine were dominated by economics: to be of any use, vaccination of the cattle had to decrease death toll of cows and sheep to below 1–2 per cent, values reached in 1886–7.

7. R. Koch, *Ueber die Milzbranddimfung*, 1882. Translation by K. C. Carter in *Essays on Robert Koch*, Greenwood Press, New York, 1987.

8. Ch. Chamberland, E. Roux and L. Pasteur, 'Le vaccin du charbon', *Comptes rendus de l'Académie des sciences*, 1881 CVII: 666.

9. Ch. Chamberland and E. Roux, 'Sur l'atténuation de la virulence de la bactéridie charbonneuse sous l'influence de substances antiseptiques', *Comptes rendus de l'Académie des Sciences*, 1883, *96*, 1088–91 and 1401–2. The existence of a difference in the procedures used in February and in June 1881 was divulged in 1938 (Adrien Loir *A l'ombre de Pasteur*, Paris, 1938). The use of chemicals to attenuate bacteria, instead of using exposure to oxygen as proposed by Pasteur, is among the first evidence of the influence of German studies on generalized usage of antiseptic substances. See J. Simon, 'Emil Behring's Medical Culture: From Disinfection to Serotherapy', *Med Hist.*, 2007 *51*: 201–18.

10. The list of publications of Pasteur, Chamberland and Roux concerning the vaccination against anthrax can be found in the *Annales de l'Institut Pasteur*, 1908 *22*: 377–80.

11. Ch. Chamberland, 'Résultats pratiques de la vaccination charbonneuse', *Annales de l'Institut Pasteur*, 1887 *1*: 301. The content of the paper is basically that of the lecture delivered at the International Congress of Hygiene, Vienna, 1887.

12. A summary of the violent debates opposing Chamberland and others to Loeffler during the International congress of hygiene, Vienna 25 September to 2 October 1887, can be found in the *Revue d'hygiène et de police sanitaire*, 1887 *9*: 910.

13. At the time of these studies, it was already known that the unseen agent of rabies was present in the brain and spinal chord of the affected animals and that brain extracts could transfer the disease to unaffected animals. Human rabies was rather rare in France.

14. G. L. Geison, 'Pasteur, Roux and Rabies: Scientific Versus Clinical Mentalities', *History of Medicine and Allied Sciences*, 1990 XLV: 341–65.

15. Although many used the name '*Institut Pasteur*', most were financially and administratively independent of the *Institut Pasteur* in Paris, which had no control over them.

16. N. Gamaleya, 'Sur les prétendues statistiques de la rage', *Annales de l'Institut Pasteur*, 1887 *1*: 289.

17. *Revue d'hygiène et de police sanitaire*, 1887 *9*: 902–9.

18. A. M. Silverstein, *A History of Immunology*, Academic Press, San Diego CA, 1989.

19. C. Gradmann, 'Maladies expérimentales. Les expériences sur l'animal aux débuts de la bactériologie médicale', in G. Gachelin (ed.), *Les organismes modèles dans la recherche médicale*, Presses Universitaires de France, Paris, 2006, pp. 75–94.

20. Neither Duclaux nor Roux ever expressed any reservations about Pasteur's work and hypotheses. The reason for keeping laboratory secrets for so long can certainly be deduced by the general policy developed by Duclaux and Roux which gradually made the *Institut Pasteur* the dominant, if not the unique, interlocutor of the French government concerning all aspects of microbiology, from hygiene to tropical medicine passing though the production and the control of therapeutic material of biological origin. The two had experienced the emotional power of Pasteurian discoveries on health and it can be assumed that the *Institut Pasteur*

could certainly not publicly acknowledge the uncertainties and risks associated to initial cures particularly faced with the growing challenge by medical universities.

21. Anonymous, 'L'Institut Pasteur', *Annales de l'Institut Pasteur*, 1889 *3*: 1–17.

22. In 1889, Roux, in addition to being the acting director of the Institute, was in charge of the 'Cours de microbie technique' offered by his own laboratory ('laboratoire de microbie technique'), which included a group of photomicrography, and the animal colony. The three other heads of laboratories were Metchnikoff, Chamberland and Duclaux. See, Sandra Legout *La famille pasteurienne: Le personnel scientifique permanent de l'Institut Pasteur de Paris entre 1889 et 1914*. DEA, Paris: EHESS, September 1999.

23. G. Gachelin, 'The Designing of Anti-Diphtheria Serotherapy at the *Institut Pasteur* (1888–1900): The Role of a Supranational Network of Microbiologists', *Dynamis*, 2007 *27*: 45–62.

24. *Archives de l'Institut Pasteur* (AIP), Fund MTC2 correspondence and AIP, Yersin's letters to his mother.

25. As a kind of family tradition Metchnikoff's brother Leon (1838–88) actively participated in the anarchist movement and was, as a geographer, the secretary of Elisee Reclus (1830–1905) himself a libertarian geographer.

26. The editors were Roux, Chamberland, Duclaux, Grancher, Nocard and Straus.

27. The first article of the first issue of the *Annales* is a letter by Pasteur describing the success of the anti-rabies vaccination. 'Lettre de M. Pasteur sur la rage', *Annales de l'Institut Pasteur*, 1887 *1*: 1–18.

28. The *Bulletin de l'Institut Pasteur*, launched in 1903, was exclusively aimed at the discussion of foreign publications.

29. E. Roux and L. Martin 'Contribution à l'étude de la diphterie (serum-thérapie)', *Annales de l'Institut Pasteur*, 1894 *8*: 609.

30. As shown for example by the creation by the *Institut Pasteur* of a commercial society for the production and distribution of the anthrax vaccine.

31. Letter from Pasteur to Grancher, 4 September 1888.

32. For more on the history of tetanus research, see Anne I. Hardy's contribution to the present volume.

33. P. Weindling, 'From Medical Research to Clinical Practice: Serum Therapy for Diphtheria in the 1890s', in J. Pickstone (ed.), *Medical Innovations in Historical Perspective*, St Martin's Press, New York, 1992, pp. 72–83.

34. A. Opinel, 'The Anti-Diphtheria Apparatus at the *Institut Pasteur* (1890–1914)', *Dynamis*, *27*, 83–106.

35. A. Opinel, 'The Emergence of French Medical Entomology: The Respective Influence of Universities, the *Institut Pasteur* and Army Physicians (1890 ca–1938)', *Medical History*, 2008 *52*: 387–405.

36. J. Weissenfeld, 'Die Veränderungen der Sterblichkeit an Diphterie und Scharlach', *Centralblatt für allgemeine Gesundheitspege*, 1900, 318, cited in the review of the press by F. H. Renaud in *la Revue d'hygiène et de police sanitaire*, 1900, *22*: 955–6.

37. The success of the anti-diphtheria serum was largely publicized in France through engravings printed on the front pages of journals in September 1894, as had happened for rabies.

38. A. M. Silverstein, *A History of Immunology*, 1989.

39. E. Roux and L. Martin, 'Contribution à l'étude de la diphtérie', *Annales de l'Institut Pasteur*, 1888, 2e année n°12, 629–61.

40. E. Roux and A. Yersin, 'Contribution à l'étude de la diphtérie. 2e mémoire', *Annales de l'Institut Pasteur*, 1889, 3e année n°6, 273–88.

41. Note de lecture from *Annales de l'Institut Pasteur*, 1893, 7, 833–7, indicating an article by Erich Wernicke 'Contribution à la connaissance du bacille diphtéri-que de Loeffler et à la sérothérapie', first published in *Archiv fur Hygiene*, 1893, *t. XVIII*.

42. Archives of the Ecole vétérinaire d'Alfort, kept at the Archives départementales du Val de Marne, Créteil, France. Some injections and bleeding may have been made in 1893–4 on the *Institut Pasteur* campus (J. Simon oral communication) and very likely at the municipal stables of Grenelle (Paris).

43. The initial name of the Annexe is *domaine de Villeneuve l'Etang*. A detailed study of the architecture and plans of the Marnes-la-Coquette production centre and of its link with the medical project of Roux and Granger is under preparation (Bottineau, Opinel, Rivoirard, Leniaud and Gachelin). For more detail about the functioning of the annexe at Villeneuve l'Etang, see J. Simon 'Monitoring the Stable at the Pasteur Institute', *Science in Context*, 2008, *21*, 181–200.

44. J. Simon, 'Emil Behring's Medical Culture', 2007.

45. Some of these records have been retained at the *Archives de l'Institut Pasteur* in Paris and at the *Musée des applications de la recherche* in Marnes la Coquette (France).

46. In a first step, Ehrlich and his followers including Roux, used a standardization of the sera based on the toxin itself (the amount of serum needed to neutralize 100 lethal doses of toxin, the lethal dose being the amount of toxin that kills a 250g guinea pig in four days). Because of the variability of the toxicity of the toxin supernatants, Ehrlich, and immediately after him the *Institut Pasteur*, turned to the use of serum standards, more stable than toxin standards. The procedure is described in detail in the course on diphtheria given by Roux and Martin at the *Institut Pasteur* (*Archives de l'Institut Pasteur*, fonds Ramon, box RAM-42)

47. E. Roux, L. Martin and A. Chaillou, 'Trois cents cas de diphtérie traités par le sérum anti-diphtérique', *Annales de l'Institut Pasteur*, 1894 8: 640–61.

48. Anti-diphtheria serum is registered in the *Codex medicamentarius gallicus* 1908, there was no Codex edition between 1884 and 1908. The law issued on 25 April 1895 defines the preparation and use of therapeutic sera (Ministère de l'Intérieur, *Sérums thérapeutique, et autres produits analogues. Législation et réglementation 1895*). A committee was created by decree of the Ministry of interior on 15 May 1895 under the auspices of the Academy of Medicine for controlling the authorizations to be granted to producers.

49. A. Hüntelmann, present issue for a France-Germany comparison of legislation concerning serotherapy.

50. G. Jorland, A. Opinel, and G. Weisz (eds), *Body Counts: Medical Quantification in a Historical and Sociological Perspective*, McGill-Queens University Press, Montreal, 2005.

51. I. Chalmers, 'Comparing Like with Like: Some Historical Milestones in the Evolution of Methods to Create Unbiased Comparison Groups in Therapeutic Experiments', *British Medical Journal*, 1998 *317*: 1167.

52. E. Roux and A. Yersin, 'Contribution à l'étude de la diphtérie. 3ᵉ mémoire', *Annales de l'Institut Pasteur*, 1890, *4ᵉ année n°7*, 384–426.

53. A. Chaillou, and L. Martin, 'Etude clinique et bactériologique sur la diphtérie. Travail du laboratoire du Dr. Roux', *Annales de l'Institut Pasteur*, 1894, 8: 449–78.

54. Roux, Chaillou and Martin, 'Trois cents cas de diphtérie traités par le sérum anti-diphtérique'.

55. *Revue d'hygiène et de police sanitaire*, 1894, *16*: 784–98.
56. J. Weissenfeld, 'Die Veränderungen der Sterblichkeit an Diphterie und Scharlach'.
57. L. Landouzy, *Les sérothérapies*, Masson, Paris, 1898.
58. J. Fibiger, 'Om serumbehandling af difteria', *Hospitalstidende* 1898, *6*: 309–25, 337–50. English translation of the lecture delivered by Fibiger at the Congress of Medicine at Moscow 30 November 1897 has been published at *Bmjjournals.com/content/317/issue 7167*. Fibiger's paper is discussed by A. Hrobjartsson, C. Gotzsche and C. Gluud, 'The Controlled Clinical Trial Turns 100 Years: Fibiger's Trial of Serum Treatment of Diptheria', *BMJ*, 1998, *317*: 1243.
59. See Axel Hüntelmann's contribution to the present volume, and A. Hüntelmann, 'Diptheria Serum and Serotherapy: Development, Production and Regulation in *fin de siècle* Germany', *Dynamis* 2007, *27*: 107–32.
60. The *Ministère d'hygiène, de l'assistance et de la prévoyance sociales*, was created by the decree of 27 January 1920. In earlier times, a *'Direction de l'hygiène publique'* was in charge of public health at the Ministry of Interior since 1889.
61. The *'Association des anciens élèves de l'Institut Pasteur'* reflects well the emotional and scientific link which persists between *Institut Pasteur* and former students of the *Institut Pasteur*.
62. For more on the local institutes in Nancy and Lyon, see Jonathan Simon's contribution to the present volume.
63. M. Kaba, 'La diphtérie à genève à la fin du XIXe siècle: l'entrée en scène de la bactériologie et l'emploi de la sérothérapie', *Gesnerus* 2004 *61*: 37–56, and Mariama Kaba's contribution to the present volume.
64. A. Guénel, 'The Creation of the First Overseas Pasteur Institute, or the Beginning of Albert Calmette's Pastorian Career', *Medical History* 1999 *43*: 1–25.
65. P. Parisot, *Revue médicale de l'Est* 1894 *26*: 641–2; 962–3.
66. J. Simon and A. Hüntelmann, 'Two Models for Production and Regulation: the diphtheria serum in Germany and France' in V. Quirke and J. Slinn (Eds), *Perspectives on Twentieth-Century Pharmaceuticals*, Oxford: Peter Lang, 2010: 37–61, p. 45.
67. See Axel Hüntelmann's contribution to the present volume.
68. E. Riestchel and J.-M.Cavaillon, 'Endotoxin and Anti-Endotoxin: The Contribution of the Schools of Koch and Pasteur: Life, Milestone-Experiments and Concepts of Richard Pfeiffer (Berlin) and Alexandre Besredka (Paris)', *J Endotoxin Res*, 2002 8: 71–82.
69. C. Debue Barazer, 'Les implications scientifiques et industrielles du succès de la Stovaïne. Ernest Fourneau (1873–1949) et la chimie des médicaments en France', *Gesnerus* 2007 *64*: 24–53.
70. J. Carroy and H. Schmidgen, 'Reaktionsversuche in Leipzig, Paris und Würzburg: Die deutsch-französische Geschichte eines psychologischen Experiments, 1890–1910', *Medizinhistorisches Journal* 2004 *39*: 27–55.
71. H. W. Paul, 'The Sorcerer's Apprentice: The French Scientist's Image of German Science, 1840–1919', University of Florida Press, Gainesville, 1972; Cl. Digeon, *La crise allemande de la pensée française, 1870–1914*, Presses universitaires de France, Paris, 1992.

5
Quality Control and the Politics of Serum Production in France

Jonathan Simon

Louis Pasteur died on 28 September 1895 at the Pasteur Institute's facility at Garches outside Paris. Today, the visitor who makes the journey to this quiet Parisian suburb is reminded of Pasteur's prominent place in the history of French science by a memorial plaque mounted on the wall of the former cavalry stables. The founder of microbiology, at least according to the French version of this history, Pasteur revolutionized the conception of infectious disease and paved the way for the major therapeutic success stories of twentieth-century medicine, from antisepsis to antibiotics. Pasteur's room at Garches, preserved in the state it was in when he died, is bathed in a calm silence broken only by the occasional passage of a train on the tracks that run behind the building, or the honking of one of the geese on the grounds of the estate. It is easy to imagine Louis Pasteur living out the end of his life in this tranquil atmosphere, with the repose of his final weeks interrupted only by visits from friends, colleagues and relatives coming to pay their last respects to the great hero of French science. But in fact, the tranquillity that envelops the buildings at Garches today is a far cry from the atmosphere that reigned at the time of Pasteur's death. In September 1895, production of the diphtheria serum, the largest project the Pasteur Institute had ever undertaken, was in full swing at the site. According to notes found in the administrative notebooks, the Institute produced some 7500 litres of blood containing serum for the treatment of diphtheria in 1895, a volume not equalled again until 1899, and then for all the sera produced by the Institute.[1] This industrial-style production of serum meant not only that there were over 100 horses living in the renovated and enlarged stables only yards away from Pasteur's deathbed, but an equally short distance away on the other side of the building there were around 3000 guinea pigs in cages. According to the caretaker, a certain Monsieur Pernin, 'These guinea pigs, shut up in long iron cages, produce a deafening sound'.[2] Indeed, the journalist from *Le Matin* who reported M. Pernin's words concluded his article with an entertaining story about what happened at the end of his visit. Upon his departure, he came across some residents of a neighbouring retirement home who asked him whether the Pasteur Institute was planning to move patients into the facilities at Garches. When the journalist enquired whether they would like that, the pensioners replied that at least sick people would make less of a racket.

With this level of noise, it is hard to see how an ailing Louis Pasteur could have managed to sleep at night, let alone pass his last days in calm tranquillity. Someone with a penchant for conspiracy theories might even imagine the situation as part of a plot by Emile Roux to dispose of the age-ing Pasteur – not only his mentor, colleague and then director at the new Pasteur Institute, but also, in certain senses his rival.[3] While Roux would not himself become director of the Institute until 1904, he had assumed the mantle of the founder's scientific heir well before Pasteur's death in 1895, rising to prominence precisely because of the public success of the Institute's treatment for diphtheria.[4] Conspiracy theories aside, the fact that Pasteur died on the site of the large-scale production of diphtheria serum is highly symbolic in terms of the history of the Pasteur Institute. The found-ing of the Institute coincided with Pasteur's decline and his withdrawal from public life, largely due to illness, ushering in a new generation of Pastorian microbiologists who would shape the future of the institution, including Chamberland, Metchnikoff, Duclaux, Roux and Calmette.

An important part of the institutional history of this period turned around the production of diphtheria serum that started in earnest at the end of 1894. Indeed, it was this serum production that would ensure the finan-cial security, if not the survival, of the Institute at the close of the nineteenth century, enabling its successful passage into the twentieth.[5] As Weindling has already argued, the French institution was able to raise a great deal of capital through its charitable fundraising drives, but regularly struggled to cover its running costs – a model that he contrasts with the financial situ-ation of Robert Koch's Institute for Infectious Diseases in Berlin where the state provided less generous capital investment but more ample long-term financing.[6] As we shall see, the dramatic discovery of an innovative treat-ment for diphtheria was mobilized to generate a very large sum of money for the Pasteur Institute (and, coincidentally, for other institutions as well) through charitable donations that flooded in from all over France as well as the rest of the world. This money was used to launch the large-scale produc-tion of the serum for treating diphtheria at the specialized plant in Garches where Pasteur died. Ironically, therefore, Pasteur's name was used to raise the funds that effectively transformed his last months into a purgatory.[7]

While the subscription, initiated by Gaston Calmette (a leading journal-ist and brother of the Pastorian Albert Calmette) and successfully run by *Le Figaro*, ensured the financial security of the Pasteur Institute, the Institute was expected to keep its side of an implicit pact made with the French people. The new hero of this cause, Emile Roux, personified the Pasteur Institute's successful treatment of diphtheria – particularly through the publicity afforded him by *Le Figaro* – and by the same account he assumed the role of the public guarantor of the quality and value of the Institute's serum. It was, therefore, of vital importance that the use of the diphtheria serum and in particular its manufacture in France passed without any hitches, especially for serum issuing from the facilities of the Pasteur Institute. Herein, I will argue, lay the principal value of the Institute's practice of evaluating their serum. The production and regulation of the diphtheria serum, as well as its perception, differed from that in Germany, so it is unsurprising that no equivalent 'Wertbestimmung' infrastructure evolved in France until much later. Nevertheless the absence of an independent institute like Paul Ehrlich's in Berlin, and then Frankfurt, and the lack of direct government control did not mean that the French producers – the Pasteur Institute foremost among them – did not impose measures for quality control; they simply operated in a different way.[8]

The present paper is organized into three sections that while in a sense chronological, more properly correspond to my analytical concerns. The first section treats the early history of serum production and the ambitions of the Pasteur Institute to establish a strict monopoly over this enterprise in France. The second section presents the subsequent developments, with government legislation and the structuring of the market for the production and distribution of diphtheria serum across the country. The last section deals specifically with the question of the evaluation and control over the quality and potency of the serum at the main production centre run by the Pasteur Institute at Garches. In this final section, I will offer an analysis of the French system of *Wertbestimmung*, or lack of it, in the context of the multiple functions of serum production and distribution in France.

Production prior to legislation – the monopoly that never was

As we have already seen in other papers in this volume, the treatment of diphtheria in humans using serum taken from animals immunized against the disease was developed in Berlin between 1890 and 1893.[9] The first commercial producers of such a serum for human use were the German pharmaceutical companies Hoechst and Schering, making their product available to the public via pharmacies in the second half of 1894. Serum had been produced by the Pasteur Institute in small quantities for use in preliminary human experiments at least since the beginning of 1894, but the serum did not become widely available until the end of that year or the beginning of 1895. France and the rest of Europe learned of the discovery of an effective treatment against diphtheria following the communication made by Emile Roux at the Eighth International Congress on Hygiene and Demography held in Budapest from the 2–8 September 1894. Emil Behring, the leading researcher in the development of the serum was not present at this conference, so there was no-one of sufficient stature in the German delegation to counterbalance the French contingent's announcement.[10] It appeared to the French, and to others, that Roux had discovered the treatment himself, and had tested it before anyone else in a trial conducted with the cooperation of Dr Jules Simon, responsible for the diphtheria ward at the *Hôpital des Enfants Malades* (a trial that Gabriel Gachelin has already discussed in his contribution to the present volume).[11] Articles in the popular press on the new cure for diphtheria, particularly the ones in *Le Figaro*, generated requests for the serum from all over France. Writing in 1895, Roux expressed his sense of being overwhelmed by the demand during this crucial period between September 1894 and the end of the year.

> Requests for serum came from everywhere, and like a rising tide threatened to submerge the bacteriologists. As for us, we would not have believed such a rapid success possible, we thought that like all good things, the serotherapy for diphtheria would only be introduced slowly, and so we only prepared enough horses to supply the hospital services, and our poor animals, even at the cost of giving up all their blood, would not be capable of furnishing one hundredth of the required quantity.[12]

By early October, the *Hopital des Enfants Malades* had to open a new ward to cope with the influx of new patients whose parents brought them to Paris in the hope of benefiting from the new treatment.[13]

Roux, it seems, had dramatically underestimated the public response to his communication in Budapest, which had received a great deal of publicity throughout Europe. He had planned a leisurely return to France that would include some tourism in the Tyrol, but had to cut this trip short to deal with the problems of serum supply in Paris. Upon his return, Roux was further embarrassed by what he saw as the eclipse of Behring in the presentation of the discovery of serotherapy by many French journalists for whom it was an essentially French affair. According to his own description of events, Roux found the general excitement over the cure for diphtheria an undesirable nuisance. Nevertheless the publicity provided an opportunity for Roux to try to ensure that the Pasteur Institute would be the sole producer of the serum in France. In an in-depth report published on the front page of *Le Figaro* at the beginning of October, Gaston Calmette offered a justification for such a monopoly.

> Two doctors – under the permanent direction of M. Roux – will be assigned to the preparation of the serum, which requires so much care and so many precautions. It is precisely this long series of precautions and the care involved that makes the absolute centralization of the services for the vaccine against diphtheria necessary, and that prevents the creation of any departmental institutes, *as the multiplication of laboratories in the absence*

of M. Roux's direct control, would take away all the guarantees concerning the
preparation of the serum and its quality.

Thus, there will be no subsidiaries of the Pasteur Institute for the diphtheria vaccine; although in a few months there will be something infinitely preferable for the public, stores of vaccine in all the large centres.[14]

The motivations behind this bid for a monopoly over serum production are a matter for speculation. While there was no doubt a concern about guaranteeing the quality of the serum, there was probably also a desire for the Pasteur Institute not to have to share the income (both direct and indirect) or any corollary benefits with other producers. In light of the French publicity that regularly identified the diphtheria serum with Roux and the Pasteur Institute, there was doubtless a fear that the revelation of poor quality or dangerous serum, even if it came from another producer, would reflect badly on the Pasteur Institute, thereby risking not only its reputation, but also its very survival. Furthermore, any such negative publicity would be a betrayal of all those who had so generously supported the Pasteur Institute through *Le Figaro*'s subscription, which had started on 20 September 1894 soon after the Budapest Congress. This fundraising exercise easily surpassed its initial optimistic goals, with Gaston Calmette fixing an initial target of 30,000 francs to pay for serum for the poor. After one month, the fund had already reached 240,000 francs, attaining 612,000 francs by the end of the year. While this money was used to buy horses and to prepare the facility at Garches for the mass manufacture of diphtheria serum, a substantial part was also invested to provide income for the Institute.

The publicity that the serum brought to the Pasteur Institute proved, however, to be a double edged sword. At first, Roux only saw the leverage it gave him to construct the French market as he saw fit, but the 'few months' indicated in the citation from *Le Figaro* quoted above would prove fatal to his dream of the Pasteur Institute's monopoly over French serum. The anticipated delay of a few months prior to the introduction of full-scale production was due to a technical constraint – the length of time it took to immunize horses against the diphtheria toxin so they could produce effective serum. Despite his awareness that diphtheria was widespread in France, as in the rest of Europe, Roux had not anticipated the demand for the serum following his intervention in Budapest. Furthermore, based on the funding practices of the Institute it is possible that he simply did not dispose of the capital that needed to be invested, as the money used to purchase around a hundred horses in September 1894 came from the subscription that took its roots in the wave of publicity following Budapest. Whatever the reasons for this delay, there were many people across France who were disappointed by the lack of serum, particularly outside Paris. Roux's notion that everyone would wait patiently for the Pasteur Institute to build up its capacity was mistaken, as we can see if we look at what happened in Lyon.

In the wake of the reports of Emile Roux's presentation at Budapest, Dr Gabriel Roux, the homonymic director of the *Bureau d'Hygiène* in Lyon, was asked by the mayor to obtain serum for the city. Roux wrote to the Pasteur Institute in Paris but received a disappointing response:

> The Pasteur Institute tersely replied that the antitoxic serum would not be sent out to the provinces within the next two months, and then would only be delivered to hospitals and patients signed up with the Welfare Centres (*Bureaux de bienfaisance*).[15]

In his report to the Mayor, Gabriel Roux suggested that Lyon should, like its smaller neighbours St Etienne and Grenoble, try to produce its own serum. Saturnin Arloing, a professor at the veterinary school as well as at the medical faculty, was chosen for the task. The project quickly grew, with Gabriel Roux conceiving an integrated microbiology laboratory that would also be able to carry out diagnostic analyses for microbial disease. Indeed, this diagnostic capacity appears to have been a common feature of these regional serum production centres, as it was, for example, also an important function of the *Institut Sérothérapique de l'Est* run by Eugène Macé in Nancy. While these provincial institutes were initially set up to produce serum for regional needs, the inclusion of a laboratory drew a number of elite provincial doctors into microbiological research. The final step taken in Nancy, and possibly in other centres as well, was to organize courses in microbiology based on the model offered by the Pasteur Institute, where many of the staff had themselves received their initial training. The irony is compounded by the fact that Pasteur Institute's celebrated *Cours de Microbie* – the model for such courses developed elsewhere – was, prior to the advent of serotherapy, Roux's most notable contribution to the Institute's activities.

Thus, Roux's initial inability to supply the provinces led to the establishment of regional centres that competed with Paris in terms of microbiological expertise and teaching. Like the Pasteur Institute, they were funded partly (but usually minimally) by sales of serum, by municipal or regional subsidies, and by substantial public donations. In Nancy, for example, the money that had already been collected for the appeal launched by *Le Figaro* was not dispatched to Paris, but was instead diverted to fund the local venture. The irony of this situation was that these regional centres found themselves in the same situation as the Pasteur Institute, needing to wait three months to have immunized horses ready to produce the serum. Thus, although he started the immunization process in November 1894, Arloing was only able to supply the Lyon hospitals with locally produced serum in February 1895, by which time the Pasteur Institute was supplying the whole of France from its site at Garches.

This kind of development was seen all over France, with the result that by the beginning of 1895 the Pasteur Institute, while by far the largest producer (most regional centres operated with only one or two immunized horses

compared to over a hundred at Garches), no longer found itself in a position
of monopoly over serum production. It is important to note, however, that
this competition was not at all on the German model, where private companies like Hoechst and Schering were competing fiercely for market share with
the aim of making a profit from the sale of serum. Although these French
regional producers were often, as in Nancy, private charitable foundations,
none were seeking to make a profit out of the serotherapy, and they seem
to have given away much more serum than they ever sold. Furthermore, the
directors and other staff were often salaried members of the local medical
faculty, which explains the facility with which such institutes could later be
integrated into these faculties as embryonic microbiology departments.

Legislation – hegemony and multiplicity

In April 1895, Félix Faure, a French President better known for dying in
office than anything else, promulgated a new law covering serotherapy in
France after six months in which anyone could in principle have produced
and sold serum for the treatment of diphtheria. According to the winning
side of the parliamentary debate, unregulated sera represented a double
danger. First, there was the risk of impure serum getting onto the market,
as an average pharmacist – untrained even in the most elementary microbiological techniques – would be incapable of ensuring the purity, let alone
the antitoxin activity of the serum. Hence, the second danger; that of ineffective serum. Very soon after the introduction of serotherapy, specialists
concluded that it was important to administer the serum as soon as possible
after the patient had been infected with diphtheria. The general consensus
among medical practitioners was that more than a week or so after the
appearance of the first symptoms of diphtheria, the use of serum could no
longer save an infected child.

This new law, which was initially intended to be appended to pharmacy
legislation under debate at the time, was introduced when the Pasteur
Institute was no longer the sole producer of serum in France. But, as there
was no competition from private companies within France, the government
was not faced with what many saw as the corrupting influence of private
enterprise and the quest for profits in the sphere of public health.[16] The
solution adopted was to institute a system of prior approval for any prospective serum producers. The law was not, however, limited to diphtheria
serum, as we can see by examining the first article:

> Art. 1. Attenuated viruses, therapeutic sera, modified toxins and analo-
> gous products that can serve as prophylaxis against or therapy for conta-
> gious diseases, and injectable substances of organic origin not chemically
> defined, applied to the treatment of acute or chronic affections cannot be
> debited, free or against payment, unless they have received a government

authorization either for their fabrication or for their origin. This authorization will be granted following advice from the French consultative committee for public health and the Academy of Medicine.

These products will benefit only from a temporary and revocable authorization. They will be submitted to an inspection carried out by a commission named by the relevant ministry.

The aim was to put some order into the burgeoning world of opotherapy, the use of extracts of animal organs for treating a range of complaints, 'medical' practices that escaped the traditional legal framework of French pharmacy. The legislation quite clearly resolved the problem faced by pharmacists in trying to evaluate the harmlessness or potency of the serum they were supposed to sell. The government relieved the pharmacist of their legal responsibility, and would (at least in principle) itself guarantee the quality of the serum – whether sold in the pharmacies or distributed via the Welfare Centres that formed the backbone of a new public health system (the *Bureaux de bienfaisance*). In turn, the government aimed to ensure the quality of the serum by allowing only approved institutions to supply the product.[17] Official approval would be granted by the Ministry of the Interior, and the law would be enforced by the police authorities under the direction of the Prefects of the individual *Départements*.

The decision concerning the official approval of a given producer was to be based on a report made by a body that came to be known as the serotherapy commission. Composed of members appointed from the Academy of Medicine and the Ministry's Consultative Committee on Public Health, this commission was to assess the prospective serum producers and transmit its opinion to the Ministry of the Interior. In this kind of technically sophisticated domain, the Minister's decision would inevitably be based on the expert opinion of the commission. The composition of the commission was in part fixed by the terms of the legislation, with the secretaries of the Academy of Medicine automatically appointed, along with those sitting on the government's Consultative Committee on Public Health.[18] Other members appointed from within the Academy of Medicine included Nocard, Duclaux, Straus and Grancher, all committed supporters of pastorian science and its institutions if not active members of the Pasteur Institute itself. With this weight of Pastorians on the committee, it is no surprise that when the first list of approved institutions appeared in January 1896, it opened with the Pasteur Institute in Paris, followed by the Pasteur Institute in Lille. Nevertheless institutes in Le Havre, Nancy, Lyon, and Grenoble were also approved, with laboratories in Bordeaux, Marseilles, Montpellier, and Rouen being approved the following year.

Thus, as we have seen with the examples of Lyon and Nancy, by the spring of 1895 when the law was introduced, a complete monopoly of the Pasteur Institute over serum production in France was no longer a realistic possibility.

Indeed, unable to supply the regions with the new treatment against diphtheria, Emile Roux had felt obliged to offer his approval if not active support of local production efforts, often undertaken by former students of his from the *Microbie* course in Paris. Furthermore, as we explained above, these were not private companies; they were aspiring Pasteur Institutes dedicated to the same selfless ends as their model, and more or less directly attached to medical faculties. Provided they met the standards of hygiene and scientific rigour imposed by the commission, therefore, they would be approved. Two interesting questions remain unanswered; whether any private companies ever applied for approval to produce diphtheria serum in France, and whether a request was ever submitted to the commission to approve the sale of imported serum.[19] We know, for example, that in the period before the Pasteur Institute's serum became generally available, the Charité hospital in Lyon used 'Behring's serum' – manufactured by Hoechst – to treat its diphtheria patients, although they replaced this costly product first with Roux's serum and then Arloing's even before the new French legislation came into effect.[20] In the absence of the relevant archival sources, however, it is impossible to say whether no such applications were ever made, or whether the commission actively excluded other actors from the French market.

Quality control – the guarantee that comes with a household name

This final section is concerned specifically with the system of quality control that was put in place by the Pasteur Institute at its main production facility at Garches, the French version of 'Wertbestimmung' as it were. Here I pose the question of how the evaluation of the quality of the serum is related to the story of its production and regulation in France recounted above. I will argue that due to the very different circumstances prevailing in the two countries, evaluation did not perform the same function in France as in Germany: in Germany, *Wertbestimmung* played an increasingly vital role in a commercial struggle for market share, assuring the consumers that they were receiving a high-quality product, while in France the main exigency was for the Pasteur Institute not to be seen to fail in its enterprise. The regional French producers had the same concern, although they were not involved in precisely the same enterprise. Thus, an important part of the history of serum production and quality control turns around the definition – implicit rather than explicit – of the 'enterprise' in question. This enterprise cannot be reduced to the simple production of quality serum; there was also the construction of the producer's image, its implication in related areas of health, its educational and political ambitions, and, necessarily, its financial viability. The stakes were evidently not the same for all the actors involved in serum production. While the production of serum at Lyon or Nancy was on a much smaller scale than at Paris, it was principally the local reputation of the product (and to a lesser extent the producer) that was at stake. After all, although he would lose face in case of a significant problem concerning the quality of the serum produced in Lyon, someone like Saturnin Arloing would not be in danger of losing his job at the medical faculty. For Émile Roux, however, the stakes included the future of the Pasteur Institute as well as his title as the legitimate successor to Louis Pasteur.

Due to the efforts of the archivists at the Pasteur Institute, we are lucky enough to have a number of registers that record the day-to-day process of production and testing of the diphtheria serum. While the series is incomplete, these registers nevertheless provide valuable information concerning the functioning of serum production in France before WWI.[21] Thanks to these documents, we can see that the French retained Behring's original technique for determining the efficacy of the serum, even after German manufacturers had adopted Paul Ehrlich's much more sophisticated toxoid neutralization process.[22] The method that served the French started with a determination of the lethal dose of the toxin – the dose that would kill a guinea pig of the appropriate weight within 12 hours – that constituted the reference for a subsequent set of tests to be performed on every batch of serum. Two tests were performed on samples of serum derived from every single bleeding of 4 or 6 litres of blood from a horse. These tests involved injecting two different guinea pigs with specific quantities of serum (1/50,000 and 1/100,000 of the weights of the guinea pigs) followed by a lethal dose of the toxin; if the guinea pigs survived this lethal dose then the serum was assigned a value of 150 or 200 units, respectively.

In a comparative assessment of the French and German sera from 1897, Thorvald Madsen, who would later head the State Serum Institute in Denmark, indicated several weaknesses in the French approach compared to the German one, reflecting a general perception that the German serum was of better quality.[23] Roux, on the other hand, was able to point to several problems associated with the German testing procedure, noting that each assessment using the German method required three different materials – the serum under test, a standard serum, and a standard toxin – rather than the two used in France (where no standard serum was used). While it was never made explicit, a major problem with adopting Ehrlich's approach was that it would undermine the independence of the Pasteur Institute, and worse still in light of Franco-Prussian relations make France dependent on Germany, as the standard serum and toxin would inevitably be supplied by Ehrlich's Institute in Frankfurt.

Indeed, as early as 1894, Roux had publicly defended the French tradition of evaluating the serum, giving a clear impression that Ehrlich's more sophisticated methods of *Wertbestimmung* were more work than they were worth:

> To tell the truth, we don't attach a great deal of importance to all these complicated definitions [Ehrlich and Behring's standard for the diphtheria

antitoxin] and we believe that the search for this degree of precision that is not found in nature is too much [unnecessary] effort. The essential thing is to understand one another: in our opinion, it was easier when one simply said of a serum that its immunizing power was 1000 or 100,000 with respect to a toxin or a virus that killed control animals in so many hours.[24]

While this attitude may seem indefensible to modern biomedical sensibilities, where precision is viewed as an end in its own right, Roux was in a certain sense right, particularly given the variability of the clinical use of the serum. Once the serum was no longer in short supply, it was not uncommon for doctors to continue injecting 20 cm^3 doses once or even several times a day until the symptoms started to abate.

Thus, the major considerations were that the product should be of sufficient potency (as determined by the guinea-pig tests presented above) and that it was 'safe'. What is striking when one looks at the registers that record the day-to-day operations at Garches is the consistency with which Louis Martin applied the Institute's quality control standards, given the small number of staff involved and the lack of external oversight. Thus, every bleeding from each individual horse was tested for its potency, and each bleeding was kept separate, allowing the traceability of the phials that were dispatched for clinical use. This contrasts, for example, with the practice developed at Lyon, where Arloing blended batches of serum to obtain a product with a uniform potency.[25] Thus, although the quality control system may not have been as sophisticated as at Frankfurt, it was nevertheless applied as methodically, and this in the absence of any legal obligation. Indeed, as we have seen, once a serum producer like the Pasteur Institute had been approved by the Ministry of the Interior, there were no longer any inspections or other requirements, and only a very vague threat that the approval could be revoked.

Of course, this constant control was in part due to the nature of the units of production – the individual horses. Microbiologists, like other biologists, knew how variable such living organisms could be, both in terms of their own variability and in comparison to the other horses, and it is not clear that such rigour would either have been demanded or respected if the process had been one of chemical synthesis rather than bio-production. Nevertheless I would argue that the stringency of the system, with its high cost in time and material (notably its consumption of guinea pigs), also reflected the fact that the Institute could not afford any suspicion that their serum was not 'good'. The control mechanisms put in place at Garches were not part of a competitive market strategy, but part of a wider approach aimed at protecting the reputation of the Pasteur Institute and its members. This, I believe, explains in part Roux's vitriolic response to Paul Moizard when the latter suggested in a presentation before his colleagues in 1895 that the serum could potentially be fatal. I want, therefore, to close this paper with a brief examination of this dispute that seemed for a short time to threaten the credibility of the Pasteur Institute's diphtheria serum.

In early May 1895, a six-year-old girl with the symptoms of diphtheria was brought to the *Hôpital des Enfants Malades* and was treated with 10 cm³ of serum from the Pasteur Institute. The child then exhibited a series of more or less dramatic symptoms before dying ten days later. In the meantime, a microbiological examination of a culture made from the child's throat suggested that although streptococcal bacteria were present, the child was not infected with diphtheria. The physician in charge of the case, Dr Bouchard, called upon Dr Paul Moizard, a colleague specialized in diphtheria and a convinced Pastorian, to help him with the case. After the girl's death, the two doctors prepared a communication presented before the Medical Society of the Paris Hospitals (*Société médicale des hôpitaux de Paris*) in early July and subsequently published in the society's journal, much to Emile Roux's displeasure.[26] The two doctors recommended not administering the serum in cases of non-diphtheric angina, a situation exemplified by this case. Nevertheless it was considered imperative to treat cases of diphtheria early and the biological diagnosis took time, particularly if one was not close to a laboratory, leading to a clinical consensus that a doctor could not afford to await the outcome of the laboratory test before starting treatment. Therefore, this case, despite the fact that in principle the recommendation should not affect the treatment of children with diphtheria, was, according to Roux at least, dissuading doctors and parents from using the serum.

Roux attacked Moizard's conclusions, pointing out how little clinical information was provided for the case, and that no autopsy had been performed. Exploiting the lack of conclusive evidence, Roux claimed that the death had been caused by the streptococcal infection identified by the laboratory analysis.[27] Moizard backed down and offered his apologies, taking the opportunity to reassert his faith in serotherapy. While Roux had quickly succeeded in defusing the situation, suppressing the suggestion that the serum might be lethal in certain circumstances, the debate did leave some traces. The way the exchange was conducted shocked the sensibilities of many doctors, with a certain Dr Variot – himself an advocate of the use of serotherapy – commenting that Roux had 'lacked that urbanity and calm that suits men of the laboratory so well when they respond to the criticisms and objections that one has the right to make of their experiments'.[28]

The tone adopted by Roux in his response to Moizard's report was not at all calm, precisely, I would claim because of what was at stake if the medical community came to suspect that the serum might be dangerous, let alone lethal. If people stopped using the serum, it would be a double catastrophe for Roux, seriously damaging his reputation and perhaps forcing the closure of the Pasteur Institute. While I certainly do not want to suggest that Roux did not genuinely believe in his own explanation, I would claim that the vehemence with which he defended his position was informed by the unthinkable consequences this kind of negative publicity could bring.[29]

Conclusion

In the absence of the competitive context that characterized German serum production, there was no obvious motivation for 'improving' French serum, or the way of assessing its quality. The increasingly sophisticated regimes that Ehrlich developed in Frankfurt responded to technical demands and were backed up by collaboration with the leading German serum producers. In France, as long as their serum remained suitably effective and was not seen to present any particular threat to public health, the Pasteur Institute could continue with its 'primitive' quality-control regimen. Already by 1895, keeping up with German standards would have meant the Pasteur Institute adapting to a German-dominated understanding of *Wertbestimmung*, providing another reason for the French not to update their system of quality control. On the other hand, the method had to work, as the idea that the serum might be ineffective, or, worse still, unsafe was unthinkable to Roux. With so much at stake for the Pasteur Institute, there was no room for any doubts concerning the quality of the serum. This was why the system for measuring the efficacy of the serum that had been established had to be consistently and rigorously applied even in the absence of any legal requirement. Of course, the fact that the serum was produced by variable living animals made the quality of each batch less predictable, but there is no hint that Louis Martin sought any routine shortcuts in the evaluation process in place at Garches.[30]

The Institute's serum, its reputation and its funding were intimately bound up together; the serotherapy service had been generously pre-financed by the subscription organized by *Le Figaro*, providing a further reason for the Institute not to drop its guard. A great deal of trust had been placed in the Pasteur Institute, and it was important, for Roux in particular, that it could justify its image as an independent institution that selflessly placed France's public health interests above all other concerns – profit in particular. One of the unintended consequences of this configuration was that the Pasteur Institute's serum was not as potent as that produced in Germany; it was, however, widely and efficiently distributed, and it was relatively cheap (particularly at first), with much being given away through administrative networks or public health offices.

I suggested at the beginning of this paper that the production of the serum contributed to Pasteur's death; but it clearly helped the perpetuation and glorification of his name. Nevertheless this was at the cost of Emile Roux eclipsing Louis Pasteur, at least for a short time, in the public eye. Quality control – *Wertbestimmung* – constituted a central part of this story, as quality was the guarantee of the good faith of a population that had generously supported the hybrid enterprise that was the Pasteur Institute by donating to the serotherapy fund. The production of diphtheria serum in France was a new venture for the Pasteur Institute both in terms of its scale and in terms of the government regulation of the product, and it therefore required the realization of a new model for production and distribution. As we have seen, the model was not the one originally envisaged by Emile Roux, as he could not, for various reasons, ensure the Institute's monopoly over the product. But a new model was put into place, in large part based on a way of functioning already established in the Institute's short history. As we can see by looking at other examples presented in this volume, this particularly French model has left its traces in a whole range of areas in the history of therapeutics, particularly in France.

Notes

1. *Archives de l'Institut Pasteur* (AIP). The notes were found in the front cover of the one of the registers used for recording the bleedings of the horses as part of the monitoring of serum production.
2. *Le Matin* 31 October 1894, 'L'Écurie du Dr Roux'.
3. For more on this rivalry, see Geison's claims concerning the attenuation of rabies. Gerald Geison, *The Private Science of Louis Pasteur*, Princeton: Princeton University Press, 1995.
4. It was Duclaux who inherited the direction of the Institute from Pasteur.
5. For an illuminating overview of the history of the Pasteur Institute in the twentieth century, see Ilana Löwy, 'On Hybridizations, Networks and New Disciplines: The Pasteur Institute and the Development of Microbiology in France', *Studies in the History and Philosophy of Science*, 1994, 25: 655–88.
6. Paul Weindling, 'Scientific Elites and Laboratory Organisation in Fin De Siècle Paris and Berlin', in Andrew Cunningham and Perry Williams (eds), *The Laboratory Revolution in Medicine*, Cambridge: Cambridge University Press, 1992, pp. 170–88.
7. Pasteur even wrote a letter of thanks that was published in *Le Figaro* on 6 October 1894: 'Je vous remercie, plus que je ne saurais dire, au nom de tous mes collaborateurs qui travaillent à la fois pour la science, la patrie, et l'humanité'.
8. See the papers by Axel Hüntelmann and Anne I. Hardy in the present volume for more information on the production and regulation of serum in Germany.
9. See the contributions by Axel Hüntelmann and Anne I. Hardy in the present volume.
10. Hans Aronson, who presented earlier tests of the diphtheria serum in Berlin, seems to have made less of an impact. It is hard to find any mention of Aronson in the French press. In a private letter to Metchnikoff, Emile Roux attributed the unexpected impact of his own communication in Budapest to the absence of Behring. Roux, correspondence, AIP.
11. According to the notes held in the archives of the *Assistance Publique* in Paris, Emile Roux, Louis Martin and Chaillou started administering the experimental treatment on 30 January 1894, *Historique des établissements*, 1894, p. 289.
12. *Archives de l'Institut Pasteur, Direction* (1888–1940) File 'Création du Service de Sérothérapie', AIP.

13. Archives of the *Assistance Publique, Historique des établissements*, 18 October 1894, pp. 292–3. 'J'ai informé M. Gallet que les enfants de la province affluaient à l'hôpital pour bénéficier du traitement de M. le Dr Roux, et qu'il fallait prévoir, à courte échéance, un encombrement du service de la diphtérie'. Paul Weindling has written about the importance of serotherapy in improving the image of the hospital as a place to be cured rather than a place to die. Paul Weindling, 'From Medical Research to Clinical Practice: Serum Therapy for Diphtheria in the 1890s', in John Pickstone (ed.), *Medical Innovations in Historical Perspective*, New York: St Martin's Press, 1992, pp. 72–83.

14. An article by Gaston Calmette, *Le Figaro* 12 October 1894, front page: 'Deux médecins seront préposés, sous la direction permanente de M. Roux, à cette préparation du sérum, qui nécessite tant de précautions et tant de soins. C'est cette longue série de précautions et de soins qui rend précisément indispensable la centralisation absolue des services du vaccin contre la diphtérie, et qui empêche toute création d'instituts départementaux, car la multiplication des laboratoires, avec l'absence du contrôle immédiat de M. Roux, enlèverait toute garantie à la qualité et à la préparation du sérum.

Il n'y aura donc en France aucune succursale *de l'Institut Pasteur* pour le vaccine de la diphtérie; mais il y aura au contraire, dans quelques mois, ce qui est infiniment préférable pour le public, des dépôts de vaccin dans les grands centres, dans les bureaux d'hygiène des villes, et ces bureaux, qui recevront un approvisionnement direct toujours au complet, délivreront des tubes à tous les médecins qui en feront la demande'. My emphasis in the English translation.

15. 'Rapport de M. le Dr Roux soumis à Monsieur le Maire', 6 November 1894, Archives Municipales de lyon.

16. While I have not come across any explicit discussion of the place of private enterprise in serum production in France, I want to suggest that there was a generalized perception that it was undesirable. This relates as much to the 'ethos' of the Pasteur Institute as the tradition of public health policy in France.

17. For more on the new system of public health care in France, see Olivier Faure *Histoire sociale de la médecine (XVIIIe – XXe siècles)*, Paris: Anthropose, 1994.

18. The Serum Commission was initially composed of the following members: Brouardel, Monod, Proust, Chantemesse, Bompard, Delaunay-Belleville, Bergeron (Secretaries of the *Académie de médecine*); Nocard, Duclaux, Straus, Grancher (ordinary members of the *Académie de médecine*); and Pouchet, Ogier, Thoinot, Netter (Members of the *Comité consultatif d'hygiène*).

19. Private companies were approved to produce other sera in France, but not diphtheria serum.

20. See Pierre Patet, 'La sérothérapie à Lyon', MD dissertation, Lyon: Faculty of Medicine and Pharmacy, 1895. For more on the German product, see Anne I. Hardy's and Axel Hüntelmann's articles in the present volume.

21. For a detailed account of the functioning of Garches based on these records, see Jonathan Simon 'Monitoring the Stables', *Science in Context*, 2008, *21*: 181–200.

22. See Anne I. Hardy's paper in this volume.

23. For more on the *Statens Serum Institut*, see Anne Hardy's paper in this volume. Thorvald Madsen, 'Ueber Messung der Stärke des antidiphtherischen Serums,' *Zeitschrift für Hygiene und Infektionskrankheiten* 24 (1897): 425–42.

24. E. Roux and L. Martin 'Contribution à l'étude de la diphtérie (Sérum-thérapie)' (3e Mémoire) *Annales de l'Institut Pasteur*, Paris 1894, 8ème année, No. 9, Septembre 1894, p. 621.

25. Paul Haushalter, *De l'application des sérums au traitement de la diphtérie et du tétanos*, Nancy: Crépin-Leblond, 1896, p. 12.
26. P. Moizard and M. Bouchard, 'Un cas d'angine non-diphtérique traité par le sérum suivi de mort', *Bulletins et Mémoires de la Société médicale des Hôpitaux de Paris*, 1985: 525–31.
27. E. Roux, 'A propos d'une observation de MM. Moizard et Bouchard sur un cas d'Angine non diphtérique traité par le sérum suivi de mort', *Journal de clinique et de thérapeutique infantiles*, 1895: 601.
28. G. Variot, 'Sur les recherches', *Journal de clinique et de thérapeutique infantiles*, 1895, p. 605.
29. Another reason for Roux's violent reaction may have been his memories of Pasteur's introduction of his new treatment against rabies a decade earlier. Roux disapproved of Pasteur's initial use of the rabies vaccine, as he felt it was not sufficiently well tested for use in human subjects. For a description of Roux's withdrawal from the rabies research, see Adrien Loir, *A l'ombre de Pasteur: Souvenirs Personnels*, Paris: Le mouvement sanitaire, 1938. This memory could only have added to Roux's discomfort when he found himself implicitly accused of similar neglect concerning the anti-diphtheria serum.
30. See J. Simon, 'Monitoring the Stable at the Pasteur Institute,' *Science in Context*, 2008, *21*: 181–200.

6

'The Geneva serum is excellent!' Autonomy and Isolation in the Swiss Cantons During the Early Years of Diphtheria Serum: The Case of Geneva

Mariama Kaba

At the end of the nineteenth century, Switzerland did not possess its own national institute equivalent to the *Institut Pasteur* in France, and nor did it promote intensive research in competitive laboratories like in Germany. Thus, the development of Swiss diphtheria serum production took place independently in each canton as a function of local resources and enthusiasm. The study of the Geneva case offers an interesting illustration of this model of diphtheria serotherapy research and production.[1] After giving some information on the context of the surge of diphtheria in the region, this paper will show how the canton of Geneva succeeded in having the product accepted and how it developed medical and social measures allowing the production, distribution and quality control of the serum in a context of total autonomy and in the absence of any national support.

The diphtheria situation in Geneva around 1890

In Geneva, the introduction of the 'anti-diphtheritic' serum in the middle of the 1890s coincided with a pronounced preoccupation with diphtheria on the part of the canton. According to the statistics from the Geneva Cantonal Hospital, during the 1880s mortality due to diphtheria in the canton had been lower than in other European cities that published similar statistics. The situation changed abruptly in 1890, despite the conditions of hospitalization and treatment remaining essentially the same.[2] The prevalence of virulent strains of the diphtheria bacillus now became comparable to that seen in England, Paris, or the rest of Switzerland, with the number of deaths from the disease rising as a consequence. The proportion of life-threatening cases of diphtheria at the cantonal Hospital grew sharply in comparison to the cases where the ailment was characterized by a local respiratory disorder usually relieved by tracheotomy.

Furthermore, between the years 1890 and at least 1897, the cantonal Hospital – where most of the population of Geneva was treated – registered a larger number of cases of diphtheria than of other diseases, and the number of deaths became preoccupying. Between 1880 and 1889, the institution recorded an average of 22 diphtheria cases per year, and 7 deaths annually, while in 1890, there were 64 cases with 16 deaths. This surge of diphtheria reached its peak the following year with 114 cases and 31 deaths (see Table 6.1).

Thus, in 1891 diphtheria accounted for 89 per cent of the mortality for the 'main acute infectious diseases' identified and recorded by the Hospital, covering essentially early-childhood diseases (smallpox, measles, scarlet fever, typhoid, diphtheria and erysipelas).[3] As can be seen in Figure 6.1, the peak in the number of deaths from diphtheria in 1891 would not be reached again in the canton of Geneva before the end of the century, and has, of course, since then never returned to this level.

The epidemic of diphtheria sweeping through the country doubtless largely explains the high mortality from this disease in Geneva. Indeed, diphtheria mortality first rose in northern Switzerland between 1876 and 1880; then, between 1881 and 1885, it reached a dozen central cantons. The increase then became apparent in neighbouring cantons and finally reached Geneva starting in 1890,[4] which is precisely the moment when the local authorities became concerned about the issue.

The fight against smallpox, regarded as the most dangerous disease in Geneva, had long been the focus of sanitary interests for the cantonal authorities. Nevertheless in January 1891, diphtheria appeared for the first time on the order for the mandatory notification of infectious diseases in Geneva. The order was drawn up by Dr Alfred Vincent (1850–1906), medical director at the cantonal Bureau of Public Health, and mainly concerned infantile diseases (measles, scarlet fever, diphtheria or croup, whooping cough, infantile cholera, dysentery, typhoid fever, erysipelas, puerperal fever, and glanders).[5] The Bureau, which dealt with issues of public hygiene, had been created at the end of 1884, among other reasons to organize the fight against the epidemics of cholera and typhoid fever that were affecting various European cities including the predominantly urban canton of Geneva. In 1891, the Bureau set up the first cantonal bacteriology laboratory, which as we will see, played a significant role in the history of the anti-diphtheria serum in Geneva. Nevertheless it was not the Bureau, which depended on the Genevan authorities, that introduced the serum into the canton.

Table 6.1   Cases of diphtheria and number of deaths between 1880 and 1891

	Cases of diphtheria	Number of deaths
1880–9	22 (average per year)	7 (average per year)
1890	64	16
1891	114	31

Source: Table based on data provided by *Rapports de la Commission administrative de l'Hôpital cantonal et de la Maternité de Genève* (1881–1892).

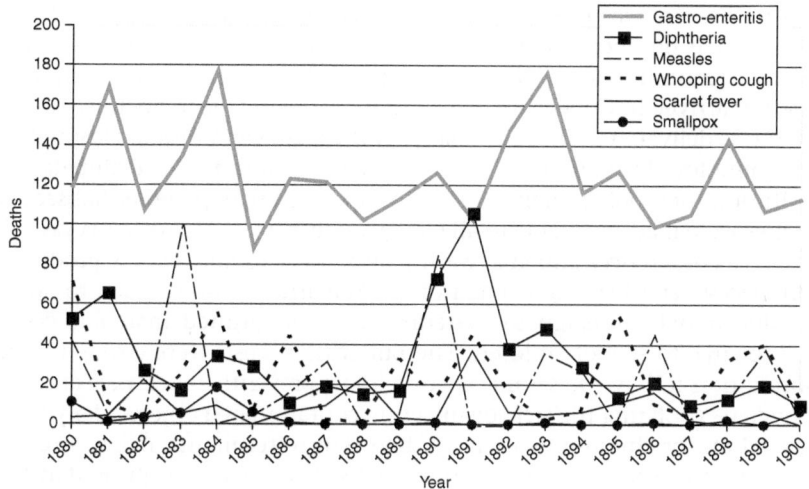

Figure 6.1 Deaths due to six infantile diseases in Geneva between 1880 and 1900
Note: This chart clearly shows the diphtheria death rate peak in 1891. In this year, diphtheria surpassed gastro-enteritis as the principal cause of death among children.
Source: Chart based on data provided by Siegenthaler and Ritzmann, *Statistique historique de la Suisse* (Zurich, Chronos, 1996), 321–9.

Introducing the serum in Geneva: An independent European network embracing Paris and Höchst

The introduction of the serum to Geneva hospitals as well as the education of the population about the new remedy was the work of the women's section of the Samaritans. This charitable organization was founded in Switzerland in 1888, and was active in fields such as voluntary first aid, the diffusion of knowledge about patient care, and the development of hygiene in general. The doctor and deputy Adrien Wyss (1856–1938) created the Geneva Samaritans Society in 1889.[6] In mid-October 1894, just a few weeks after Emile Roux's announcement about the serum at the International Hygiene Congress in Budapest and its subsequent publication in the *Annales* of the Pasteur Institute, the women's section of the Samaritans launched a public subscription in order to buy the serum. The Geneva population responded very positively and three days after it opened, the subscription already counted 2500 Swiss Francs (CHF). By the time the operation came to an end on 3 November 1894, the amount had reached CHF 3400.[7]

This amount, already qualified as 'considerable' at the time, allowed a few Geneva doctors to obtain the serum. They had previously been in contact with laboratories abroad: the Pasteur Institute in Paris to obtain Roux's serum, and the pharmaceutical company in Höchst near Frankfurt where

Behring's serum was produced. Indeed, starting in the autumn, the French and German producers had supplied their anti-diphtheria serum to several different places in Europe.[8] On 17, 20 and 31 October 1894, three shipments of serum arrived in Geneva for distribution free of charge to doctors. The first batch was delivered to those in charge of the Geneva Cantonal Hospital and the Home for Sick Children (*Maison des enfants malades*) in Plainpalais, an institution that took in many of the Hospital's younger patients. Subsequent batches were made available to all Geneva's doctors by Dr Adrien Wyss, who also gave conferences aimed at raising awareness about the new product.[9]

In December 1894, the women of the Samaritans also decided to launch an educational campaign for parents, based on printed materials distributed for free to school children. The publication costs were covered by the subscription to buy the serum. Beyond a few preventive measures and rules of hygiene (isolating the sick, disinfecting contaminated objects, daily throat inspections for children by their mothers, salt water mouth rinses, strengthening children's health with outdoor exercise and walks), the printed instructions called for 'serum vaccination, without delay, for all the children in a school or family where a case of diphtheria occurs'.[10]

The freedom with which Adrien Wyss and the women's section of the Samaritans were able to introduce the serum in Geneva was remarkable. They succeeded in having doctors as well as the general population accept the product thanks to a privately financed public campaign that mobilized the press and the schools. At this initial stage, the cantonal Bureau of Public Health never once intervened, although the Genevan authorities had placed it in charge of all anti-epidemic measures. Meanwhile, behind the scenes, political rivalries and divergent opinions opposed Adrien Wyss and Alfred Vincent, the director of the Bureau of Public Health. Although both were radical deputies at the local parliament, Vincent was re-elected until he died, whereas Wyss had only a short three-year career.[11] Following his political failure, Wyss dedicated himself exclusively to the Samaritans Society, trying to recruit members from the middle and lower classes in Geneva, thus distancing himself from the mostly bourgeois philanthropy movements of the time, such as the Red Cross. The anti-diphtheria serum apparently allowed the young institution and its founder to become quite popular within Genevan society.

It did not take very long, however, for the management of the anti-diphtheria serum to pass from private to public hands. At the beginning of 1895, Adrien Wyss ran into difficulties with the financial management of the Samaritans Society. Furthermore, he had helped himself to what remained of the 1894 subscription for the acquisition of the serum, which led to his dismissal from the direction of the Society.[12] Probably due to inadequate funding and in light of the large scale of the serotherapy operation, the Samaritans Society had to hand over the project to those responsible for public health at the cantonal level. The Bureau of Public Health then charged the director of its bacteriology laboratory, Léon Massol (1838–1909), with the task of obtaining the serum. The French-born Massol had trained as a railway engineer in Paris, and only became interested in bacteriology at around the age of 50, studying the science at the Pasteur Institute, where he became a member of the administrative board. A remarkable researcher, constantly keeping up with new developments, he was called to Geneva in the fall of 1891 to become the director of the very first cantonal bacteriology laboratory set up by the Bureau of Public Health, an appointment that reflects the lack of trained personnel available at the time.[13] Massol's close ties with the Pasteur Institute made it easy for him to import French serum, although it was not long before Geneva, like other cities of comparable size, decided to start producing its own serum. This local production of anti-diphtheria serum did not take place at the laboratory of the Bureau of Health, but instead at a new municipal laboratory created specifically for the purpose.

Producing the serum in Geneva: The Municipal Bacteriology and Serotherapy Laboratory

Indeed, in addition to the Bureau of Health's cantonal laboratory, which had been operating since the end of 1891, the Geneva city authorities opened a Municipal Bacteriology and Serotherapy Laboratory in January 1896. Léon Massol was then hired by the new laboratory (see Figure 6.2), leaving his position as director of the Bureau's cantonal laboratory to another researcher. Because of the exceptional demand for scientific expertise created by the emergence of the serum, the Geneva authorities set up an unprecedented infrastructure to supply the product. The Geneva faculty of medicine had been created in 1876, first and foremost as a professional school, and less than a third of the positions there were in basic or fundamental medicine.[14] The few laboratories that had allowed bacteriology to develop since the 1880s were funded by the researchers who had set them up. This situation is typical of the lack of visibility and scientific isolation of the institutions of individual Swiss cantons at an international level. Geneva tried to escape this kind of phenomenon by hiring a Frenchman, Léon Massol, who could use his contacts with the Pasteur Institute to give the project a higher profile, thereby adding a further link to the France-Geneva scientific network. This relationship dated from at least the sixteenth century, when many French Protestants fled catholic France for Calvin's Protestant Geneva, and in the course of the nineteenth century, many doctors from Geneva had trained in Paris.

From 1896 onwards, the Municipal Bacteriology and Serotherapy Laboratory served to fill many gaps in the field of bacteriology in Geneva: it provided the medical community with experimental research and shed light on problematic diagnoses. In addition to the anti-diphtheria serum, the municipal laboratory distributed anti-tetanus and anti-streptococcus serum.

Figure 6.2   Léon Massol, the first director of the Geneva Municipal Bacteriology and Serotherapy Laboratory (*La Patrie Suisse*, 9. June 1897)

These three products were mainly distributed to the cantonal Hospital, to the Home for Sick Children in Plainpalais and to private clients. The sera were also supplied to the Hospice du Prieuré in Geneva (a private institution), to the cantons of Vaud and Neuchâtel, and abroad, as well as to veterinarians in the canton who were interested in the anti-tetanus serum, although the quantities concerned were much smaller.[15]

From the beginning, diphtheria dominated the horizon of the municipal laboratory. Whereas the other two sera were supplied by the Pasteur Institute, the anti-diphtheria serum was produced directly by the Geneva municipal laboratory, using a horse immunized at the Pasteur Institute that provided 'a serum of truly extraordinary curative power'.[16] But in May 1896, the horse fell ill, and Léon Massol had difficulties in finding another animal. It was then that the Swiss national government intervened for the first time in the history of serotherapy in Geneva. The federal administration in Bern offered Massol old horses at a price between CHF 800 and 900 for the municipal laboratory in Geneva.[17] Given the age of the horses, Massol did not consider this a very good deal, but would have had to accept it had not two citizens of Geneva donated two vigorous horses. Massol immunized these horses himself between July and December 1896, taking all the necessary precautions: at each bleeding, the curative power of the serum was evaluated at the laboratory, and a sample was sent to the Pasteur Institute for control. As the first horse from the Pasteur Institute recovered, Geneva now had a three-animal stable with all the horses prepared for serum production by the municipal laboratory.

Geneva, like a number of other European cities, had received the initial doses of serotherapy from the Pasteur Institute in Paris or the Behring serum from Höchst, before quite rapidly establishing a serotherapy centre where it could prepare its own.[18] However, in the case of Geneva the involvement of the public sector was slight, and the fight against diphtheria through serotherapy was the work of a handful of actors. Indeed, it was Adrien Wyss and the women of the Samaritans who took the initiative of informing professionals and the public as well as distributing the first doses of serum that came from abroad. Moreover, it was only thanks to the generosity of two Geneva horse-owning philanthropists that Léon Massol – who was a bacteriologist and not a doctor – was able to continue and then develop the local production of anti-diphtheria serum.

Due to the system of federalism that left each canton with considerable political and economic independence, there were many different ways of managing serotherapy in Switzerland. In the canton of Vaud bordering on Geneva, for example, the initial reaction came from doctors at the Medical Society, who were particularly panicked when the disease reappeared in 1894, after being practically absent for the preceding 16 years. They solicited the Department of Interior, who promised to obtain the Behring serum at cost price.[19] By contrast, in the canton of Fribourg the Health Commission only very reluctantly acquired the serum during the month of December 1894. Then, following some unsatisfactory trials, the Commission decided to abandon the serum in January 1895, deeming it too dangerous, and leaving the administration of the treatment to doctors' discretion, and, more particularly, leaving the patient's family to bear the cost of the serum. In April 1897, under pressure from the Fribourg Medical Society and confronted with a large-scale diphtheria epidemic, the Commission reversed its decision and ordered a large quantity of serum from the Häfliger laboratories in Bern. The State now paid for half the cost of the remedy, but the remaining half was still borne by the patient's family.[20]

The absence of centralization of production or distribution of the anti-diphtheria serum in Switzerland did not, therefore, make it easy to institute any form of control of the product at a national level. In addition to the diversity of the products, doctors' use of the serum depended on their appreciation of not only the severity of the disease and its susceptibility to respond to serotherapy but also the potency of the particular product available for administration.

The early use of the serum in Geneva and its results

As soon as the serum was introduced to the medical community in Geneva, the problem arose of knowing which dose to prescribe for the desired curative effect. In the absence of any official standards concerning doses to administer, it is interesting to note the hesitancy of the first endeavours in the field of therapeutic experimentation. Before the end of 1897, the doses prescribed at the Geneva Cantonal Hospital for the first days of treatment did not exceed a maximum of 30 cc of serum, but by 1898 and 1899, the doses had already climbed to 50 cc. After 1901, much higher doses became standard: at the cantonal Hospital, 60 to 80 cc were injected in two or three days, with up to 150 cc for adult patients, though at the Home for Sick Children in Plainpalais the dose was no more than 30 cc in the first 24 hours.[21] When considering these varying dosages, we also need to take into account the fact that the serum had itself evolved since 1894–5, with a wide range of potencies.

Nevertheless according to statistics published by the principal users of the serum in Geneva – the cantonal Hospital and the Home for Sick Children in Plainpalais – the beneficial effects of the product were noticeable from the first months of serotherapy, even when only small doses were used. For diphtheritic angina, the administration of the serum resulted in the disintegration of the false membranes, and tracheotomy appeared to become rarer. Furthermore when a tracheotomy for croup was needed, the operation seemed to be more successful when serotherapy had been administered than when it had not. Comparing the years 1889 to 1894, preceding the introduction of the serum, with the years 1895 and 1896 (first trimester) when the serum was used, the average mortality fell from 37 per cent (126 deaths out of a total of 338 incoming patients) to 5 per cent (3 deaths out of 53 cases). The proportion of cases that had to undergo tracheotomy went down from 34 per cent to 13 per cent, and mortality after tracheotomy (the most severe cases), dropped from 62 per cent to 14 per cent.[22]

Results were equally satisfying at the Home for Sick Children in Plainpalais, where deaths due to diphtheria fell from a rate of 35 per cent (103 deaths out of a total of 289 cases of diphtheria) in the five years preceding the introduction of the serum treatment (1890 to September 1894) to a rate of 6 per cent (4 deaths out of 62 cases of diphtheria) for the first year that the product was used (October 1894 to 1895).[23] The statistics published by the Health Bureau of the canton of Vaud in May 1895 suggested a similar improvement, with a mortality rate of 64 per cent in 1893, reducing to 36 per cent in 1894 (148 deaths out of 404 cases) and attaining 16 per cent for diphtheria treated with the serum (14 deaths out of 85 diphtheria cases) in 1895.[24]

However, we need to be cautious in our interpretation of this data, as the criteria for diagnosing the disease changed when bacteriological examinations became a regular practice. As the cantonal Hospital intern in charge of the first serotherapy treatments noted, among the patients previously diagnosed with diphtheria, there must have been quite a few who did not have the Löffler bacillus,[25] since many doctors only took into account the clinical diagnosis and not the bacteriological one. In these circumstances, it is almost impossible to know whether the patients recorded in the statistics dating from before bacteriological analysis really died because of the diphtheria bacillus. Furthermore, it is also possible that the decreasing mortality credited to the serum starting from the end of 1894 was a result of more general factors such as the attenuation of the waves of epidemics, the reduced virulence of the disease, or the growing resistance of individuals thanks to better nutrition, or something else.[26]

Notwithstanding certain doubts concerning the quality of the medical examinations that lay behind these statistics, there was much interest in the results of the use of the serum, although the analyses remained local. In Geneva, the cantonal Medical Society created a commission in February 1895 specifically to collect information on anti-diphtheria serotherapy, centralizing and analyzing all the relevant observations. Members of the commission included doctors from the cantonal Hospital and the Home for Sick Children in Plainpalais, the primary users of the serum, as well as Alfred Vincent, the director of the cantonal Public Health Bureau. This commission represented the introduction of a cantonal collaboration around serotherapy, as the Bureau was responsible for making the serum available to doctors and giving them a form to fill in that supplied data for cantonal statistics.[27] At the national level, a single survey on diphtheria was conducted between 1896 and 1898 across the whole of Switzerland, to assess the evolution of the disease. The federal authorities then gave subsidies to laboratories in various Swiss cantons so as to disseminate the use of bacteriological tests in order to complete national epidemiological statistics. This national survey was not, however, concerned with either promoting the serum or even simply knowing the results of its use by means of collecting data from the different cantons.

The serum at the dawn of the twentieth century

At the very end of the nineteenth century, the medical community's enthusiasm for the novel serotherapy was clear. At the Twelfth International Medical Congress held in Moscow in August 1897, it was unanimously acknowledged that there had been no effective remedies to fight diphtheria before the introduction of the serum. Nevertheless, this enthusiasm did not mean that there were no controversies concerning its effectiveness and harmlessness. At a November 1898 meeting of the French-speaking Medical Society of Switzerland, doctors from Vaud and Geneva shared their experiences.[28] The testimonies from the hospitals in these cantons, as elsewhere, underlined that the serum was unable to overcome the most severe cases of diphtheria poisoning, and that it was powerless against mixed infections (such as diphtheria anginas associated with streptococci). Allergic effects due to the product, at times severe, were also observed in some of the patients treated. But the Geneva representatives at the meeting seemed confident in the serum. They underlined the importance of the bacteriological diagnosis, which worked better than the clinical diagnosis in identifying 'true' cases of diphtheria (containing the Löffler bacillus) that required treatment using the serum. In June 1901, at a Geneva medical Society meeting, Edouard Martin (1844–1931), the head doctor at the Home for Sick Children in Plainpalais, pointed out that while all sera did not operate with the same intensity, 'the Geneva serum is excellent'.[29]

It is clear that confidence in the serum's efficacy was reinforced in Geneva by the decreasing mortality observed following the introduction of this treatment (see Figure 6.1). No more peaks in mortality occurred on the scale of those seen in 1890 and 1891, with 73 and 106 deaths respectively. Between 1895 and 1901, Geneva recorded around 20 deaths per year due to diphtheria, far from the annual death rate of between 30 and 60 that existed before 1894.[30] Subsequently, the death rate would stabilize at around 10 per year until the 1930s with no deaths reported for 1924 and 1926. The disease itself did not decline rapidly with the annual number of cases of diphtheria in Geneva varying between 100 and 250 from 1892 to 1920, and an average of 160 cases per year.[31] The last serious outbreaks in the canton occurred in 1921, 1929, and 1944–5. While the serum also started to be used in the context of prevention, it was not used in any systematic way, which explains the continuing presence of the disease. The medical literature had little to say about such prophylactic treatment, as the debate focused on the great novelty of the curative use of diphtheria serotherapy. The principal concern in such journals was to establish the outcome of what was considered an experimental treatment. The optimistic and progressive attitude of the authors writing on the subject probably contributed to the success of the serum on the market, inspiring enthusiasm in the private sector as well as among doctors and health authorities.

Conclusion

Geneva's scientific community, and some notable individuals like Wyss and Massol, successfully networked with foreign institutions, facilitating the importation of the new serum from Paris and Höchst and its distribution in Geneva starting in October 1894. The question remains, however, concerning what might have occurred in Geneva with respect to the development of bacteriology in general, and serotherapy in particular, without the virulent diphtheria epidemics at the beginning of the 1890s. It is almost certain that this difficult time, which coincided with the recent discovery of the bacillus, encouraged Geneva's political and medical communities to become involved in research and anti-epidemic sanitary measures that led to the creation of the cantonal Health Bureau's laboratory in 1891 and the municipal serotherapy laboratory five years later. In the end, thanks to the close relationship established with the *Institut Pasteur*, Geneva followed the French model for serum production when it came to set up its own institute.

Finally, it should be noted that Switzerland introduced a system of official control of serum at a national level in 1926.[32] In 1932, the Geneva government was the first in the world to make anti-diphtheria vaccination mandatory for all children before they entered the school system, with Poland following suit in 1936 and France in 1938. The last diphtheria epidemic recorded in Geneva was during the difficult period of 1944–5. After WWII there were only a few isolated cases of the disease before it disappeared entirely from the canton.[33]

Notes

1. This paper was elaborated on the basis of a thesis for a Masters (DEA) in 'Social and Cultural History of Health Knowledge and Practices', defended in July 2002 at the University of Geneva (Faculties of Liberal Arts and Medicine). See also Mariama Kaba, 'La diphtérie à Genève à la fin du XIXe siècle: L'entrée en scène de la bactériologie et l'emploi de la sérothérapie', *Gesnerus. Swiss Journal of the History of Medicine and Sciences*, 61 (2004), 37–56.
2. Hector Maillart, 'La diphtérie à l'Hôpital cantonal de Genève, du 1er janvier 1890 au 1er avril 1892', *Revue médicale de la Suisse romande (RMSR)* (1892), 443–5.
3. *Rapports de la Commission administrative de l'Hôpital cantonal et de la Maternité de Genève* (1881–92).
4. Robert Nadler, *Statistischer Beitrag zu dem Verlaufe der Mortalität an Diphtherie, Keuchhusten, Scharlach und Masern in der Schweiz in den Jahren 1876–1900* (Bern, Stämpfli & Cie, 1903).
5. Registre du Conseil d'Etat 467, 13 January 1891 (Archives d'Etat de Genève).
6. Charles Heimberg, *'L'œuvre des travailleurs eux-mêmes ?' Valeurs et espoirs dans le mouvement ouvrier genevois au tournant du siècle (1885–1914)* (Genève, Slatkine, 1996), 263–7; *Statuts de l'Alliance suisse des Samaritains* [1948].
7. *La Tribune de Genève*, 14, 15 October 1894; *Le Journal de Genève*, 16, 31 October 1894.
8. See Gachelin's contribution to the present volume.
9. *Le Journal de Genève*, 19 October 1894.
10. Instruction publique Q 171 (1894–5), 13 December 1894, no. 1212 (Archives d'Etat de Genève).
11. When Wyss returned to politics, he turned his back on the radical party and became the leader of the Geneva Labour Party (1901–13). (Marcel Godet (dir.), *Dictionnaire historique et biographique de la Suisse* (Neuchâtel, Administration du DHBS, 1933) vol. 7.

12. Charles Heimberg, *'L'œuvre des travailleurs eux-mêmes ?*, 270.

13. Marc Cramer, Jean Starobinski and Marc-A. Barblan, *Centenaire de la Faculté de médecine de l'Université de Genève (1876–1976)* (Genève, Médecine & Hygiène, 1978), 132; Constant Picot and Emile Thomas, *Centenaire de la Société médicale de Genève, 1823–1923* (Genève, Sonor, 1923), 103–4.

14. In comparison, two-thirds of the ordinary tenures at the faculty of medicine in Strasbourg, founded in 1872, were in fundamental medical science. (Philip Rieder, *Anatomie d'une institution médicale: La Faculté de médecine de Genève (1876–1920)*, Lausanne, Editions BHMS; Genève, Médecine & Hygiène, 2009, 145).

15. Léon Massol, *Rapport au Conseil administratif de la Ville de Genève sur les travaux du Laboratoire municipal de bactériologie et de sérothérapie pendant l'année 1896* (Genève, 1897), 3 ff.

16. Massol, *Rapport au Conseil administratif de la Ville de Genève*, 10. Thus the Geneva laboratory, like several of its French counterparts, produced its own serum under the supervision of people trained and approved by the Pasteur Institute. See Gachelin and Simon's papers in this volume.

17. Massol, *Rapport au Conseil administratif de la Ville de Genève*, 10.

18. On the beginning of serum production abroad, see Paul Weindling: 'From Isolation to Therapy: Children's Hospitals and Diphtheria in *Fin De Siècle* Paris, London and Berlin', in Roger Cooter (ed.), *In the Name of the Child: Health and Welfare 1880–1940* (London, Routledge, 1992), 124–45; Heinrich Zeiss and Richard Bieling, *Behring: Gestalt und Werk* (Berlin, Bruno Schultz Verlag, 1940), 170 ff.

19. André Guisan, *Centenaire de la Société vaudoise de médecine, 1829–1929* (Lausanne, Imprimerie La Concorde, 1929), 95–6.

20. Alain Bosson, *Histoire des médecins fribourgeois (1850–1900): Des premières anesthésies à l'apparition des rayons X* (Fribourg, Université de Fribourg, 1998), 98–9.

21. Emile Thomas, 'Note sur l'emploi des fortes doses de sérum antidiphtérique', *RMSR* (1901), 552–6; Société médicale de Genève, séance du 5 juin 1901, *RMSR* (1901), 442–3.

22. *Rapports de la Commission administrative de l'Hôpital cantonal et de la Maternité de Genève* (1896), 63.

23. Eugène Revilliod, 'Note sur quelques cas de diphtérie traités par la sérothérapie', *RMSR* (1896), 73–7.

24. Congrès des médecins suisses à Lausanne, 3, 4, 5 May 1895, *RMSR* (1895), 283.

25. Arnold Vallette, *La sérothérapie de la diphtérie à la clinique médicale de Genève du mois d'octobre 1894 au mois de juin 1895* (Genève, Georg, 1895), 4.

26. Gachelin claims that 'An overall two- to five-fold decrease in mortality rates was observed through most countries in Europe during the following two years and was attributed to the use of antiserum, although the effects of new isolation rules in hospitals may have contributed to the decline in mortality'. See Gachelin's contribution to the present volume.

27. Adolphe d'Espine, 'Rapport sur les cas de diphtérie traités à Genève par la sérothérapie, d'octobre 1894 à la fin de mars 1895', *RMSR* (1895), 177; Eugène Revilliod, 'Note sur quelques cas de diphtérie traités par la sérothérapie', *RMSR* (1896), 72–3.

28. Société médicale de la Suisse romande, session of 3 November 1898, *RMSR* (1898), 629–35.

29. Société médicale de Genève, session of 5 June 1901, *RMSR* (1901), 443; see also the session of 4 May 1904, *RMSR* (1904), 443.

30. Hansjörg Siegenthaler (dir.), Heiner Ritzmann-Blickenstorfer (éd.), *Statistique historique de la Suisse* (Zurich, Chronos, 1996), 324.
31. Paul Bouvier, 'Soixante ans de vaccination contre la diphtérie à Genève', *RMSR* (1993), 147–53, 149.
32. 'Circulaire du Département fédéral de l'Intérieur aux gouvernements cantonaux concernant l'institution d'un contrôle officiel des sérums et vaccins (du 20 novembre 1926)', *Bulletin du service fédéral de l'hygiène publique* (Berne, 1926), 393 ff.
33. Bouvier, 'Soixante ans de vaccination contre la diphtérie', 148–50.

7

The State, the Serum Institutes and the League of Nations

Pauline M. H. Mazumdar

At the turn of the twentieth century, serological treatment of disease was the most powerful weapon in the therapeutic armamentarium. At the heart of its successful diffusion lay the standardization process, carried on by serologists in serum institutes all over the world. The founding generation all looked to Paul Ehrlich and his procedure of *Wertbestimmung,* and as Hüntelmann shows in an earlier chapter, to Ehrlich's transformation of immune horse serum into a reliable pharmaceutical, produced and certified under state control. The involvement of the state from the outset meant that by the side of the technical process of standardization, lay the political.[1] The political and economic element in many ways came before the technical: the groundwork for international standardization had been laid during the second half of the nineteenth century in the field of communications and weights and measures. Standardization agreements were commonplaces of international trade.[2] The economist Charles Kindleberger calls them international public goods, like peace, an open trading system and fixed exchange rates.[3]

The International Sanitary Conferences began in 1851, a full decade before the first agreements on communications, but they took much longer to reach any firm conclusion. They were an attempt to standardize port quarantine regulations as part of the international trading system. Each of the states taking part sent one sanitary expert and one government representative. But although agreements were drafted and signed they were never ratified, and so never went into effect. By 1885, 28 states were represented at the Conference. Agreement collapsed on the question of quarantines at Suez, opposed by British trading interests, the argument being that quarantines would force shipping to abandon the Suez Canal route from India into Europe. A limited agreement followed the conference at Venice in 1892, which established a quarantine station at El Tor on the Sinai Peninsula. It was soon famous for its virulent cholera organisms.[4]

The first general agreement on shipping quarantines did not come till the Eleventh International Conference of 1903; it was followed by the formation of the first permanent international health organization, the *Office international de l'hygiène publique,* set up by the Rome Agreement of 1907, mainly to track the flow of epidemic disease through the ports.[5] Twelve states signed and this time ratified it, including the US, with another 30 adhering later, though neither Germany nor Austria-Hungary ever joined.[6] The officials of the Office were in most cases senior bureaucrats in the health ministries of their states, rather than technical experts. Neville Howard-Jones thinks this was because the International Sanitary Conferences had found it so difficult to get agreement between different scientific points of view, though it seems more likely that the theoretical disagreements on the effectiveness of quarantines had been a smokescreen for interests of state. Standardization as an interest of state was included in the Office's mandate.[7]

When meetings started up again after WWI, the British Ministry of Health proposed that the Office support free venereal disease clinics in all ports, along with a standardized health card which would allow sailors diagnosed with syphilis to carry on with treatment from port to port.[8] This proposal led to the Brussels Agreement of 1924.[9] For this to work, the Wassermann test for syphilis would have to be standardized. The Danish delegate to the Office, Thorvald Madsen, was commissioned to look into it since his institution, the *Statens Serum Institut* of Copenhagen, had been doing Denmark's syphilis testing since 1909.

The League of Nations came into existence under the Treaty of Versailles in January 1920. At first, it was expected that the *Office international* would amalgamate with the League's Health Organisation.[10] But that plan fell through, and the Office continued to exist.[11] The Health Organisation's Medical Director was the Polish bacteriologist Ludwig Rajchman, who worked at the League's Secretariat in Geneva. Its President was Thorvald Madsen, now Director of the *Statens Serum Institut*, who was also Chair of one of its subcommittees, the Standardisation Commission. (Figure 7.1).

In the post-WWI world, it would be the *Statens Serum Institut*, working through the Standardisation Commission, that replaced Frankfurt as the central laboratory for serum and standards. As Anne Hardy shows in another chapter of this book, it was in many ways the natural successor: its leaders all looked to Frankfurt for their standardizing techniques.

The first meeting of the League of Nations Health Organisation held in 1921 decided to work on the standardization of sera and the serological tests for syphilis.[12] Madsen and Rajchman's list of workers and laboratories for the first Conference was a list of friends and colleagues from the little world of the Serum Institutes.

The knotting of their ties with each other through the League's Standardisation Commission only formalized old relationships. Except for the group at the *Institut Pasteur*, the serologists were like-minded: they had mostly been trained in Frankfurt under Paul Ehrlich, and they used Ehrlich's technique for measuring the antibody content of a serum and the toxicity of a toxin. The idea of state control of standardized sera had originated in

Figure 7.1   A friendly relationship: Thorvald Madsen, Director of the Danish State Serum Institute, with Ludwik Rajchman, Medical Director of the League of Nations Health Organisation. Undated photograph from the Archives of the State Serum Institute. My thanks to Dr Jørn Lyng, then Head of the International Laboratory of Biological Standards, State Serum Institute

Robert Koch's Berlin Institute, where Ehrlich was running the *Kontrolstation* for diphtheria antitoxin from 1895. Ehrlich and the *Kontrolstation* together moved to the Frankfurt laboratory in 1899. Madsen, trained in Frankfurt, had been using the Frankfurt methods since the 1890s. (Table 7.1).

Germany and the League of Nations

Germany, however, was not allowed to join the League. In fact, a recent account suggests that the League was expressly designed as an organization of the victors for 'keeping Germany down'.[13] It certainly seemed like that in Germany.[14] The hostility extended to individuals: the boycott of German scientists organized by the Royal Society of London and the Belgian Royal Academy of Sciences in October 1918 had been confirmed in July 1919.[15] In theory, that should not have mattered. Members of the League's technical committees, including Economic, Financial, Intellectual, Transit and Health, were not to be regarded as delegates of a particular state, but as technical experts making purely technical decisions: national interests were to be put aside.[16] We have many contemporary statements on that ideal: Harold Greaves writing in 1931 on the technical committees saw them as creating a 'disinterested viewpoint wherever national interests conflicted'.[17] When there is technical regulation to be done, the work should be carried on by a committee of expert technical administrators. The

Table 7.1 A conference of expert serologists, almost all of whom were trained in the Frankfurt tradition: bold type shows Frankfurt connexion. Adapted from report of Second International Conference on the Standardisation of Sera and Serological Tests, held at the Pasteur Institute in November 1922 (1923)

Countries:	Institutes and workers:	Projects
Austria:	University of Vienna: R. Müller	Syphilis test
Belgium:	Institut Pasteur du Brabant, Brussels: J. Bordet	Syphilis test
Denmark:	**Statens Serum Institut, Copenhagen: Th. Madsen, K. A. Jensen**	**Anti-diphtheria, -tetanus, -meningococcal, -pneumococcal serum, syphilis test**
France:	Institut Pasteur, Paris: A. Calmette, I. Cotoni, Chas. Dopter, L. Martin S. Mutermilch Institut Prophylactique, Paris: A.Vernes	Anti-diphtheria, -tetanus, -meningococcal sera, -pneumococcal serum, syphilis test
Germany:	**Inst f. Infektionskrankheiten: R. Koch Berlin: F. Neufeld, J. Morgenroth**	**Anti-meningococcal, -pneumococcal sera**
	Inst. f. exper. Therapie, Berlin: A. von Wassermann	**Syphilis test**
	Staatl. Inst f. exper. Therapie P. Ehrlich, Frankfurt: W. Kolle	**Anti-diphtheria, -tetanus sera, syphilis test**
	Inst. f. Cancer Research: Heidelberg H. Sachs	**Syphilis test**
Japan:	**Kitasato Institute, Tokyo: O.Hida**	**Anti-diphtheria, -tetanus, -dysentery sera, -meningococcal, -pneumococcal sera**
Poland:	**State Epidemiological Institute, Warsaw: L. Hirszfeld**	**Anti-dysentery serum, syphilis test**
Rumania:	**University of Bucarest: J. Cantacuzene**	**Anti-dysentery serum**
Russia:	**Institute for Public Health, Moscow: L. Tarassevitch**	**Anti-diphtheria, -tetanus, -meningococcal, -pneumococcal serum, syphilis test**
Switzerland:	**Institute of Hygiene, University of Basel: R. Doerr**	**Anti-dysentery serum**

(Continued)

Table 7.1   (Continued)

Countries:	Institutes and workers:	Projects
UK:	Medical Research Council, London: S. R. Douglas, H. H. Dale, M. H. Gordon	Anti-dysentery, anti-meningococcal sera
	Pathological Institute, University of Oxford: G. Dreyer	Syphilis test
	Ministry of Health, London: Griffith, L. W. Harrison, E. J. Wyler, G. Buchanan	Anti-pneumococcal serum, syphilis test
US:	Hygiene Laboratory, Washington DC: McCoy (absent)	
	Rockefeller Institute, New York: A. B. Wadsworth, S. Flexner (absent)	Anti-diphtheria, -tetanus, -meningococcal, -pneumococcal serum

Note: Source, LoN Archives, C.H. / S.S. / 19.

League's committees were only advisory, but national sovereignty itself, he thought, was put in question by a committee's mere existence.[18]

It was a highly idealized view of the independence of science and technology, and one that our present historiography finds naive.[19] But it was the official view of the League, confirmed in 1922; Madsen and Rajchman could argue that whatever the political situation, they must have the German serologists, if that was where the expertise lay.[20]

As to *who* should come, diplomacy was required. Gottfried Frey of the Reichsgesundheitsamt had already refused an invitation. Nothing embarrassing had happened at a Health Committee meeting so far, but at a big conference, there could be an incident.[21] Bernhard Nocht, Director of the Hamburg Institute for Tropical Diseases, a port medical officer who had been at the Paris International Sanitary Conference of 1911, was the next choice.[22] But Nocht was uncomfortable: everything must be purely scientific and technical, not official or even semi-official. The Foreign Office must give its permission.[23] In the end, Nocht did not come, but four other German serologists were invited: Wilhelm Kolle, Director of the Frankfurt Institute and successor to Ehrlich; Fred Neufeld, Director of the Robert Koch Institute; serologist Hans Sachs of the Heidelberg Cancer Research Institute, and August von Wassermann of the Institute for Experimental Therapy, Berlin.

Wilhelm Kolle felt he had a right to be part of the team: he was after all Ehrlich's appointed successor. It was reported that he had been furious at not being invited.[24] (Figure 7.2).

But he still felt uneasy about going abroad: he had heard stories of how Germans were being snubbed and insulted. But Paris was not so bad after all. Madsen's famous diplomatic skill, he wrote after his first visit, has helped to heal a Europe sick unto death.[25] But Kolle himself was not healed. His letters to Madsen of the early twenties are a mixture of reports about the serological work he was doing for the Standardisation Commission and distress about the French occupation.[26]

You cannot imagine the hatred and bitterness caused by the French robbery of a people made utterly defenceless by the Entente – a truly cowardly act of the French. This is the *grande nation* that fought for freedom and justice among men, for the freedom of nations. Arbitrary brutality, looting of private and public property, ill-treatment of civil servants who are doing their duty, of families who are forced out of their house with all their goods and chattels at three days' notice, nightly abduction of high functionaries, arrest of everyone who does not submit to the arbitrary orders of the French – that is the fame and glory of Poincaré and Marshall Foch.

I wonder whether the world, whether the 50 million neutrals in Europe, whether America and England – proud but now powerless Albion – will calmly accept this punch in the face of all civilised nations, by those who were once leaders in the struggle for the rights of nations – I am afraid they will. Everyone just watches while we are robbed, looted and raped at will.

It has had one good effect, though: the whole of Germany is united, everything that divided us is put aside and everyone is calling for revenge.

Figure 7.2 Wilhelm Kolle, Director of the Paul Ehlich Institute from 1916 to1935, in succession to Ehrlich. Photo courtesy of Wcllcome Images

Dies irae, dies illa – the day of wrath must and will come, perhaps sooner than they think. Starvation and misery are getting rapidly worse. It is a torture to live in this place.

Dr Mørch [from the *Statens Serum Institut*] has been working hard here; he is an intelligent young man.[27]

On 11 February 1923, three weeks before this letter was written, France had occupied the Ruhr. As British economist Maynard Keynes understood, the combination of economic distress with patriotic rage was making Germany desperate.[28] His version of events corroborates Kolle's:

If ... the French insist on ... a military occupation of the Ruhr, ... such occupation cannot be compatible in practice with the financial and economic integrity of the German empire, if the French military authorities are to retain the power of daily interference which they are now exercising in the Palatinate and the Rhinelands. ... The occupation must be limited to the presence of French troops in barracks, with no administrative powers in normal circumstances.[29]

A few weeks later, Kolle wrote again to Madsen, telling him that a dried sample of dysentery toxin was now ready to send off, and giving its parameters in Ehrlichean terms of its *Dosis letalis*. He ends his letter pathetically:

I hope that you are pleased with our work. What it means I need not explain to anyone who reads the newspapers, to manage to keep on thinking and doing scientific work. I might just mention that on the first of October, the dollar stood at 1 million Marks, and today it has gone up to 1 billion, a 1,000 times fall in the value of our money in four weeks. We are standing on a knife-edge.[30]

As Keynes wrote in 1922, Germany was near to a nervous breakdown: prices in the shops changed every hour: the effect of the crashing fall of the Mark on the general public was terrifying and disintegrating.[31] Keynes, like Kolle, saw the Rhineland occupation as a sword in Germany's side that wounded Europe, and did France no good either.[32]

The *Institut Pasteur*

Germany was the obvious problem for the diplomats of the Standardisation Commission, but there was also the relationship with the *Institut Pasteur*. The *pastoriens* had had a long-running but variably hostile relationship with German bacteriologists, as Gachelin explains elsewhere in this book, and as Simon tells us, the *Institut Pasteur* had been independently producing antiserum in commercial quantities since 1895, the same year as the German group. It used its own system of normalization and it sold its own products over its own distribution networks, reaching out, as Anne-Marie Moulin has shown, from Paris to the French colonies of Asia and Africa.[33] It generally tended to favour vaccines over anti-sera, and prophylaxis over therapy, and it preferred colloidal precipitation tests to the Ehrlich *Dosis letalis* technique. [34]

It also preferred a flocculation test to the (German) Wassermann reaction as a test for syphilis. The Wasserman was the mainstay of clinicians. But many technical workers would have liked a simpler, more reproducible test, since the 1924 Copenhagen Conference on the standardization of syphilis tests, and particularly since the international agreement on seamen's syphilis was soon to be signed. Surprisingly the new flocculation tests did not after all look as good as the original Wassermann test.[35] But Stefan Mutermilch, the *Institut Pasteur*'s representative at the conference on the standardization of syphilis testing of 1923, could not accept this. He told Rajchman scornfully that he had learned nothing new from the meeting. It was useless, he said, to try to standardize the old Wassermann test when new flocculation methods were being published every day. The *Institut Pasteur*'s flocculation test was the most sensitive test, although perhaps a little too sensitive. He was personally in touch, he said, with some of the other participants and was sending them his protocol.[36] The *Institut Pasteur* did not wish to work through the central laboratory. It *was* the central laboratory.

The standardization of tuberculin created problems too. Koch had introduced it as a remedy for tuberculosis in 1890. But since the Graz pediatrician Theodor Escherich recognized that the reddening at the injection site was an immune reaction to tuberculin, followed by the flow of publications on it by his protégé Clemens von Pirquet from 1907 onwards, tuberculin had been used not as therapy, but as a skin test for the disease.[37] Frankfurt already had a standard, but it turned out that the *Institut Pasteur* had one too, and a different procedure for measuring its potency. Rajchman foresaw trouble: Albert Calmette, Director of the *Institut Pasteur*, should be warned in advance that the proposal was on the Agenda of the next meeting of the Standardisation Commission. Sir Henry Dale of the British Medical Research Council, placed in charge of the standardization of biologicals other than anti-sera in 1923, had no problem with Frankfurt's tuberculin standard.[38] But he agreed that the *Dosis letalis* method was clumsy where the material was to be used for skin testing. He felt, he said, some sympathy for Calmette's 'open door' as regards method, as long as the standard was accepted.[39] Calmette however, rejected both the Frankfurt standard and the method.[40] He showed his hand in his next letter:

> I would like to go along with Dale. But why choose as a standard the Frankfurt tuberculin rather than the one from Paris or from the Bureau of Animal Industry in Washington? If there were a scientific argument for choosing Frankfurt, I would gladly bow to it; but I am still waiting to hear

what it is, *and if it doesn't exist, why should you inflict yet another German standard on us, as if the Reich had the monopoly on bacteriological science, which does in fact, after all, owe something to Pasteur?*[41]

Rajchman copied this letter on to Dale, but diplomatically cut out the last three lines as unnecessarily provocative. Dale, a Frankfurt loyalist, wrote back:

> I do not think that I can be expected to give a *scientific* reason in favour of the adoption of the Frankfurt Standard Tuberculin rather than that of the Pasteur Institute, or of the Bureau of Animal Industry of Washington. Scientifically, any one of the three would be acceptable. Our only concern would be to ensure that the Standard was accepted by everybody, and was kept at an Institute in which the whole scientific world would have confidence. ... I think there would be a strong reason for choosing the Frankfurt Institute on practical and historical grounds.[42]

Rajchman passed on Dale's reasons to Calmette, though not, he said, his actual text.[43] Calmette agreed that Koch had prepared it first, but he kept it secret until the *Institut Pasteur* repeated the experiment and published it. He suggested that everything could be solved 'rationally', by giving Madsen's Copenhagen laboratory the task of standardizing and issuing the tuberculin standard.[44] Madsen agreed, and, as Rajchman told Dale, would probably arrange with Kolle to get it from Frankfurt.[45] This carefully edited indirect correspondence appears to have led finally to a diplomatic success. Paris, Copenhagen, Geneva, London and Frankfurt were all satisfied.

In 1924, with the development of a diphtheria vaccine by Gaston Ramon at the *Institut Pasteur's* facility at Garches, a new element entered the serological picture.[46] Soon after the first publication of Ramon's work, Albert Calmette wrote triumphantly to Rajchman that the days of antitoxin treatment of diphtheria (unchallenged since 1894) might well be coming to an end. 'The revolution is on the march', wrote Calmette, 'This is not the moment to have some theory crystallised under the aegis of the League of Nations'.[47] Over the next decade, the Ramon anatoxin began a general shift away from serotherapy to vaccination. The *Institut Pasteur* was soon at the centre of a web of users trying out the new vaccine on a grand scale.[48] And as Calmette immediately pointed out, vaccines should not be standardized like toxins, using the classical Ehrlich assay method of finding a lethal dose for guinea pigs: vaccines were not lethal. The *Institut Pasteur* used a flocculation test to predict the immunizing effect of the anatoxin. Ramon resented the attempt of Madsen and his colleagues to butt into the *Institut Pasteur's* affairs by trying to adapt the Ehrlich assay to the standardization of its vaccine.[49] The *Institut Pasteur* did not want to work through another laboratory, or some group of miscellaneous experts.

The League of Nations' mandate was to work towards world peace, and standardization fitted it very well. Standardization contributed to 'collective security based on a cooperative universality', in the words of Georg Schwarzenberger, writing in 1936.[50] Collective security was the slogan of the times. The term itself came into use in the early 1930s, but it went back in principle to the League's Covenant: the League was set up as an international order that would resolve disputes between nations without going to war. As a failed attempt at a guarantee of peace, it had a bad press in hindsight after the end of WWII. According to Roland Stromberg writing in 1956, the concept of collective security was fabricated by 'journalists, moralists, popular politicians', rather than by the more cynical diplomats and statesmen at Versailles in 1919.[51] The Canadian historian George Egerton regarded it in 1980 as a myth: 'one of the most potent ideals in attempts to structure a peaceful international order', but a myth just the same.[52] In the interwar period, however, it was taken very seriously. The International Institute of Intellectual Cooperation, an offshoot of the League's Committee on Intellectual Co-operation, was established in Paris in 1928, with a mandate to develop a system of intellectual and technical collaboration among its affiliated institutions and to organize collective research.[53] The Institute's International Studies Conferences of 1933 and 1935 put 'Collective Security' at the top of their agenda.[54]

The League's technical committees were composed of experts making decisions independently of national origins and interests of state.[55] But the further development of the system of standardization, if it was to contribute to collective security between nations as Schwarzenberger suggested, meant that states had to get involved. In the light of this hopeful rhetoric, it looked as though the nation states of the League must now play their part. Madsen's plan was to set up an international collective with parallel institutions in each national unit, on the model of the network of serum institutes, but reaching more deeply into the formal administrative structure of the states. An Intergovernmental Conference on Biological Standardisation, with official state representatives, was something that the Standardisation Commission had never suggested before. It completely contradicted the principle of independent technical expertise, which had guided it up till that time. A series of letters between Sir Henry Dale and Raymond Gautier, Secretary of the Health Committee, shows how difficult it was to accept this change of direction.

When he received the Agenda for the Conference in April 1935, Dale wrote a startled letter to Gautier at the Geneva secretariat: he had not realized that this was to be an international conference of government representatives:

> With regard to the further additions proposed by Dr Madsen ...
> I venture, with all respect, to think that he is proceeding on the wrong
> lines. I understood from the beginning that the Commission was not

to be officially representative, but to consist of those people from any country who, at the time, were regarded as most likely, on account of their personal qualifications, to contribute to the organisation of the International work. I believe it was for that reason that the Commission was never called 'International' as it was desired to avoid any implication that it was officially representative.[56]

The Intergovernmental Conference was held in Geneva in October 1935; 23 countries were represented.[57]

The 27 delegates – they were national representatives this time – were not the Commission's usual serologists. Only one of them had ever attended a standardization meeting before. They were administrators, professors of medicine and directors of their national institutes. Some, from the smaller countries, were diplomats and *chargés d'affaires*. The *Office international de l'hygiène publique* and its veterinary counterpart, the *Office international des épizooties*, were both represented: this type of intergovernmental meeting was very much in their style. There was, of course no delegate from Nazi Germany: it was an official occasion and the German directors could not attend. Listed separately were just three 'experts'. One of them was Sir Henry Dale, responsible since 1923 for the growing numbers of non-serum standards, held mostly at the Hampstead laboratory (see Table 7.2).[58] The meeting broke up into two subcommittees, one on sera chaired by Madsen, the other on remedies, including pharmaceuticals such as digitalis and strophanthin, and vitamins and hormones, chaired by Dale. As the leader in non-serum standardization, Dale's stature was growing. (Figure 7.3).

It is noticeable that by this time, there were many more non-serum than serum standards, though standardization was still broadly based on the Ehrlichean approach, as we can see in the contributions to this book by Jean-Paul Gaudillière, and, more particularly, by Christian Bonah on digitalis and strophanthin.

The key meeting of the conference was on the 'national centres'. Thorvald Madsen opened with a statement:

> It is desirable that each country should have a recognised national center to take charge of the national and international standards intended for biological assay, and that every such center should have a qualified staff to control the application of the international standards in its own country and thus to serve as the national scientific authority in this field.[59]

The Standardisation Commission was moving from the technical to the political, from the standardization of sera to the standardization of scientific authority internationally. However, there was no agreement signed for later ratification by governments, as the *Office international* would probably

Table 7.2 List of League's Standards sent out with questionnaire, reported on September 23 1935

List of Products for which the Permanent Commission on Biological Standardization has Adopted International Standards

1. Sera and Bacterial Products:
Anti-diphtheria serum
Anti-tetanus serum
Anti-dysentery serum (S. Shiga)
Anti-pneumococcus sera (Types I and II)
Anti-gas gangrene sera (V. oedematiens, septitique, perfringens)
Tuberculin

2. Glandular Products:
Insulin
Pituitary extract (posterior lobe)
Female sexual hormone (oestrus-producing)

3. Vitamins:
Vitamin A
Vitamin B$_1$
Vitamin C
Vitamin D

4. Therapeutical Substances:
Digitalis
Strophanthus (Ouabain)
Arsphenamine
Silver-salvarsan
Sodium-salvarsan
Neo-arsphenamine
Neo-silver-salvarsan
Sulpharsphenamine

Note: Note that the antitoxin standards shown in bold are held at the *Statens Serum Institut*, Copenhagen, under Thorvald Madsen. All the other biologicals, the hormones, vitamins, plant products and arsphenamines are under the control of Sir Henry Dale at the National Institute for Medical Research, Hampstead. It is striking that most of the standards by this time were for hormones, vitamins and therapeutic substances, rather than sera. Adapted from LoN Archives, C.H. / C.P.B. / 28.

have wanted: the Standardisation Organisation drafted no treaties, but only passed resolutions. But it was moving in that direction. At the Conference, nobody saw any problem in setting up what amounted to a system of government-controlled standards officials in each member country answering to an International Standardisation Commission. Publicly at least, Dale welcomed the proposal. Standardization clearly promoted both the progress of science and collective security (Table 7.3).

Figure 7.3   Thorvald Madsen, Director of the State Serum Institute, standing, left and Sir Henry Dale, Director, National Institute for Medical Research, Hampstead, London, standing, right, with Ludwik Rajchman, Medical Director of the League's Health Organisation, in front. Photo taken in the park at the League of Nations, Geneva (1937) possibly at the same time as Figure 7.1. My thanks to Dr Jørn Lyng, then Head of the International Laboratory of Biological Standards, State Serum Institute

Raymond Gautier's Report on Biological Standardisation to the Health Committee was enthusiastically positive, so much so that he made it sound as though the proposals of the Intergovernmental Conference were more or less agreed upon and all ready to be put into effect. In 1937, the Health Committee was able to say that standardization had always received support from the League's Council as a fine example of international cooperation; by this time 19 countries in Europe plus 11 in Asia and the Americas

Table 7.3 Provisional List of Delegates to the Intergovernmental Conference, 1st to 4th October 1935. President: Thorvald Madsen, *Statens Serum Institut*, Copenhagen

Country	Delegates	Institutions represented
Australia	Frank T. Wheatland	Senior MO, Commonwealth Serum Laboratories
Austria	Prof. Bruno Busson	I/C State Control, Vienna Serotherapeutic Institute
	Prof. Ernst P. Pick	I/C State Control, Vienna Serotherapeutic Institute
Belgium	M.R. Willems	Dir., Veterinary Res. and Diagnostic Laboratory, Brussels
Britain	M.T. Morgan	MO, Ministry of Health, London
China	Prof. Robert Lim	Peiping Union Medical College, Peiping (Beijing)
Czechoslovakia	O. Schubert	Chief, Biological Control, State Hygiene Inst., Prague
Ecuador	Alejandro Castelu	Consul General, Republic of Ecuador at Geneva
Estonia	H. Peterson	Dir., State Serotherapeutic Institute, Tartu
France	**Louis Martin**	**Dir., Institut Pasteur, Paris**
	Prof. M. Tiffeneau	Deptt Pharmacology, Faculty of Medicine, Paris
Guatemala	Louis Willemin	Consul of Guatemala, at Geneva
Hungary	Prof. Joseph Tomcsik	Dir., Institute for Hygiene, Szeged
India	Lt. Col. A. J. H. Russell, IMS	Public Health Commissioner, Govt. of India, Simla
Ireland	Prof. W. D. O'Kelly	Bacteriologist, Local Govt. Deptt Public Health, Dublin
Japan	M. Tsurumi	Central Bureau of Public Health, Tokyo
Lithuania	Karlis Kalnins	Sec., Lithuanian Delegation to League of Nations
Netherlands	Prof. U. G. Bilsma	University of Utrecht
	W. Aeg. Timmerman	Dir., State Institute for Public health, Utrecht
Norway	Prof. Klaus Hansen	University of Oslo
Poland	**Ludvik Hirszfeld**	**Dir., State Institute for Hygiene, Warsaw**
Rumania	C. Ionescu-Mihaesti	Dir., Institute for Experimental Medicine, Bucarest
Sweden	Prof. Hans v. Euler-Chelpin	Faculty of Medicine, Stockholm
	Prof. Carl A. Kling	Dir., State Bacteriology Laboratory, Stockholm
	Prof. Göran Liljestrand	Faculty of Medicine, Stockholm

(Continued)

Table 7.3   (Continued)

Country	Delegates	Institutions represented
Switzerland	Otto Stiner	P.H.S., Chief, Control of Serums & Vaccines, Berne
Yugoslavia	Marján Banić	State Institute of Hygiene, Zagreb
International:	Professor Carl A. Kling	Office international d'hygiène publique
International:	Professor E. Leclainché (absent)	Office international des épizooties
Experts:	**Sir Henry H. Dale**	**Director, National Laboratory for Med. Research, UK**
	Percival Hartley	**Chief, Standards Deptt., Nat. Lab. for Med. Research, UK**
	Claus Jensen	**Chief, Biological Standardisation, Statens Serum Institut**

Note: Note that the members are national delegates this time, mainly directors of the national institutes, and that Germany is not represented. Only two of the national delegates (in bold) had ever been at one of the League's standardisation meetings before. The last three people on the list, Dale, Hartley and Jensen, are the only non-national 'experts'. From LoN Archives, C.H. / C.P.S.B. / 32.

had adopted the standards that they sent out.[60] But privately Dale was still doubtful:

> I hope that the list you give … will not mislead the Health Committee into the conclusion that the arrangements advocated by the International Conference have to a large extent already taken effect. … You will understand that I only desire that the Health Committee should not sit back with mistaken satisfaction, on the idea that your lists represent what has been accomplished in reality, whereas there is a good deal, as you will agree, which is still only on paper.[61]

Gautier wrote back to say that his list was really only a promotional attempt to encourage the formation of national centres. He had just launched a fresh offensive in their favour: so far, twelve countries had been attacked.[62] We hear of him next organizing a visit to be made by the Directors of the two central institutes, Dale and Madsen, accompanied by Dr Gautier himself, to the main national centres in Europe in 1938.[63]

But Dale had found another objection: it was his laboratory that was responsible for the new hormonal, vitamin and therapeutic substance standards and he said they were not ready yet to send out. Dale tried to get Gautier to add something to his report to dampen expectations and warn that more work by experts was going to be needed before governments could get involved.[64]

However, even those standards were in fact available before WWII actually broke out. In 1939, with war coming closer, Percival Hartley, as director of the Department of Biological Standards, sent out extra-large stocks of his non-serum standards to a list of 35 officially approved National Control Centres for the Drugs, Hormones and Vitamins. The USSR was now on the list, but not Germany.[65] The serum standards were sent for safety to Ottawa: two separate packages of four sealed ampoules of each of the International Standards went to the Department of Pensions and National Health. A few months later, in April 1940, Denmark was invaded.[66] Geneva asked the British Medical Research Council to authorize the Hampstead Institute under Sir Henry Dale to take over as the sole central laboratory for the duration of WWII. Records and lists of National Centres from the *Statens Serum Institut* went to Geneva, and from there to Dale in Hampstead.[67] Wartime conditions seem to have firmed up Dale's belief in governmental cooperation, but made Gautier more cautious: perhaps Geneva knew more than Hampstead about governments and their weaknesses, and perhaps about the weaknesses of the League itself.[68]

The issue of whether the Standardisation Commission's members were national representatives remained unclear. The death of G. J. Fitzgerald of the Connaught Laboratory in Toronto in 1940 was a case in point: if he was on the Commission as an expert, his replacement too should be an expert; if he was a Canadian representative, his replacement should be appointed by Ottawa, and need have no experience of standardization. It was not clear whether he had been invited to join because Madsen knew him well as a serologist, or as a Canadian, and Madsen was unreachable in Copenhagen.[69]

Conclusion

This paper has focused mainly on the political aspects of the League of Nations Standardisation project. The League's technical committees were expected to be groups of technical experts who did not represent their countries, and whose opinion should be independent of national interests, free of all values except those of science. In some ways, the serologists found that easy: they were already a community of like-minded experts that owed its existence as much to the Frankfurt tradition as to the League. Although historians of science tend to see the ideal of a science free of all national, political and personal values as utterly naïve, the Standardisation Commission made good use of it as an argument for collaboration between German and French workers in the politically tense conditions of the 1920s. Without it, there would have been no German serologists at the Commission's meetings.

As time passed, the importance of sera declined, and that of standards for vitamins, hormones and pharmaceuticals (and later on, antibiotics) increased. Along with the sera, the power of the Ehrlich method as the technical bond that kept the group together tended to diminish. At the same time, there came a change of direction, away from the principle of the independence of technical matters from national interests. The fading cohesion of the group may have had something to do with this, added to the influence of the hopeful rhetoric about the collective security of nations. Later historians have again called this naïve idealism, but collective security was the slogan of the age, and it made technical cooperation itself a national interest. The Inter-governmental Conference of 1935 invited the national representatives to form themselves into a system of government-approved National Centres under the two central laboratories, Copenhagen and Hampstead, to be responsible for receiving and passing on the standards to acceptable users. The arrangements were hurried into force when war broke out in 1939.

No treaty was actually drafted, but this change of direction represented a return to the government-oriented attempts to create a standardized international system of health regulations, as favoured by the Paris *Office international*, and before that, by the International Sanitary Conferences of the nineteenth century.

Notes

1. Carola Throm, *Das Diphtherie-serum: ein neues Therapieprinzip, seine Entwicklung und Matkteinführung*, Stuttgart: Wissenschaftliche Verlagsgesellschaft, 1995.
2. Samuel Kryslov, *How Nations Choose Product Standards and Standards Change Nations*, Pittsburgh, PA: University of Pittsburgh Press, 1997, pp. 26–52.
3. Charles P. Kindleberger, 'Standards as Public, Collective and Private Goods', *Kyklos*, 1933 36: 377–96; Kindleberger, 'International Public Goods without International Government', *American Economic Review*, 1986 76: 1–13.
4. Neville M. Goodman, *International Health Organisations and Their Work*, Philadelphia, PA: Blakiston, 1952, pp. 39–66.
5. C. W. Hutt, *International Hygiene*, London: Methuen, 1927, pp. 3–64; Hutt writes as a hygienist, author of several handbooks: he explains the work of the *Office international de l'hygiène publique* from a contemporary point of view; G. Abt, *Vingt-cinq ans d'activité de l'Office international d'hygiène publique: 1909–1933*, Paris: *Office international de l'hygiène publique*, 1933; Goodman, *International Health Organisations*, 1952, 82–94: Appendix II: Text of the Rome Agreement, 95–8; Goodman was a member of the League of Nations Health Organisation as well as the World Health Organisation. He writes at the point of transition between the two, tracing their history back to the sanitary movement of the XIXth c.
6. René Lacaisse, *L'hygiène internationale et la Société des Nations*, Paris: Editions 'Mouvement Sanitaire', 1926, p. 38. The 30 later adherents were: Argentina, Bolivia, Britain (signing for her colonies Australia, Canada, India, New Zealand and South Africa); Bulgaria, Chile, Denmark, France (signing for her colonies Algeria, Indo-China, French East Africa, and Madagascar); Greece, Japan, Mexico, Monaco, Morocco, Norway, Peru, Persia, Poland, Rumania, Serbia, Sweden, Tcheko-Slovakia, Tunisia, Turkey, Uruguay.
7. Report, *Bulletin mensuel de l'Office international de l'hygiène publique*, 1912.
8. *Office international d'hygiène publique*, Comité international permanent, *Session extraordinaire d'avril 1920, Première Séance 26 avril 1920* Paris, 1920, pp. 201–3.

9. Foreign Office. Treaty Series No. 20. *Agreement Respecting Facilities to be Given to Merchant Seamen for the Treatment of Venereal Disease.* Signed at Brussels, December 1 1924, PP 1926 (Cmd 2727).

10. M. D. Dubin, 'The League of Nations Health Organisation', in Paul Weindling (ed.), *International Health Organisations and Movements 1918–1939,* Cambridge: Cambridge University Press, 1995, pp. 56–80; Dubin gives a comprehensive overview of the Health Organisation and its various sections.

11. Madsen to Velghe, ltr d. 22 February 1923. SSI, Madsens Papirer. SSI, Madsens Papirer1923 File I Folkevorbund.

12. League of Nations. Health Organisation, *Reports on Serological Investigations Presented to the Second International Conference on the Standardisation of Sera and Serological Tests, held at the Pasteur Institite, in November 1922,* 1923, p. v.

13. Margaret O. MacMillan, *Paris 1919: Six Months that Changed the World,* New York, NY: Random House, 2001, pp. 166–79; her chapter title is, 'Keeping Germany Down'.

14. Eduard Bernstein, *Volkerbund oder Staatenbund: eine Untersuchung,* 2nd edn., Berlin, 1919.

15. Brigitte Schroeder-Gudehus, *Deutsche Wissenschaft und internationalen Zusammenarbeit 1914–1928. Ein Beitrag zum Studium kultureller Beziehungen in politischen Krisenzeiten.* Thesis No. 172, Geneva: University of Geneva, 1966, pp. 90–6; Karl Kerkhof, 'Die internationalen naturwissenschaftlichen Organisationen vor und nach dem Weltkrieg und die deutsche Wissenschaft', *Internationale Monatsschrift für Wissenschaft Kunst und Technik,* 1921 15: 226–42.

16. Cited by Harold Richard Goring Greaves, *The League Committees and World Order: A Study of the Permanent Expert Committees of the League of Nations as an Instrument of International Government,* Oxford: Oxford University Press, 1931, p. 42.

17. Greaves, *League Committees,* 1931, pp. vii–viii.

18. Greaves, *League Committees,* 1931, p. 42.

19. Pauline M. H. Mazumdar, '"In the silence of the laboratory", the League of Nations Standardizes Syphilis Tests', *Social History of Medicine,* 2003 16: 437–59.

20. League of Nations. Health Committee, Minutes of the Fourth Session, Geneva August 14th to 21st, C. 555, M. 337, 1922, III, p. 4; cited in Rajchman to Madsen ltr d. Geneva, 26 December 1922. LoN Archives, 12 B 26213x/11346.

21. Gottfried Frey, Reichsgesundheitsamt, Berlin to Rajchman, ltr d. Berlin 1 September 1922. Archives LoN. R 838. 12B 26311x / 20109.

22. Madsen to Rajchman, ltr d. Copenhagen, 3 October 1922, p. 1. LoN. 12B 26213x/11346, File #3. Nocht had been working on ship's diseases, disinfection and questions of quarantine since the turn of the century. His most recent publication was Nocht, *Die Malaria: Einführung in ihre Klinik, Parasitologie und Bekämpfung,* Berlin, 1918.

23. Gottfried Frey, Reichsgesundheitsamt, Berlin to Rajchman, ltr d. Berlin, 6 August 1923. LoN. R 852. 12B / 26773x / 26249, File #2.

24. Albert Calmette, Director, *Institut Pasteur,* to Madsen, ltr rec'd October 1922, enclosed in Madsen to Rajchman, ltr d. 3 October 1922. LoN. 12B 26213x/ 11346.

25. Wilhelm Kolle to Madsen, ltr d. 29 November 1921. SSI, Madsens Papirer 1921–2 File Folkevorbund. Cited by Mazumdar, 'In the Silence of the Laboratory', 2003, p. 450.

26. Frankfurt was on the eastern edge of the French-occupied Rhineland: see map in MacMillan, *Paris 1919,* 'Germany and Europe in 1920', n.p., front matter, 2003.

27. Kolle to Madsen, ltr d. Frankfurt, 5 February 1923. SSI. Th. Madsens Papirer. I 1923. File Folkeforbund 1923. My thanks to Ms Magdalene Duckwitz, erstwhile

Cultural Secretary, German Embassy, New Delhi, for her help in transcribing the angry handwriting of this letter. Raymond Poincaré was Premier of France 1922–4, 1926–9; his repeated statements that German War debts must be repaid in full contributed to the fall in value of the Deutschmark. Marshall Foch was the supreme Allied Commander during WWI. Johan R. Mørch was a serologist from the *Statens Serum Institut* who at this time was travelling round the European laboratories working on the standardization of syphilis tests, trying to get some uniformity in the Wassermann protocols for comparison with the new colloidal tests. Later in the same year, he was in Warsaw and Vienna. See Mazumdar, 'In the Silence of the Laboratory', 2003, p. 451.

28. John Maynard Keynes to *Westminster Gazette*, ltr d. 1 January 1923; cited in in Elizabeth Johnston (ed.), *Collected Writings of John Maynard Keynes. XVIII: Activities 1922–1932: the End of Reparations*, London: Macmillan, 1978, pp. 105–8.

29. John Maynard Keynes, review in *The Nation and Athenaeum* 12 April, 1924; cited in Johnston (ed.), *Collected Writings XVIII*, 1978, pp. 235–41, 236.

30. Kolle to Madsen, ltr d. Frankfurt-am-Main, 3 November 1923. SSI. Th. Madsens Papirer I 1923. File: Volkeforbund 1923.

31. John Maynard Keynes, 'German People Terrified by Uncertainty', *Manchester Guardian* dispatch d. 28 August 1922; in Johnston (ed.), *Collected Writings XVIII*, 1978, pp. 28–30.

32. Keynes, *Manchester Guardian* dispatch d. 28 September 1922, in Johnston, (ed.), *Collected Writings XVIII*, 1978, pp. 32–43, 36.

33. Anne Marie Moulin, 'The Pasteur Institutes between the Two Wars: The Transformation of the International Sanitary Order', in Paul Weindling (ed.), *International Health Organisations and Movements 1818–1939*, Cambridge: Cambridge University Press, 1995, pp. 244–65.

34. Pauline M. H. Mazumdar, 'Antitoxin and *Anatoxine*: The League of Nations and the *Institut Pasteur*, 1920–1939', in Kenton Kroker, Jennifer Keelan and Pauline M. H. Mazumdar (eds), *Crafting Immunity: Working Histories of Clinical Immunology*, Aldershot: Ashgate, 2008, pp. 177–200.

35. League of Nations Health Organisation, Permanent Commission for the Standardisation of Sera, Serological Reactions and Biological Products (1923), *Investigations on the Serodiagnosis of Syphilis: Report of the Technical Laboratory Conference*, Copenhagen, C.5.1 (1924) iii; (C.H. 148) p. 71.

36. Stefan Mutermilch to Rajchman, extract from ltr d. 31 January 1924. LoN. R 828. 12B/26156x/7692.

37. Clemens Freiherr von Pirquet, 'Tuberculindiagnose durch cutane Impfung', Verhandlungen ärztliche Gesellschaften, Sitzung von 8 Mai 1907, *Berliner klinische Wochenschrift*, 1907, No. 20: 644. von Pirquet argued against Calmette's suggestion that tuberculin testing could be done (painfully!) on the conjunctiva instead of the skin: v. Pirquet, 'Kutane und konjunktivale Tuberculinreaktion', in R. Kraus and C. Levaditi (eds), *Handbuch der Technik und Methodik der Immunitätsforschung*, 2 v. +*Supplement*, Jena, 1907, pp. 1035–62; see Gabrielle Dorffner and Gerald Weippl, *Clemens Freiherr von Pirquet: ein begnadeter Arzt und genialer Geist*, Vienna: 4/4 Verlag, 2004, pp. 85–92.

38. Record of a Conversation [between Rajchman and] Dr Dale, 6 October 1923. Following the Conference on the Standardisation of Biological Products held at Edinburgh, July 23 1923, 'Dr Dale agreed to act as Director of Research for these products in the same way as Professor Madsen does for the Serological Standards'.

39. Sir Henry Dale to Rajchman, ltr d. Hampstead, London, 15 November 1927. LoN. R899. 12B 55189x / 29418, File # 1; Rajchman to Dale, ltr d. Geneva, November 21 1927. LoN R899 12B 55189x / 29418 File # 1; Dale to Rajchman, ltr d. Hampstead, London 25 November 1927. LoN R899 12B 55189x / 29418 File # 1.
40. Calmette to 'Mon cher ami', [Rajchman] ltr d. 1 December 1927. LoN R899 12B 55189x / 29418 File # 1.
41. Calmette to Rajchman, ltr d. 12 December 1927. LoN R899 12B 55189x / 29418 File # 1. My emphasis.
42. Dale to Rajchman, ltr d. 23 December 1927. LoN R899 12B 55189x / 29418 File # 1. Note also *The Collected Papers of Paul Ehrlich in Four Volumes including a Complete Bibliography.* Compiled and edited by F. Himmelweit, with the assistance of the late Martha Marquardt, under the editorial direction of Sir Henry Dale, London: Pergamon, 1957. The fourth volume with the bibliography never appeared.
43. Rajchman to Dale, ltr d. Geneva, 27 December 1927. LoN R899 12B 55189x / 29418 File # 1.
44. Calmette to Rajchman, ltr d. Paris, 14 January 1928. LoN R899 12B 55189x / 29418 File # 1.
45. Rajchman to Dale, ltr d. Geneva, 10 March 1928. LoN R899 12B 55189x / 29418 File # 1.
46. Gaston Ramon, 'Sur la toxine et sur l'anatoxine diphthériques: pouvoir floculant et propriétés immunisantes', *Annales de l'Institut Pasteur*, 1924 38: 1–10; Ramon, 'L'anatoxine diphthérique', *Annales de l'Institut Pasteur*, 1928 42: 959–1009.
47. Albert Calmette, Director, *Institut Pasteur* to Rajchman, ltr d. Paris, 17 January, 1924, LoN. 12B/ 30923x / 2941, cited in Mazumdar, 'Antitoxin and *Anatoxine*', in Kroker, Keelan and Mazumdar, *Crafting Immunity*, 2008, pp. 185–6.
48. Mazumdar, 'Antitoxin and *Anatoxine*', in Kroker, Keelan and Mazumdar, *Crafting Immunity*, 2008, p. 188. An adaptation of the classical Ehrlich method did in fact continue to be used, at least in Britain: see Percival Hartley, 'State Control of Diphtheria Prophylactic in Great Britain', *Wissenschaftliche Woche zu Frankfurt a.M. Probleme der Bakteriologie, Immunitätslehre und experimentelle Therapie* (2–9 September), 3: 81–9; Madsen however came to prefer Ramon's flocculation method: League of Nations, 2008, Inter-governmental Conference on Biological Standardisation, Committee to Review International Standards Established for Sera and Bacterial Products, Second Meeting, 3 October 1935, p. 6. LoN. Conf. S.B. / S.P.B. / P.V.2.
49. Mazumdar, 'Antitoxin and *Anatoxine*', in Kroker, Keelan and Mazumdar, *Crafting Immunity*, 2008, p. 189.
50. Georg Schwarzenberger, *The League of Nations and World Order: A Treatise on the Principle of Universality in the Theory and Practice of the League of nations*, London: Constable, 1936, pp. 149–51.
51. Roland N. Stromberg, 'The Idea of Collective Security', *Journal of the History of Ideas*, 1956 17: 250–63.
52. George W. Egerton, 'Great Britain and the League of Nations: Collective Security as Myth and History', in *The League of Nations in Retrospect: Proceedings of the Symposium Organized by the United Nations Library and the Graduate Institute of International Studies Geneva, 6th to 9th November, 1980,* Berlin: de Gruyter, 1983, pp. 95–117, 95.
53. Greaves, *League Committees*, 1931, pp. 111–38, 122.
54. Maurice Bourquin (ed.), *Collective Security: A Record of the Seventh and Eighth International Studies Conference,* Paris: International Institute for Intellectual Co-operation, 1936.

55. Greaves, *League Committees*, 1931, p. 42.
56. Dale to Raymond Gautier, ltr d. Hampstead, 4 April 1935. LoN. 8E / 8391 /1060.
57. League of Nations Health Organisation, Intergovernmental Conference on Biological Standardisation, 1 to 4 October, 1935. C.H. / C. P.S.B. / 32.
58. Dale's responsibility for the non-serum standards dated back to the First International Conference on Biological Standardisation of Certain Remedies, held in Edinburgh in 1923. At the Second Conference held at Geneva in 1925, under Dale's chairmanship, the list of remedies included pituitary extract, insulin, digitalis, thyroid gland, ergot, the antihelminthics *Felix mas* and oil of Chenopodium, vitamin A, then known as 'growth factor', as well as salvarsan and the other arsenobenzene derivatives, used in syphilis. These last were held by Wilhelm Kolle at Frankfurt until 1935, since that was where Paul Ehrlich had invented them. After Kolle's death in 1935, when Germany had left the League, they were taken over by Dale. Although the arsenobenzes were not biologicals, each batch was tested biologically for toxicity and therapeutic effectiveness on mice and rats infected with pathogenic typanosomes, in a modification of the *Dosis letalis* method. LoN C.532. M. 183. 1925. III (C.H. 350).
59. League of Nations. Health Organisation, Intergovernmental Conference on Biological Standardisation. Third Plenary Meeting, Geneva, 3 October. Advisability of Establishing in Each Country a National Centre for the Distribution of International Standards and Possibility of Establishing International Standards for Certain Standardised Preparations, 1935, p.1. LoN. Conf. S.B. / P.V.3.
60. League of Nations. Health Organisation, *Report to the Council on the Work of the Twenty-Fourth Session of the Health Committee* Geneva, 5 to 9 February. C.148.M.96.1937.III., 1937, pp. 18–22.
61. Dale to Gautier, ltr d. Hampstead, 16 April 1937. LoN. 8E / 8391 / 1060.
62. Gautier to Dale, ltr d. Geneva, 23 April 1937. LoN. 8E / 8391 / 1060.
63. Gautier to Dale, ltr d. Geneva, 11 February 1938. LoN. 8E / 8391 /1060.
64. Dale to Gautier, ltr d. Hampstead, 14 September 1938. LoN. 8E / 8591 / 1060.
65. Percival Hartley, 'Note on the International Standards for Drugs, Hormones and Vitamins: August 1940', in Raymond Gautier, 'The Health Organisation and Biological Standardisation (Second Memorandum)', *League of Nations: Bulletin of the Health Organisation*, 1945–6, 12: 1–110, 98–110.
66. According to Gautier, the serum standards were still being sent from the *Statens Serum Institut* to 15 European countries and 11 overseas ones, including the US, China and Japan, and five in South America. Hampstead supplied the British Empire. Gautier to Dale, ltr d. Geneva, 20 May 1941. LoN. 8E / 13635 /x(3).
67. Dale to Gautier, ltr d. Hampstead, 8 June 1940. LoN. 8E / 13635 / x (3).
68. Gautier to Dale, ltr d. Geneva, 14 November 1945: Dale seems to have suggested appointing the heads of the control centres to a renewed post-War Commission; Gautier, now more reserved about the international project than Dale, replies that 'In certain countries I could name, the heads of these centres do not attain a high standard, and are merely concerned with the practical aspect of biological standardisation. Would the presence of such members not result in its leaning too heavily on the administrative side? It is not in every country that one finds a Percival Hartley combining in one person scientific and administrative capacities'. LoN. 8E / 8391 /1060.
69. Dale to R. E. Wodehouse, Deptt of Pensions and National Health, Ottawa, ltr d. Hampstead, 15 October 1940. UK Public Record Office, ED1 /2383; J no. 11797.

8
Questions of Quality: The Danish State Serum Institute, Thorvald Madsen and Biological Standardization

Anne Hardy

The opening of the Danish State Serum Institute (SSI) in Copenhagen on 9 September 1902 was a festive occasion, attended by renowned figures from the wider bacteriological community including the German scientists Paul Ehrlich, Carl Weigert, and Julius Morgenroth, future Nobel prize-winner Svante Arrhenius from Sweden, Ole Malm and Armauer Hansen from Norway, and William Bulloch and German Sims Woodhead from England.[1] Established as a national resource for the production of diphtheria antitoxin, the SSI was from its inception concerned to deliver a quality product at a minimum price, and to link pharmaceutical production with research into, and further development of, biological products. In the course of the twentieth century, the institute acquired an international reputation for the quality of its products and its cutting edge research, and, in the 1920s, achieved international authority as the League of Nations Health Commission's central laboratory for the preservation and distribution of all standard sera and bacterial products.[2] The rise of the SSI to international prominence came about through a combination of factors, personal, scientific and political, but above all, perhaps, from its early association with questions of quality in the production of the new generation biological medicines, of which diphtheria antitoxin was the first to emerge.

Diphtheria and the development of the Danish State Serum Institute

The creation of the SSI was largely due to the energy and determination of Carl Julius Salomonsen (1847–1924), the 'father of danish bacteriology',[3] for Danish medical culture at that period was largely traditional with a focus on hospital medicine and general practice, rather than on research, let alone bacteriology.[4] Late nineteenth-century Denmark was a small state on the European periphery, with a population of just over two million in 1890, which was then beginning to adopt progressive social welfare policies in emulation of the German welfare model elaborated from the mid-1880s – a political context which helped Salomonsen's promotion of several medical projects.[5] Like almost every European state at this period, the country suffered severely from epidemic diphtheria in the years between c. 1880 and 1895; at the peak in 1893 a total of 23,695 cases were noted.[6] Public concern over the domestic tragedies resulting from this epidemic was considerable, and was not helped by press publicity. In 1890, for example, one of the leading Danish newspapers carried an article which graphically described the diphtheria wards at Copenhagen's isolation hospital where 'tragic children struggle against powerful death'. The article concluded:[7]

> We understand mothers' terror of this dreadful disease, that sometimes kills at once, sometimes when the child is convalescent. Diphtheria is nearly always followed by paralysis in the throat or the heart. When one believes a little child has recovered, it is suddenly overtaken by a heart attack, and tumbles over on the floor in its play, dead. (Author's translation)

In this context, Emile Roux's announcement in the summer of 1894 of the successful diphtheria antitoxin trials generated, as elsewhere, significant interest from a wide section of the general public as well as within the small medical community.

The significant mortalities and vivid popular anxieties which still surrounded several major infectious diseases in the 1880s generated perilous hopes of the new knowledge and new techniques that were beginning to emerge from the bacteriological laboratories. Although, as Jonathan Simon notes in his contribution to this volume, Roux was surprised by the avid public interest in the new anti-diphtheria serum in 1894, his reaction seems odd, even naïve. Given the intense popular anxieties that surrounded the disease, such a reaction seems predictable – more especially since precisely such an overwhelming public response had greeted Robert Koch's announcement of an apparent cure for tuberculosis in 1890.[8] The new therapies quickly altered public expectations and popular practices; the introduction of serum therapy and laboratory diagnosis, for example, reconciled the middle classes to hospital treatment for their sick offspring.[9] On the one hand anxiety generated popular interest and support for these new treatments, but on the other the treatments could also generate new concerns. Deaths associated with the new treatments, such as that of Ernst Langerhans described by Axel Hüntelmann, indicated that these novel therapies were not without risks, and could not be accepted as an unconditional good.[10]

The introduction of these new therapies was in fact far from straightforward. Public acclaim and public suspicion, scientific rivalry, financial considerations and ethical issues surrounded their introduction into clinical practice. In the last two decades of the nineteenth century, the scientists who pioneered these techniques learned, faute de mieux, to negotiate conflicting sets of interests. They also learned, gradually and sometimes painfully, that to be successful the new techniques required precision of method, the setting of standards, and the implementation of hygienic practices. It is often forgotten that the young science of bacteriology emerged in a scientific world where standards of scientific practice were rudimentary. Koch's postulates were only one small step on the road to modern science; the success of both scientific experiment and commercial manufacture could depend crucially on the physical environment of the laboratory. One of the most striking features of the memoirs by distinguished Danish practitioners written about the early years of bacteriology is the lasting impression of habitual chaos in the laboratories of the period 1870 to 1900. The Danes were proud of having been the first to adopt antiseptic practices,[11] and were somewhat shocked by failures to recognize the importance of the practice in both Germany and Paris, while the filthy conditions prevailing in many prestigious laboratories were also a source of remark. As one observer later noted: 'The laboratories then were uniformly depressing to look at, grey or brown and often terribly dirty. Apparatus and bottles filled the tables, racks of half-full test-tubes were the order of the day, and there was generally a prohibition on moving anything'.[12] Pasteur and Ehrlich's labs were no exception, and when the League of Nations Health Commission in the 1920s held a laboratory conference at the *Institut Pasteur* in Paris, the designated rooms were so dirty that it took several days' cleaning before they were useable.[13] Asked if these distinguished bacteriologists had no sense of order and cleanliness, this witness replied that laboratory equipment was then less highly regarded, and that researchers were perhaps also under the impression that if people like Pasteur could make ground-breaking discoveries in a filthy room in the Rue d'Ulm, then lesser spirits were in no position to demand better conditions.[14]

More scrupulous laboratory practice and protocols did, however, develop with, sometimes bitter, experience. The new biological therapies that began to emerge after 1880 were of a very different nature from the traditional galenic medicaments. They were not compounded from inert organic and inorganic substances but prepared with and from living organisms and live body fluids. As live preparations, these biological products were more volatile than the drugs derived from the pharmacopeia; they could vary unpredictably in strength, and in their reactions with other living material. They contained in themselves the unpredictable essence of nature. Late nineteenth-century researchers were still learning not only how to make and handle such substances but also the importance of precision if effective replication was to be achieved. As Gabriel Gachelin points out in his chapter of this book, early descriptions of the anthrax and rabies vaccines published from the Pasteur Institute were 'rather imprecise' as to the technical procedures used, while Robert Koch notoriously withheld any details of the manufacture and testing of tuberculin in 1890. By contrast, Emile Roux and Louis Martin were quick to publish detailed descriptions of the protocols for the isolation and inactivation of diphtheria toxin in 1894.[15] That publication, together with the generosity which the French researchers showed in welcoming visitors to the Pasteur Institute and instructing them in their methods of production, ensured that manufacture of the antitoxin serum became a possibility even in laboratories on the European periphery.[16]

National responses to the possibilities of diphtheria antitoxin varied. While the excitement generated by the promise of the new therapy spread widely, different communities took the initiative towards implementing introduction and manufacture in different places. In Geneva and in London, the initiative came from the private sector: in Geneva from the women's branch of the Samaritans, as Mariama Kaba shows in her contribution to this volume, while in London the privately funded British Institute for Preventive Medicine (later the Jenner, then the Lister), then under the direction of Pasteur protégé Armand Rueffer, set up a Serum Department as early as August 1894.[17] In Denmark, the incentive came from Copenhagen University's professor of microbiology, Carl Julius Salomonsen.[18] Patterns of manufacture and distribution also varied, as they became established. In Britain and in Germany, production passed to commercial companies.[19] In France, locally funded provincial serotherapy facilities, acting under the umbrella of the Pasteur Institute in Paris, supplemented the activities of the Paris centre, which operated as a quasi-monopoly of serum production in France (Gachelin; Simon). In Denmark, serum production began in a university facility but became formalized under state patronage and transformed into a national facility.[20]

The history of the Danish State Serum Institute, as previously noted, is of much more than purely national significance. Under the leadership of its second Director, the suave, cosmopolitan Thorvald Madsen (1870–1957), the SSI achieved an international reputation for the quality of its products and the status of arbiter of international standards in the production of biological medicines. Although microbiology in late nineteenth-century Denmark was a minute enterprise compared with the programmes and personnel active in the bacteriological heartlands of France and Germany, it was fronted by powerful, well-connected personalities possessed of a clear sense of scientific and human priorities. Carl Julius Salomonsen and Thorvald Madsen were between them the architects of the SSI and its subsequent reputation.

Carl Julius Salomonsen had initially trained as a doctor in Copenhagen, but was seriously attracted to the study of microbiology, and spent an enjoyable and profitable summer working under Julius Cohnheim's instruction at Breslau in 1877.[21] Here he made the acquaintance, among others, of Paul Ehrlich and William Welch, and completed an important study of tuberculosis in the eye of the rabbit.[22] On his return home he was appointed to teach microbiology at Copenhagen University, the first such lecturer to be appointed in any university. His small laboratory was located in a basement under the

Botanic Gardens Museum, and here he conducted the first ever taught course in bacteriology.[23] Salomonsen was among the many fired with enthusiasm for Roux's new therapy, and in September 1894 he travelled to Paris to study it. He was given every facility to familiarize himself with the processes of production, remarking particularly on the generosity of the French researchers in this respect.[24] Back in Copenhagen, Salomonsen successfully sought financial support from the Ministry of Education for a serum production facility and the training of two young co-workers. In November 1894 he established a serotherapy department in two small rooms within the university Medical Bacteriology Laboratory on Ny Vestergade. His team consisted of Thorvald Madsen, who had previously taken his bacteriology course, and who had qualified as a doctor the previous year; Miss Louise Hoeg, as under-assistant, and a laboratory technician, N. Rasmussen. Salomonsen's ambition for this department was two-fold. First, that it should produce diphtheria antitoxin for free distribution to Danish doctors and hospitals; and second, that it should engage in research into the processes of immunity.[25]

It was soon discovered that serum production was a lengthy procedure with unpredictable outcomes. Just as the facility at Nancy began the immunization process in November 1894 but was unable to supply the local hospitals until February 1895, so it was not until June 1895 that the first home-produced serum reached Copenhagen's Blegdamshospital.[26] It was not just that the immunized horses took a couple of months for their serum to ripen. The Danes also experienced much greater difficulty in developing a satisfactory product than Salomonsen had anticipated, and assistance was not forthcoming from elsewhere. 'It must be remembered', he recalled, 'that at that time the few existing serum manufactories which possessed a greater experience in the matter than ourselves, had, in the interests of their business, preserved an absolute silence with regard to their methods and results'.[27] The reticence of the successful manufacturers greatly increased the difficulties experienced by new would-be producers: they had to negotiate the same learning curve as everyone else. They learnt the hard way that the strength of strains of diphtheria bacillus attenuated with time; that the strength of serum produced by individual horses varied, and also depended on the strength of the toxin used, the mode of immunization, and the time of bleeding; and 'upon other circumstances as well, of which we possess as little knowledge as of the varying toxigenic power of the diphtheria bacillus'.[28] The importance of measuring the strength of a given serum was identified early, and remained a constant concern. Writing on the occasion of the opening of the SSI in 1902, Salomonsen noted, 'It is this enigmatical uncertainty on all the chief points that makes the production of the serum both very expensive and very precarious'.[29] In his report for 1903–8, Thorvald Madsen echoed these sentiments: 'Serum production is still from day to day a very difficult and uncertain business'. The sentence was to appear practically unaltered some 30 years later, in his report for 1940.[30]

Despite all the difficulties, the Danish serum project was taken steadily forward, encouraged by disappointing experiences with imported German serum in the Blegdamshospital in the autumn of 1894.[31] Salomonsen's ambition to produce enough serum for the whole country was underpinned by a determination not to spend national resources on expensive German commercial imports of variable quality.[32] In the end, cooperation with the British Institute for Preventive Medicine, who donated a powerful diphtheria strain, resulted in a satisfactory outcome. The production method eventually adopted followed the BIPM's very closely, but the methods of measurement were, 'of course', those of Paul Ehrlich.[33] Ehrlich had been a very constant friend to the Danish project.[34] Salomonsen had met him in Breslau in 1877, and he and Thorvald Madsen visited Ehrlich in 1899. Moreover, Madsen's doctoral dissertation, completed in 1896, endorsed Ehrlich's methods of measuring serum quality as superior to the French.[35] Ehrlich's work remained a powerful influence within the SSI for many years, and Madsen long maintained warm sympathies for Germany and his German colleagues.[36] Cay-Rüdiger Prüll has noted the importance of Ehrlich's influence on Henry Dale as a formative contribution to Dale's role in achieving international standards for insulin in the 1920s (Cay-Rüdiger Prüll, 'Paul Ehrlich's Standardization of Serum; *Wertbestimmung* and its Meaning for Twentieth-Century Biomedicine'), but Ehrlich's influence on Thorvald Madsen, who was to chair the LNO Health Commission for much of its existence, and who was a prime mover in establishing the Commission's Committee on Biological Standards, was perhaps of greater significance to the interwar biological standardization project in general.

Thorvald Madsen, diphtheria and the SSI

Thorvald Madsen was, as we have seen, appointed director and researcher of Salomonsen's small serum production facility in 1894. He was then a very newly qualified doctor, who had come to Salomonsen's attention as one of only two students attending the microbiology course a few years earlier.[37] Moreover, he came to the post with a particular personal interest: he himself had been hospitalized with diphtheria in 1889, and his youngest sister had died of the disease at the same time.[38] Although Madsen himself never refers to this incident in any of his memoirs, it is generally assumed that this experience was not unconnected to his later scientific interests.[39] While Johannes Fibiger, also a former student of Salomonsen, was conducting his more famous study of the outcomes of antitoxin use in the Blegdamshospital, Madsen was conducting a landmark comparative study of the French and German methods of serum production and measurement for his doctoral dissertation, completed in 1896.[40] Although Gachelin has emphasized the importance of knowledge transfers in respect of serum evaluation between France and Germany and the similarity of their methods, Madsen concluded that the German methods – those of Paul Ehrlich – were superior to the

French. Significantly, the dissertation ('Experimental Investigations into Diphtheria Toxin') included the following observation:[41]

> It would be of the greatest significance for these measurements, if agreement could be reached on an international unit for determining the strength of the antidiphtheria serum.

When Madsen visited Ehrlich as part of his bacteriological 'Grand Tour' in 1899, he found himself in high favour with Ehrlich as a result of his doctoral work, and he clearly felt very much at home with Ehrlich and his colleagues.[42] By contrast, a visit to the *Institut Pasteur* was less congenial: 'It was one of those places where you dumped down in a corner, and one could stay there without ever learning anything about the rest of it'.[43] It seems apparent that, from the beginning, Paul Ehrlich stood as patron and godfather to the Danish serotherapy enterprise, and his presence as an honoured guest at the inaugural festivities for the new Serum Institute in 1902 reflected that fact.

The translation from a small university facility to a state-funded, purpose-built institute was achieved through a combination of circumstances. In the first place, diphtheria remained an acute concern for the Danish people, and medical demand for the new serum was rising; secondly, the Danish state was already embarked on a programme of infectious disease control.[44] Finally, Salomonsen was a man of some determination when he had identified a project, as his achievements of the initial serum production unit, and later of Denmark's first Institute of Pathology, prove; and Madsen was well-connected politically, his father being Minister of War between 1901 and 1904. When, in 1898, Denmark's hospital physicians united in calling for the further development of the serum facility so that all the country's doctors could be supplied with diphtheria antiserum free of charge, Salomonsen took action and approached the Ministry of Education.[45] While the Ministry and the University's Medical Faculty considered how best to establish a new and extended facility, Salomonsen and Madsen identified a suitable site for the new facility on War Ministry property on the island of Amager, lying just across a narrow strait south of the mainland city.[46] Delayed by various political complications, the State Serum Institute Act was finally passed in spring 1901.

The first clause of the Act provided for the establishment of an Institute for the production of diphtheria antiserum,[47]

> under the jurisdiction of the Ministry of Justice. It shall undertake to supply such serum on the largest possible scale to the medical men of this country and its colonies, on application to the said Institute.

The sum of 172,800 Danish kroner was set aside to provide for the new buildings on Amager Common, and the sum of 23,000 kroner per annum was provided to meet running costs. Government priorities for the new

Institute were reflected in its governance provisions: the Institute Director was to receive a salary of 1000 kroner a year, but the Laboratory Director responsible for serum manufacture, the work connected with it, and for directing research, was to receive an annual 3200 kroner, with an incremental increase of 600 kroner every 5 years, to a maximum of 5600 kroner.[48] Thus Salomonsen as Director of the Institute was considered a far less important figure than Madsen as Director of the Laboratory with responsibility for the day-to-day functioning and production.

The original intention had been for the SSI to deliver its products to the medical community free of charge, but the 1901 Act actually specified that the product be sold. The Ministry of Justice managed to circumvent this provision with the consent of parliament, setting the price so low as to cause minimal inconvenience to existing supply agreements. At a price of 25 øre (100 to the krone) per dose, the SSI price compared very favourably to the 6 kr 25 øre charged by the German commercial firms.[49] Where the small University serum facility had been stretched to produce 6305 doses of antiserum in 1901–2, the SSI in its first full year of production (1903) produced 8800 doses.[50] At this time it staff was barely larger than when in Ny Vestergade: it consisted initially of Madsen, Louise Hoeg, a watchman and a stable-master. The first research assistant arrived six months after the building opened, by which time the SSI was already making a name for itself. The first foreign visitor was Hideyo Noguchi, who came to study under Madsen in 1903, but eminent scientific guests over the years came to include Theobald Smith, Elie Metchnikoff, William Welch, Robert Koch and Jules Bordet, besides a constant stream of lesser luminaries.[51] The SSI had quickly come to represent a gold standard in the production of biologicals.

In the years that followed, the SSI began to accrue additional responsibilities for bacteriological and serological diagnoses, as the field developed, and as anxieties about diseases other than diphtheria came into existence. Such work in turn generated related research projects. A new small laboratory for the study of plague and cholera was approved in 1908, just as it was realized that the Wasserman test had important consequences for the diagnosis and medical control of syphilis. The perceived necessity of giving private practitioners and hospitals access to the new procedure precipitated plans for a major expansion of the SSI's remit and facilities.[52] The newly established government Health Commission, which had the brief to organize Denmark's civilian health administration, recognized the SSI as the country's central laboratory for epidemiological and serological research in support of its own remit, and supported the proposed expansion. The law enabling the SSI's physical and scientific expansion was enacted in 1910.[53] At the same time, the Health Commission articulated its expectations for the newly expanded establishment:[54]

> The Health Commission strongly emphasizes that the Institute is established as a humane and socially useful institution, and it therefore

regards any attempt to make this facility a directly profitable undertaking as highly regrettable... The Commission considers this desirable not only on humanitarian grounds, but also because by the reduction in the costs of hospitals and epidemics, and in the losses incurred by commerce and transport resulting from epidemics, it will achieve a much greater compensation for its outlay. (Author's translation)

A clear recognition of the economic benefits to be obtained by supporting the production of and research into biological medicines underpinned the Danish government's continuing financial support for the expansion of the SSI.

The scientific expansion of the SSI beyond the initial remit of diphtheria anti-serum production was presided over by Thorvald Madsen, initially as Director of the Laboratory and, from 1 April 1909, on Salomonsen's retirement from the post to concentrate on the University's new Ordinary Pathology ('Almindelig patologi') facility, also as Director of the Institute.[55] In taking on these new responsibilities, Madsen's aim remained clear: the development and production of high-class biological medicines for the prevention and treatment of human disease.

The SSI and biological standardization

Quality in the production of biologicals was an issue for the Danish serotherapy community even before the establishment of the SSI. Madsen's doctoral thesis, as we have seen, marked an early interest in issues of comparability, quality and standardization, and this proved an enduring preoccupation in the context of the SSI's ambition to provide the best possible products at the lowest possible price. With the development of an international trade in antitoxin after 1894, concerns over the relative strength of the different products on the market became rife. Physicians needed to know how many units or millilitres of serum were needed for individual patients. German, French, Danish and English sera came in varying strengths and qualities, but these differences might not be immediately apparent, unless the place of production was carefully noted. Although the 'unit' (the smallest amount of serum that would neutralize the smallest lethal dose of toxin in a mouse or guinea pig) as the standard measurement of strength was adopted early, it became apparent that different national units were not one and the same thing. Differing biological characteristics between populations of laboratory animals significantly affected the comparative quality of the final product. As historian Karl Jensen observes:[56]

The Germans used some very sensitive mice, the French mice were sound as bells and difficult to kill, while the English mice were pure rubbish. For this reason the French antitoxin unit was stronger than the German which in turn was stronger than the English. (Author's translation)

The differences were significant enough for the English *Lancet* Commission on diphtheria antitoxin of 1896 to conclude that the English products did not work.[57]

The biological problems of standardization quickly became central to the research conducted at the SSI. In 1897, Madsen and Salomonsen demonstrated that there were significant differences in the ability of various sera to neutralize toxins. Some horses produced poor antitoxin, others a high quality product.[58] In that year, however, Paul Ehrlich produced his ingenious solution to these problems as regards the diphtheria antitoxin in developing a standardized antitoxin preparation, which prepared the way for the later standardization of a wide range of biologicals from toxins and antitoxins to insulin and penicillin.[59] From 1897 until the outbreak of WWI, Ehrlich's preparation was internationally accepted as the standard for all diphtheria antitoxins. Problems of unit comparability resurged, however, in the political and scientific fragmentation associated with war. Within weeks of the outbreak of hostilities, tetanus had become a serious problem among the wounded of all combatants on the Western Front. Both British and German military authorities sent urgent requests for supplies of anti-tetanus serum across Europe, and production was stepped up in many areas outside the military zones. In Denmark, neutral in this war, but economically heavily dependent on both England and Germany before 1914, production was hurriedly expanded.

The level of wartime demand for the tetanus antiserum may be illuminated by the British experience. From August to October 1914 there were no definite instructions on the administration of antitoxin to the wounded, and many were not given it; in this period tetanus deaths stood at 8 per 1000 wounded. From the middle of October every wounded man was given a dose of 500 units; from June 1917 dosage was increased to four doses of 500 units at weekly intervals.[60] This reduced the incidence of tetanus to 1 per 1000 wounded. Given the number of casualties, vast quantities of antiserum were required to meet demand. Denmark's response to military demands may not have been typical, but is suggestive of the efforts that were made across Europe to meet the military requirements. At the outbreak of war, the SSI maintained just two 'tetanus horses' – enough to meet the national need in peacetime. As the calls began to come through from the front, Madsen sent his stable-master out to buy horses, managing to acquire a further 46 animals, in the teeth of aggressive German buying to meet demand on the battlefield. By the end of the war, the SSI's stable of tetanus horses numbered around one hundred animals.[61]

As a result of extensive laboratory efforts at expanded production, death rates on the battlefields began to fall by early 1915, but with the pan-European and American involvement in production, issues of standardization soon became apparent. One German anti-tetanus unit was discovered to be the equivalent of 67 American units, or of around 3000 French units.

As Madsen later observed: 'The unfortunate doctors, who ... had to use teta-nus serum from many different countries, were often terribly dis-oriented when it came to dosage'.[62] René Gautier, Secretary of the League of Nations Health Organisation's Permanent Committee on Biological Standardisation, phrased it rather more trenchantly:[63]

> Many deaths could have been averted if the sera used during the war had been assayed in relation to a unique standard. Doctors would not have been betrayed by the unitage given on foreign ampoules into injecting quantities of serum which they had good reason to regard as sufficient, but which were in fact inadequate, since the assay had been effected in terms of a unit of lesser potency than that to which they were accus-tomed.

Wartime experience thus dramatically reinforced the realization of the necessity for internationally accepted biological standards being agreed upon and implemented.

Biological standardization was one of the first issues taken up by the League of Nations Health Committee after the war, at least in part because Thorvald Madsen had been appointed its President – a position he was to hold until 1937. At the Committee's second session, held in Geneva in 1921, Madsen proposed international collaboration on the question of biological standardization.[64] As Pauline Mazumdar shows in her chapter, such inter-national collaboration, on a topic which necessitated the involvement of French and German scientists as leading authorities in the field, was by no means easy in the initial aftermath of the war. Madsen's diplomatic skills, the networks of scientific association he had built up in the pre-war years, the standing which he and the SSI had already achieved in this new field, and his own recognition of the importance of the standardization project to future developments in international health, were critical in enabling the achievement of the Health Committee's international collaborative effort.

Biological standardization in context

The interwar biological standardization project was not, therefore, primarily initiated in response to problems with diphtheria antitoxin preparations. None the less, Paul Ehrlich's pioneering work in standardizing the diphtheria antitoxin had established the scientific basis from which work in respect of other pathogens could go forward in the interwar period when the principle found general application in pharmacology, physiology and bacteriology.[65] Writing in the early 1940s, the Danish bacteriologist and SSI researcher Johannes Ipsen, whose own work was to put biological standardization on a new plane after 1945, recorded that the adoption of the procedure was the result of intensive international co-operation, sponsored in particular by the League of Nations Health Committee under the direction of Thorvald Madsen. Three countries, Ipsen noted, were principally concerned in this work, although many institutions had shared in the establishment of standard preparations for series of hormones, vaccines and sera. Those three countries were Germany, where the traditions created by Paul Ehrlich had been maintained by the Frankfurt Institute; England, 'the home of modern biometry'; and Denmark, which 'has contributed greatly to the biological standardization under the leadership of Aug. Krogh and Th. Madsen'.[66] In other words, Germany had set the standards; England provided the mathematical means for calculation; and Denmark the bacteriological and physiological underpinning of the standardization project.

Modern histories of the early decades of bacteriology have tended to focus on contributions from the two great pioneering nations of the young science, Germany and France, as well as on such leading figures as Louis Pasteur, Robert Koch and Paul Ehrlich. But bacteriology rapidly became an international science, whose fascination attracted world-class scientists across the globe, many of whose contributions remain as yet under-explored and undervalued by historians. In the case of Denmark, a handful of scientists working initially in primitive conditions in a small country on Europe's Nordic fringe, played a critical role in the eventual establishment and implementation of internationally accepted standards for the groundbreaking biological therapeutic innovations that derived from the new science. Biological standardization was a core interest of the SSI from its earliest incarnation as a centre of production for the anti-diphtheria serum, and it was a particular interest of Thorvald Madsen, who developed into one of the subtlest medical politicians operating on the international stage during the interwar years. While Germany and France were the colossi that bestrode the world of bacteriology before WWI, it was Denmark, with its particular concern for biological standards, that can be seen as a crucial connecting thread, especially in the pulling together of the international standardization project of the interwar period.

Notes

1. T. Madsen, *Statens Seruminstitut* Copenhagen: Staten's Serum Institut, 1940, p. 21.
2. R. Gautier, 'The Health Organisation and Biological Standardisation', *Quarterly Bulletin of the Health Organisation* 4 1935: 499–554, 501, 504.
3. T. Madsen, 'Carl Julius Salomonsen (1847–1924)', *Journal of Pathology and Bacteriology* 28 1925: 702–8, 702.
4. See, for example, T. Madsen, 'Dr Thorvald Madsen fortaeller', *Medicinsk Forum* 6 1953: 197–205, 198–202; C. J. Salomonsen, 'Reminiscences of the Summer Semester, 1877, at Breslau', in C. L. Temkin (ed.), *Bulletin of the History of Medicine* 24 1950: 333–51, 344.
5. For Danish medical organisation and culture, see J. Lehmann, J. Carlson and A. Ulrik (eds), *Denmark: Its Medical Organisation, Hygiene and Demography* Copenhagen and London: Gjellerup, 1891.

6. J. Lehmann et al. *Denmark: Its Medical Organisation*, 1891, pp. 319–20.
7. 'Mellem dødsyge. Et besøg ved midnat', *Politiken* 15 December 1890, cited in K. Jensen, *Bekaempelse af infektionssygdomme* Copenhagen: Nyt Nordisk Forlag/ Arnold Busk, 2002, p. 18.
8. See C. Gradmann, *Krankheit im Labor: Robert Koch und die medizinische Bakteriologie* Göttingen: Wallstein Verlag, 2005; T. D. Brock, *Robert Koch: A Life in Medicine and Bacteriology* Madison WI: Science Technology Publishers, 1988, pp. 201–5.
9. P. Weindling, 'From Medical Research to Clinical Practice: Serum Therapy for Diphtheria in the 1890s', in J. V. Pickstone (ed), *Medical Innovations in Historical Perspective* Basingstoke: MacMillan, 1992, pp. 72–83.
10. Axel Hüntelmann, 'Das Diphtherie-Serum und der Fall Langerhans', *MedGG* 24 2006: 71–104.
11. Salomonsen, 'Summer Semester', 1950, p. 345.
12. T. Madsen, 'Dr Thorvald Madsen Fortaeller. II', *Medicinsk Forum* 9 1956: 161–77, p. 161.
13. Madsen, 'Madsen Fortaeller. II', 1956, p. 162.
14. Madsen, 'Madsen Fortaeller. II', 1956, p. 162.
15. E. Roux and L. Martin, 'Contribution a l'étude de la diphtérie (sérum-thérapie)', *Annales Institut Pasteur* 8 1894: 609–39.
16. Jensen, *Bekaempelse*, 2002, p. 22.
17. R. Church and E. M Tansey, *Burroughs Wellcome & Co: Knowledge, Trust, Profit and the Transformation of the British Pharmaceutical Industry, 1880–1940* Lancaster: Crucible, 2007, p. 203.
18. Jensen, *Bekaempelse*, 2002, p. 22.
19. Church and Tansey, *Burroughs Wellcome*, 2007, Chapter 7.
20. Jensen, *Bekaempelse*, 2002, Chapter 1.
21. See Salomonson, 'Summer Semester', 1950.
22. Madsen, 'Salomonsen', 1925, pp. 704–5.
23. 'Salomonsen', 1925, pp. 704–5.
24. Jensen, *Bekaempelse*, 2002, p. 22.
25. Madsen, 'Salomonsen', 1925, pp. 706–7.
26. Jensen, *Bekaempelse*, 2002, p. 25; For the situation in Nancy, see Jonathan Simon's contribution to the present volume.
27. C. J. Salomonsen, *Contributions from the University Laboratory of Medical Bacteriology to Celebrate the Inauguration of the State Serum Institute* Copenhagen: Carlsberg Fund, 1902, p. 5.
28. Salomonsen, '*Contributions*', 1902, pp. 5–6.
29. Salomonsen, '*Contributions*', 1902, p. 6.
30. Jensen, *Bekaempelse*, 2002, p. 34.
31. *Politiken*, 23 October 1894, cited in ibid., p. 26.
32. Jensen, *Bekaempelse*, 2002, p. 31.
33. Salomonsen, *Contributions*, 1902, p. 6.
34. Salomonsen, *Contributions*, 1902, p. 6.
35. E. Schelde-Møller, *Thorvald Madsen. I videnskabens og menneskehedens tjeneste* Copenhagen: Nyt Nordisk Forlag, 1970, p. 41.
36. Madsen's strong pro-German sympathies were reported by a French observer at the League of Nations Health Commission in the early months of WWII. I am grateful to Iris Borowy for this information.
37. Madsen, 'Madsen Fortaeller I', 1953, p. 198.
38. Schelde-Møller, *Madsen*, 1970, p. 29.

39. Schelde-Møller, *Madsen*, 1970, p. 29; see also Jensen, *Bekaempelse*, 2002, p. 22.
40. T. Madsen, 'Experimentelle undersøgelser over difterigiften' (unpublished PhD thesis), Copenhagen, 1896.
41. Cited in Schelde-Moller, *Madsen*, 1970, p. 41.
42. Madsen, 'Madsen Fortaeller I', 1953, pp. 203–6.
43. Madsen, 'Madsen Fortaeller I', 1953, p. 208.
44. See J. Lehmann et al. *Denmark: Its Medical Organisation*, 1891, pp. 65–128.
45. Salomonsen, *Contributions*, 1902, p. 10.
46. Salomonsen, *Contributions*, 1902, p. 13.
47. Salomonsen, *Contributions*, 1902, p. 13.
48. Salomonsen, *Contributions*, 1902, p. 14.
49. Jensen, *Bekaempelse*, 2002, p. 34.
50. Salomonsen, *Contributions*, 1902, p. 7; Jensen, *Bekaempelse*, 2002, p. 34.
51. Madsen, *Statens Seruminstitut*, 1940, pp. 14–15; see also T. Madsen, 'Hideyo Noguchi', *Medicinsk Forum* 13 1960: 112–13.
52. Madsen, *Statens Seruminstitut*, 1940, pp. 22–5.
53. Jensen, *Bekaempelse*, 2002, p. 36.
54. Madsen, *Statenens Seruminstitut*, 1940, pp. 26–9.
55. Schelde-Møller, *Madsen*, 1970, p. 83.
56. Jensen, *Bekaempelse*, 2002, p. 66.
57. *Lancet* 1896, ii: 182–95, 196.
58. Jensen, *Bekaempelse*, 2002, p. 66; See Salomonsen, *Contributions*, 1902, pp. 8–10 for bibliographical references.
59. Jensen, *Bekaempelse*, 2002, p. 67.
60. J. Boyd, 'Tetanus in Two World Wars', *Proceedings of the Royal Society of Medicine*, 53 1958: 109–10, p. 109.
61. Schelde-Møller, *Madsen*, 1970, pp. 86, 110; Jensen, *Bekaempelse*, 2002, pp. 51–2.
62. *Madsen*, 1970, p. 110.
63. Gautier, 'The Health Organisation', 1995, p. 500.
64. Gautier, 'The Health Organisation', 1995, p. 500; Jensen, *Bekaempelse*, 2002, p. 68.
65. J. Ipsen, *Contribution to the Theory of Biological Standardization* Copenhagen: Nyt Nordisk Forlag/Arnold Busk, 1941, p. 11.
66. Ipsen, Theory of Biological Standardization, 1941, p. 11; August Krogh (1874–1949), Nobel Prize winner, 1920. See B. Schmidt-Nielsen, *August and Marie Krogh: Lives in Science* Oxford: Oxford University Press, 1995.

9

'The Wright Way': The Production and Standardization of Therapeutic Vaccines in Britain, 1902–13

Michael Worboys

Vaccine therapy was the medical sensation of the 1900s in Britain.[1] Serum therapy, the breakthrough of the previous decade, was by then well established for diphtheria and tetanus, though bacteriologists had found few other obviously successful applications of the principle.[2] In contrast, the promoters of vaccine therapy promised specific cures for a range of infectious diseases and the details of their exploits were promoted and debated at medical meetings and in medical journals throughout the decade.[3] The fluctuating fortunes of vaccine therapy were covered in the popular press and it was a central theme in George Bernard Shaw's play *The Doctor's Dilemma*.[4] The treatment was pioneered by Almroth Wright and his colleagues at St Mary's Hospital in London, but it was soon taken up by physicians, surgeons and general practitioners across the country and overseas, particularly in the US, who were supplied by laboratories producing vaccines on a commercial scale.[5] The basis of vaccine therapy was to extend the principle of preventive vaccination to those already suffering from an infectious disease. It was based on the assumption that with many infections, especially chronic and localized ones, the full immune system had not been alerted to, and mobilized against, the pathogenic bacteria. Given that most infections were self-limiting because the powers of bodily immunity overcame those of infection, the aim of vaccine therapy was to accelerate this natural process by boosting immunity qualitatively and quantitatively. In Wright's words, the aim of vaccine therapy was to 'exploit in the interest of the infected tissues the unexercised capacities of the uninfected tissues', by 'calling into action the forces of resistance that lie latent in the organism'.[6] Vaccine therapists mainly used graduated doses of killed vaccines to boost the immune system, which they believed stimulated the production of so-called opsonins – bacteriotrophic chemicals in blood serum – that attracted phagocytes (other terms used were white blood cells or leucocytes) to ingest the infecting bacteria. A popular analogy was that opsonins were like relishes – substances that improved the attractiveness of food. The 'fighting power' of the immune system, in terms of its ability to produce opsonins and stimulate phagocytosis, was termed its opsonic index (OI).

Wright stated in 1903 that 'the physician of the future will, I foresee, take upon himself the role of an immunisator'.[7] More specifically, he anticipated that 'the physician of the future' would be: first, a maker of vaccines using killed bacteria taken from the patient's own infected tissues; second, an administrator of these killed vaccines with the hypodermic syringe replacing the stethoscope as the icon of medicine; and third, a monitor of the effects of vaccines through laboratory estimations of the immune status of the patient's serum. Standardization was obviously important in ensuring the correct strengths of the graduated doses of vaccines, and also in monitoring the immune status of the patient and their response to vaccines, through estimations of their OI. The standardization of therapeutic vaccines was said by its advocates to be even more critical than with the more widely used serum therapy for two reasons. First, because Wright's vaccines, although produced from killed bacteria, generated active rather than passive immunity, and second, because Wright maintained that the immune system was unstable during infection and that the first response to any infection was to lower the serum's immune power – the so-called negative phase, a phenomenon observable in a low OI.[8] Hence, it was vital that the first dose of a vaccine was relatively weak and that subsequent doses were increased gradually to allow the immune system to move from the negative to the 'positive phase' of enhanced immune power.

The standardization of sera had focused on an assay of the strength of the therapeutic sera – a *product* – against a fixed standard; 'the Wright way' with therapeutic vaccines was quite different.[9] He emphasized the standardization of the *processes* of production and monitoring, because each laboratory estimation was unique to an individual patient and their disease. In other words, what was crucial was the technique and morality of the vaccine therapist, such that a skilled, reliable man could be depended upon to produce precisely graded vaccines and make accurate estimations of opsonic indices. Moreover, the calibrations of both vaccine strength and opsonic indices were not against a fixed, agreed standard, as with Wertbestimmung in serum therapy, but against 'healthy' blood taken anew from healthy people each time a test was made. Essentially, Wright's view was that clinical practice needed to become based on more laboratory-based methods, but he acknowledged that certain margins of error were inevitable, but that acceptable tolerances were best guaranteed by standardized processes carried out by individuals following rigorous professional and moral standards in their laboratory work.

In this article, I first discuss Wright's career and then review Rosser Matthew's discussions of his conflicts with Karl Pearson and Major Greenwood over mathematical precision in OI measurements, to highlight key features of Wright's view of laboratory medicine. I then discuss in detail the methods of vaccine therapy showing their focus: on *processes* rather than *products*; on relative rather than absolute standards; and on the morality of the laboratory worker. Finally, I look at contemporary challenges to vaccine therapy's methods and at the response of Wright and his supporters. Standardization was critical to the fate of vaccine therapy because other doctors found it hard to achieve the outcomes reported by Wright's men and those trained as St Mary's, whose success was attributed, by themselves and by observers, to the fact that they were 'masters of the opsonic art'.[10]

Almroth Wright: Innovator and controversialist

In the early decades of the twentieth century Wright was often styled as the British Pasteur, the man who had done for therapy what Pasteur had done for disease prevention. The epithet also implied that he would eventually enjoy equal status and that future historians would write the great era of medical bacteriology in terms of the successive eras of Pasteur, Koch and Wright. No less modestly, his supporters suggested that he might enjoy equal status in immunology with Metchnikoff and Ehrlich, as the man who brought cellular and humoral theories together through the discovery of opsonins – the humoral component that regulated phagocytosis. Wright was a contentious figure and a controversialist throughout his career, a status that he seems to have enjoyed and cultivated.

Wright was born in 1861 and qualified at Trinity College Dublin in 1883; he then worked in Germany, Cambridge, Sydney and London, before taking up the Chair of Pathology at the Army Medical School at Netley in 1892.[11] It was a surprise appointment as Wright had no Army background, was young and had mainly worked in physiology. At Netley, he continued to work and publish on physiological topics, especially coagulation of the blood, but increasingly moved to bacteriology and the emerging field of immunology.[12] He came to vaccine therapy from work on preventive vaccinations for typhoid fever. Pasteur had used live bacteria in his protective vaccines, altering or attenuating them by exposure to air; however, Wright's work was based on heat-killed vaccines. This was a type of vaccine that was associated with doctors working with the British military at home and abroad; indeed, Wright was influenced by Haffkine's use of killed vaccines against cholera and the plague in India in the mid-1890s. After experimentation on himself and those in his laboratory, Wright first tried his anti-typhoid vaccine on patients in an asylum at Maidstone in Kent in 1897, before offering it, without authority, to troops in India when he was travelling as a member of the Indian Plague Commission in 1899. Both trials seemed successful, so Wright offered the vaccine to the War Office at the outbreak of the South African War and it was given to 14,000 troops.[13] However, the Army was reluctant to offer official endorsement without a full enquiry, a decision that eventually led to Wright's resignation from Netley. He became embroiled in debates with various authorities over his methods and their value, most notably Karl Pearson who challenged the statistical basis of his claimed success. None the less, the trials continued and eventually in 1909 Wright was vindicated, as after a prolonged five-year investigation, led by Sir William Leishmann, the vaccine was shown to be effective and was adopted by the British military for overseas posting and with great success in WWI.[14]

Wright's claims about 'the physician of the future' came soon after he took up the post of Professor of Pathology at St Mary's Hospital, London, and signalled a switch from preventive to therapeutic vaccines. At St Mary's, he built up a large Inoculation Department, which in the period 1900–14 was Britain's highest profile bacteriological research laboratory. Wright provocatively championed laboratory over clinical methods, at a time when most of his colleagues saw these approaches as complementary, though with the laboratory supporting the clinic. He is perhaps best known today for advancing the claims of Alexander Fleming for the discovery of the antibacterial powers of penicillin, as against those of Howard Florey and Ernst Chain in the 1940s. This claim was and remains controversial.[15] While there is no doubt that Fleming identified the antibacterial powers of penicillin in 1928, he did not see or explore its clinical potential for very long, and abandoned work on the substance in the early 1930s.[16] Indeed, Fleming's last major publication before the announcement of penicillin therapy in 1940 was a large volume on vaccine therapy, in which the value of antibiotic chemotherapy was doubted.[17] Wright was also a controversialist on social issues, being a vocal opponent of women's suffrage and holding unconventional views on language and philosophy. He was undoubtedly a charismatic figure, especially in the 1900s, when he enjoyed the fierce loyalty of a group of doctors known as 'Wright's Men', some of whom defended him against a growing body of sceptics, long after his ideas and practices had been abandoned by most doctors. The critics often referred to Wright as 'Sir Almost Right'.[18]

Wright's attitude to quantification, an important feature of standardization, has been discussed by James Rosser Matthews in a number of articles that explore his clashes with Karl Pearson and the emerging group of Biometricians.[19] Matthews portrays these disputes as clashes between different types of disciplinary expertise, with Wright, the clinician, claiming to have a 'feel' for figures based on 'experience', whereas the Biometricians worked with mathematical and statistical methods that were based on transdisciplinary principles. Indeed, Karl Pearson and his followers maintained that the application of rigorous, externally validated methods was the best test of the significance and robustness of data in any discipline. Matthews portrays Wright as a pathologist-clinician, while acknowledging that there were differences between the two roles. Indeed, he points to a three-way split in British bacteriology between physician, bacteriologist and statistician, suggesting that there was an accommodation between the former, as a reliance on experience and skill remained at the core of medical identity.

Most historians now accept that any conflict between bedside and bench around 1900 was more rhetorical than real, with knowledge from both sources being seen as complementary and each counting differently in different contexts. However, Wright was no ordinary doctor combining the roles of pathologist and clinician; he wished to accentuate the differences to define a new type of clinician. In 1912, he stated that 'It is not, at this hour of the day, arguable that the verdict of the bacteriologist stands in need of confirmation by the clinician. In blunt language, "the boot is on the other foot"'.[20] He told of a history of germs in Britain where, first surgeons, and then physicians, had adopted antiseptics without bothering to learn any bacteriology, and hence had used them indiscriminately and uncritically. This was also part of his argument about the historic weakness of medical research in Britain, and the need for state and voluntary support of laboratory research.[21] Wright styled ordinary physicians and surgeons as 'plain, practical men' who were 'empirical' and hence easily misled by superficial observations.[22] In comparison, the laboratory man could work with controlled conditions, where key variables could be isolated. He could make crucial experiments working with exact methods, and he understood basic principles, such as mechanisms of the immune response. I will argue that it was in part Wright's failure to live up to his own exacting ideals and standards that was a major impediment to the acceptance and adoption of vaccine therapy in Britain.

Vaccine therapy methods and standardization

Two laboratory processes were at the core of vaccine therapy: (i) the measurement of the OI of the serum of the infected person; and (ii) the production of graduated doses of vaccine containing specified numbers of bacteria. The OI was a measure of the immune power of serum, based on a comparison between the serum of the patient and a healthy person. Its origins were in the work of William Leishmann, which Wright linked to his theory of opsonins and then associated it with therapeutics.[23] The phagocytic power (i.e. OI) of a healthy person was taken by Leishmann and Wright to equal 1, while that of an infected person was taken to be less than 0.8, indicating depressed immunity and vulnerability to infection, or to be over 1.2, indicating enhanced immune activity as the patient was 'fighting' the infection with stimulated phagocytes. To determine an OI, blood was taken from the patient and a healthy volunteer – originally from Wright or one of his men. Both samples were spun in a centrifuge to separate serum, phagocytes, and red blood cells. The determination of the OI required parallel estimations of the number of bacteria ingested by phagocytes in healthy serum and the patient's serum.

To make this estimation the investigator would set up two pipettes A and B: into A they would draw 'healthy' serum, 'healthy' phagocytes, and an emulsion of bacteria taken from the patient's infection; into B would go the patient's serum, 'healthy' phagocytes and emulsion of bacteria taken from the patient's infection. The contents of both pipettes would then be mixed and after removing the rubber teat both ends would be sealed. Both 'tubes' were then incubated at 37°C for 15 minutes. After opening, a sample was then taken from each pipette and spread as a film on a microscope slide, allowed to dry, then fixed and stained. The investigator would then count the number of bacteria in a set number of phagocytes in the two samples, usually between 20 and 50 cells, though sometimes hundreds. Thus, if such a count showed 10 bacteria in 40 phagocytes in the healthy serum (A) and 5 bacteria with the same count in the patient's serum (B) then an OI of 5/10 or 0.5 would be indicated. The variable in the parallel procedures was the serum, while the count of bacteria ingested was a proxy of the phagocytic power of the serum sample. An OI of 0.5 would have implied that the patient's immune system was deficient, which both explained their illness and indicated that vaccines needed to be given to raise the level of opsonins, or as Wright would have termed it, to raise the 'bacteriotrophic pressure' of the blood.[24]

The key techniques in these estimations were physiological and bacteriological: (i) the micromanipulation of serum, phagocytes, and bacteria in capillary pipettes; (ii) taking accurate samples, ensuring uniform mixing and incubation; (iii) the preparation of slides ensuring the consistency of the film and staining; and (iv) the accurate counting of the numbers of ingested bacteria, which required careful 'reading' of the slides. The vaccine therapist also needed to have skills in venesection, pipetting, culturing, and microscopy, not to mention access to the necessary equipment and reagents, and the time to perform intricate procedures, often repetitively. So-called all night sittings were a regular occurrence at St Mary's in the early 1900s.[25]

Very similar techniques were central to the standardization of therapeutic vaccines, where their strength was defined by the number of killed bacteria per cubic centimetre (cc), usually between 500–1000 million per cc. The ideal was to use bacteria from the patient's active infection to produce so-called autogenous vaccines, however, the same methods were used by individuals and companies to produce 'stock vaccines', which were made from pooled cultures of the same bacteria, or of mixed bacteria (so-called polyvalent vaccines, also known as phylacogens). While Wright strongly advocated the use of autogenous vaccines, as being more specific and more active, as they were freshly prepared from specific bacteria, his Department at St Mary's quickly moved into the production of stock vaccines for the Parke Davis Company. Other companies, such as Martindales, also began to produce stock vaccines.

Therapeutic vaccines were made by first growing bacteria on solid media plates in Petri dishes to produce pure cultures, which were identified by microscopy and staining if necessary. The cultures were then washed off the plate to produce an emulsion that had to be carefully mixed to avoid the clumping of the organisms. It was essential to have an even distribution to give the phagocytes an equal chance of 'finding' bacteria to ingest. The next stage was counting. Here, Wright maintained that direct counting was too difficult because of the numbers involved, the problems with movement, and accurate volumetric measurements. His answer was to ascertain the number of bacteria in a given mixture by counting against a known standard; the one he chose was that there were 5 million red cells per cc of healthy blood. Hence, the vaccine therapist drew into a pipette equal volumes of blood, the bacterial emulsion and 1.5 per cent sodium citrate (to prevent clotting) using the method described above. The components were then thoroughly mixed in the pipette, or on the slide, before the mixture was spread on a microscope slide, fixed and stained ready for counting. Counts of red blood cells and bacteria were made in a number of sample fields, defined either by the total field of view, or by areas taken from an etched grid. Working from simple ratios, Wright calculated the number of bacteria against the 5 million per cc standard. Hence, if the average number of red cells per field was 500 and the average number of bacteria was 400, the result was 4 million bacteria per cc; figures of 100 red cells and 300 bacteria would give a strength of 15 million bacteria per cc. The production of the vaccine required the killing by heat of the bacteria in the emulsion, and then its sterilization, before serial dilutions gave vaccines of known strengths.

The key factors in standardization of both the opsonic index and vaccine were the laboratory techniques of manipulating blood and bacteria, and of counting. Wright and his followers styled themselves as virtuoso laboratory scientists who had developed unique skills in the micromanipulation of blood and bacteria. (See Figure 9.1) In 1912 Wright published a volume of nearly 400 pages entitled the *Technique of the Teat and Capillary Glass Tube being a Handbook for the Medical Research Laboratory and the Research Ward*, which set out in the style of a manual all of the methods developed in his laboratory and which was complemented by his 1906 treatise *Principles of Microscopy*.[26]

Vaccine therapy required many samples of blood from patients and from healthy people. To ease the problem of obtaining multiple samples and overcome patient resistance to hypodermic syringes, Wright introduced the technique of taking blood samples in capillary tubes from small puncture wounds on the fingers. This meant that vaccine therapists had to work with small and variable quantities, which meant that using measured volumes was difficult, if not impossible. Initially, Wright used calibrated micropipettes of a near uniform diameter, but abandoned these as unreliable and unnecessary. The alternative was to use relative volumes, accurately measured, to give precisely proportioned mixtures and dilutions. His favoured tool was the long capillary pipette with a mark on the stem, say, a centimetre or two from the tip. (See Figure 9.2) To produce the mixture of equal volumes of serum, phagocytes and emulsion, the vaccine therapist would: first draw serum up to the mark, then introduce a small air gap, then draw phagocytes to the mark, and then the bacterial emulsion to the mark. The three equal parts would then usually be expelled on to a slide and mixed mechanically by repeated taking up in the pipette and expelling, or they might be shaken if the pipette had a bulb.

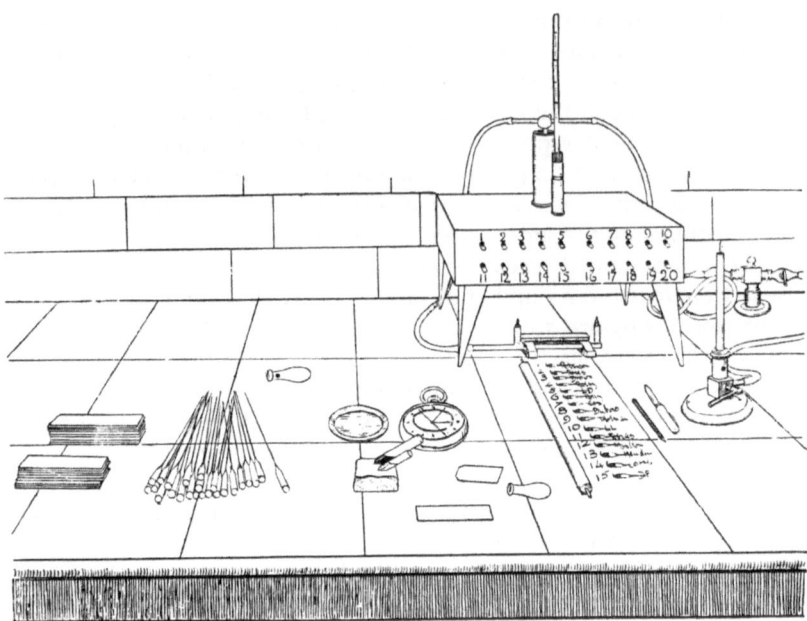

Figure 9.1   The bench of the vaccine therapist, showing from right to left, capillary pipettes, slides in preparation, a set of slides ready for reading, and at the back an incubator for holding sealed tubes horizontally
Source: A. E. Wright, *Handbook of the Technique of the Teat and Capillary Glass Tube and Its Application in Medicine and Bacteriology*, 1912, 228.

There was no agreed ratio of volumes, some doctors used equal parts, but it was not uncommon to choose other multiples, which was relatively easy to do. Norman Leeming wrote in the *Guy's Hospital Gazette* in 1909 that, 'Every worker has his own favourite proportion, but providing the same is used for the patient and for the normal slide, the results will be concordant'.[27]

Other features of opsonic methods depended on the 'experienced eye'. For example, the quality of the bacterial emulsions was judged by 'inspection', with the immunizator looking for evenness to ensure that bacteria had not clumped and that they were uniformly distributed. The films from which the counts were made usually had a gradient of thickness, meaning the vaccine therapist had to judge the best place along the slide to count in relation to depth of field, and when estimating OIs they had to try and ensure that the slides of the patients and healthy blood were equivalent. (See Figure 9.3)

Reading the slides also meant determining how many fields to count, and what to do about boundary objects: did they count a cell that was partially in the field as 'in' or 'out', or as a fraction? There were no rules, the vaccine therapist had to decide and apply their own standards consistently.

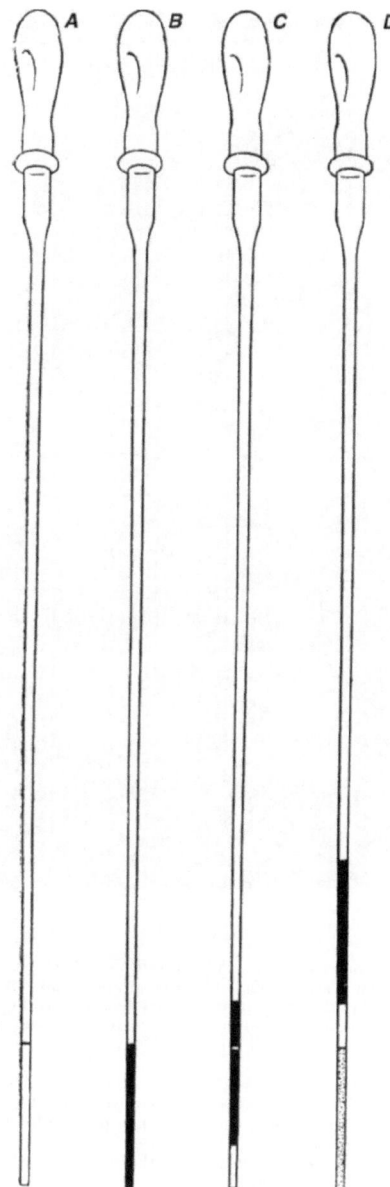

Figure 9.2 The technique of the capillary pipette, showing from left to right: a marked pipette; one unit of blood; blood and air gap; and one unit of serum, air gap and one unit of blood

Source: A. E. Wright, *Handbook of the Technique of the Teat and Capillary Glass Tube and Its Application in Medicine and Bacteriology*, 1912, 60.

162

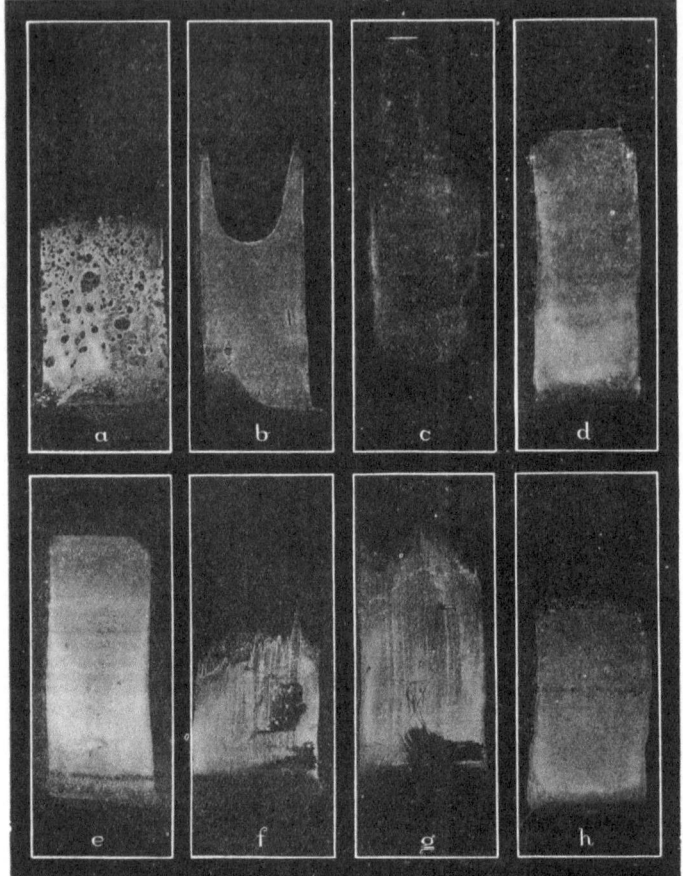

PLATE II.

This plate shows what has to be avoided and what is to be aimed at in making
film-preparations of blood, and in particular film-preparations for phagocytic counts.

a. An extreme example of the unsatisfactory honeycomb film which is obtained
when the surface tension of the blood-fluid strips it off from some regions of the slide,
leaving the glass bare, and piles it up on other regions so that the corpuscles lie
there many layers deep.

b. An example of a film with a concave end. Such a film is produced when the
spreader is held too upright, or its working end is too much hollowed out.

c. An example of a film drawn out into a tongue by a rectangular spreader which
has also been held too obliquely.

d. Example of a film which is open to criticism only in the respect that it is
carried a little too far down the slide.

f and *g.* Examples of unsatisfactory films.

The frilled edges and striation marks which are the characteristic features of
these films show that the slide which was used for spreading them had an edge

Figure 9.3 Illustrations of poor slide preparation, only slide 'h' is satisfactory

Source: A. E. Wright, *Handbook of the Technique of the Teat and Capillary Glass Tube and Its Application in Medicine and Bacteriology*, 1912, 104.

Wright termed the discrepancies that might emerge from such matters as 'functional errors', similar to what would later be called experimental errors. These came from the properties of the materials being worked upon, the limitations of the apparatus and reagents being used, and the skills of the laboratory worker. It was the latter that seemed to distinguish St Mary's men from the 'ordinary practitioner'; however, according to Wright, it was not simply technical prowess that explained their success, it was also their 'Code of Morality'; the ideal vaccine therapist had to show judgement, resolve and be conscientious.

> The proper morality in all these cases is to make up our minds as to how they [counting criteria] may best be dealt with and then to abide consistently by what we have decided.[28]

Moreover, 'functional errors' could be controlled by 'Introspection and Self-Examination'.

> Every man who sets up before himself any standard of efficiency will know the feeling of having done his work well .... And he will also know how it feels to have failed to exercise sufficiently strict supervision of himself, to have been guilty of momentary lapses of attention, and to have come short in the matter of keeping intellectually alive when doing his work.[29]

Wright was fond of quoting Carl Ludwig's dictum that 'The whole of science lies in technique' and believed that the best scientist was the 'kind of man' who knows that the mastery of laboratory methods requires investment of time. Wright also wrote that 'no one can hope to acquire proficiency in billiards or golf unless he gives up years to the task'. His choice of sporting parallels was interesting; obviously excellence in both required commitment, good hand-eye coordination and the right equipment, and both sports were well known for their codes of honour.

Wright maintained that the only way the necessary skills could be learned was by apprenticeship at St Mary's and indeed, that is how most doctors did learn 'the opsonic art'. None of Wright's men produced a manual or textbook on vaccine therapy; doctors who could not attend St Mary's had to learn the methods from journal articles or manuals produced by other enthusiasts such as Richard Allen, whose book on the subject went through four editions from 1907 to 1914.[30] The only book to emerge from St Mary's was Wright's *Studies on immunisation and their application to the diagnosis and treatment of bacterial infections* in 1909, which was a collection of journal articles. Indeed, publications remained scattered across medical, bacteriological and scientific journals. The *Journal of Vaccine Therapy* only began publication in 1912 and closed in September 1913, so there was little opportunity for systematic reporting and refinement of methods amongst aficionados or for building a specialism. Yet, for many years, the Inoculation Department was a Mecca for young doctors and a centre that attracted medical scientists from across the world. Its reputation ensured that it was the first British laboratory to receive compound 606 (Salvarsan) from Paul Ehrlich, and Wright was also visited by Robert Koch and Elie Metchnikoff. The Department was also famous for its camaraderie based on long days and nights exercising and refining laboratory skills, its shared commitment to Wright's doctrines, and its collective sense of being in the vanguard of a revolution in medicine. Wright's style seems to have inspired great loyalty, while the mounting criticisms of his ideas and methods seem to have served only to solidify the group.

Criticisms

Vaccine therapy was controversial from the outset. This was in no small measure due to Wright's reputation and style, and to his antagonism towards clinicians. However, it was also due to disappointing results experienced by many doctors and patients. Most doctors found it impossible to replicate OI estimations, or to produce standardized vaccines; also, therapeutic outcomes were mixed. Some laboratory scientists also questioned the validity of OI determinations.[31] There were few doctors and scientists who publicly doubted the 'theory' of the treatment, Some suggested that the claims about the importance of cellular immunity were unproven, but there is little or no evidence of anyone who doubted the 'reality' of opsonins until the 1910s.[32]

The standing of vaccine therapy was not helped by its association with the Tuberculin treatment of tuberculosis; indeed, this may have been its most common application given the prevalence of the disease, its high public profile in the early 1900s, and the availability of many varieties and strengths of off-the-shelf Tuberculins. Of course, Tuberculin had a chequered history, though British doctors had been quite sceptical from its introduction in 1890.[33] Wright adopted Tuberculin as a form of vaccine therapy for tuberculosis as early as May 1903, arguing that while the original Tuberculins were based on 'toxi-therapeutics' (i.e. antibacterial therapy), the new Tuberculins 'have been definitely invested with the character of therapeutic inoculations of a tubercle vaccine designed to call forth an antibacterial reaction in the organism'.[34] This support attracted a lot of attention and reports of many trials were soon published, mainly on disease localized in joints and other organs, but Wright's work gave further impetus to the many doctors still using Tuberculins with pulmonary disease.

Wright was soon confronted with powerful opponents for his support of Tuberculins, none more so than W. Watson Cheyne, a close associate of Joseph Lister, a pioneer bacteriologist and by then senior surgeon at King's College London. Cheyne attacked Wright's ideas on the treatment and his notions of immunity; many would have enjoyed the pun when he described the notion of opsonins as 'deliciously simple'. [35] But Cheyne focused especially on the OI stating that,

> As to the accuracy of the opsonic index I do not presume to judge, but it has struck me from what I have seen that the result of the examination of the blood depends a good deal on the personal equation and has not the strict mathematical accuracy that one would desire. Individuals will differ in the way in which they will prepare the dilutions, more especially of the bacilli, and so on. Nevertheless, judging from the reports there is a remarkable correspondence in the results obtained by different observers. The chief difficulty seems to be in the cumbersomeness of the method which puts it outside the power of the busy practitioner'.[36]

Indeed, the measurement of what became known as the Tuberculo-opsonic index proved to be the most difficult of all vaccine therapy procedures as Tubercle bacilli, more than any other micro-organisms, clumped together in emulsions.[37] This problem led to many innovations in the preparation of emulsions, in mixing emulsions, sera and phagocytes, and in the use of substrates.[38] (See Figure 9.4) However, the main result was the abandoning of OIs and a switch to the use of clinical indicators, principally body temperature, to monitor the effects of Tuberculin injections.

The use of clinical indicators to monitor vaccine therapy developed rapidly, especially when stock vaccines became available, and there were three reasons for this. Firstly, changes in the clinical presentation of disease were what doctors were used to assessing, so it was not so much a switch away from laboratory monitoring, but an unwillingness or inability to adopt the complex methods. Second, doctors seem to have made a cost-benefit analysis: was the time, effort and cost of obtaining repeated OI estimations worth it? Were there quicker, easier and cheaper alternatives? In many, many instances doctors concluded that OIs were uneconomic, unreliable and often simply unobtainable, yet the promise of vaccine therapy meant that it had to be tried. Given the high profile of the treatment, doctors also had to respond to patient demands. Third, clinical monitoring seems to have been adopted by those who were sceptical of opsonic theory, but who had tried and found vaccine therapy valuable empirically. While Wright and his acolytes continued to maintain that opsonic theory and vaccine therapy were inextricably linked, they aided the decoupling of theory and practice through their production of stock vaccines.

For those doctors who pursued the laboratory version of vaccine therapy, there remained technical problems with the preparation and management of cultures, emulsions, sera, leucocytes, and with counting.[39] First, the use of relative measures for mixtures and for dilutions using micropipettes led to complaints that these were approximations rather than controlled scientific measurements, and that they would differ between investigators. Second, there were questions about bacteriological technique. Many of the cultures

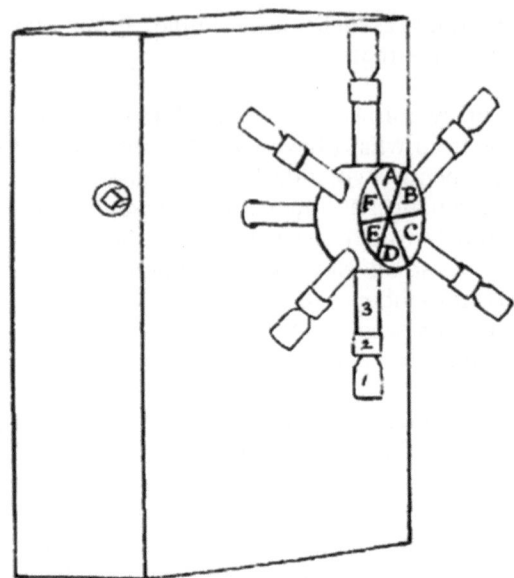

The " opsonic mill." This clockwork-driven machine, standing in the incubator, rotates the opsonic mixture and
prevents sedimentation of its constituents. It also con-
tinues the mixing of bacteria and white corpuscles.
1, The shortened opsonic pipette. 2, An indiarubber collar.
3, A copper tube packed with fine copper wires surrounding
the pipette. Speed of rotation = 1¾ minutes per revolution.
The brake is not illustrated.

Figure 9.4 The opsonic mill
Source: C. Russ, 'An Improved Method for Opsonic Index Estimations', *Lancet*, 1912, i: 1463.

grown from infected lesions contained mixed bacterial colonies, hence, how
did the investigator know which to use in determining OIs or for the vaccine?
And, were vaccine therapists certain that they were using pure, single organism samples? Third, the manipulation of emulsions led investigators to vary
substrates and mixing methods to prevent clumping; reports showed that
many of these methods were individualistic and used in an ad hoc manner.

Fourth, in determining OIs Wright had originally used 'normal' serum
taken freshly from healthy individuals in his laboratory, however, repeat tests
showed that there were wide variations between individuals; hence, vaccine
therapists moved to using 'pooled serum' to give a more average level. This
shift changed the meaning of 'normal' from 'healthy' to 'average', and raised
the questions of how large a pool needed to be, and of whether such serum,
which usually had to be stored, lost potency over time as opsonins degraded?
This pointed to the need for OIs to be performed promptly, which suggested
a need for large, dedicated laboratories and for such facilities in every large

hospital and available to general practitioners.[40] A number of alternative methods of determining OIs were proposed to meet the objection that it was 'a very delicate and somewhat laborious and difficult matter'.[41] In America, Simon, Lamar and Bispham proposed simply mixing bloods in varying dilutions with bacterial emulsions, and then counting the percentage of phagocyting leucocytes. This method was simpler technically, but took just as long. Another similar method was to use whole blood and emulsion mixtures, but to count and estimate the levels in the Wright way.

The issues around counting raised both technical and statistical issues of standardization. One of the main objections of Major Greenwood and the biometricians was that using the average of the number of bacteria observed in a set number of leucocytes would lead to errors, as there was never a normal distribution of bacteria. Rather the distribution was usually skewed. For while most leucocytes had ingested two to four bacteria, there were usually a few leucocytes having 20 or more bacteria, which meant that a few atypical observations raised the average and could radically alter the OI. Furthermore, the reason for such large numbers in one phagocyte seemed to be the ingestion of clumped bacteria. There were also questions about the accuracy of the counting of large numbers of bacteria in a single phagocyte, and how evenly they might be stained. However, a key issue remained the reliability of counting and the differences between observers, and with the same observer at different times. There was no norm for the optimum thickness of slide preparations, or controls over their evenness, when the depth of an emulsion would determine the numbers of cells seen and how they were seen. In the counting area, should observers strive to produce a slide with cells in a single plane, or only observe in one plane, or observe on all planes? By the late 1900s, Wright was recommending counting 100 leucocytes, though it is clear from reports that individual investigators varied widely, from 600 downwards. Two Danish workers, Reyn and Kjer-Petersen, commented unfavourably on errors in 1908, combining both technical and statistical objections.[42] Leonard Noon and Alexander Fleming replied from Wright's laboratory, suggesting that the Kjer-Petersen work 'lacks accuracy' and that because he delegated the counting of slides to another, he had thrown away a valuable check on the previous stages of the technique.[43] They cautioned that 'a high level of perfection in making and staining films is only to be maintained by that constant self criticism which is incidental to the counting of films one has oneself prepared'.[44] However, criticisms of its methods, if not vaccine therapy itself, continued to mount. An investigation by Hort, published in 1909, showed very large variations in the OIs produced by different observers. His summary chart (See Figure 9.5), shows the range for each of 13 determinations, with the length of the line indicating the range or degree of error.[45]

One alternative that was widely canvassed, but rarely used, was to use haemocytometers, a technology of counting developed in haematology. The centrality of questions of counting was somewhat ironic, as Wright had first

168

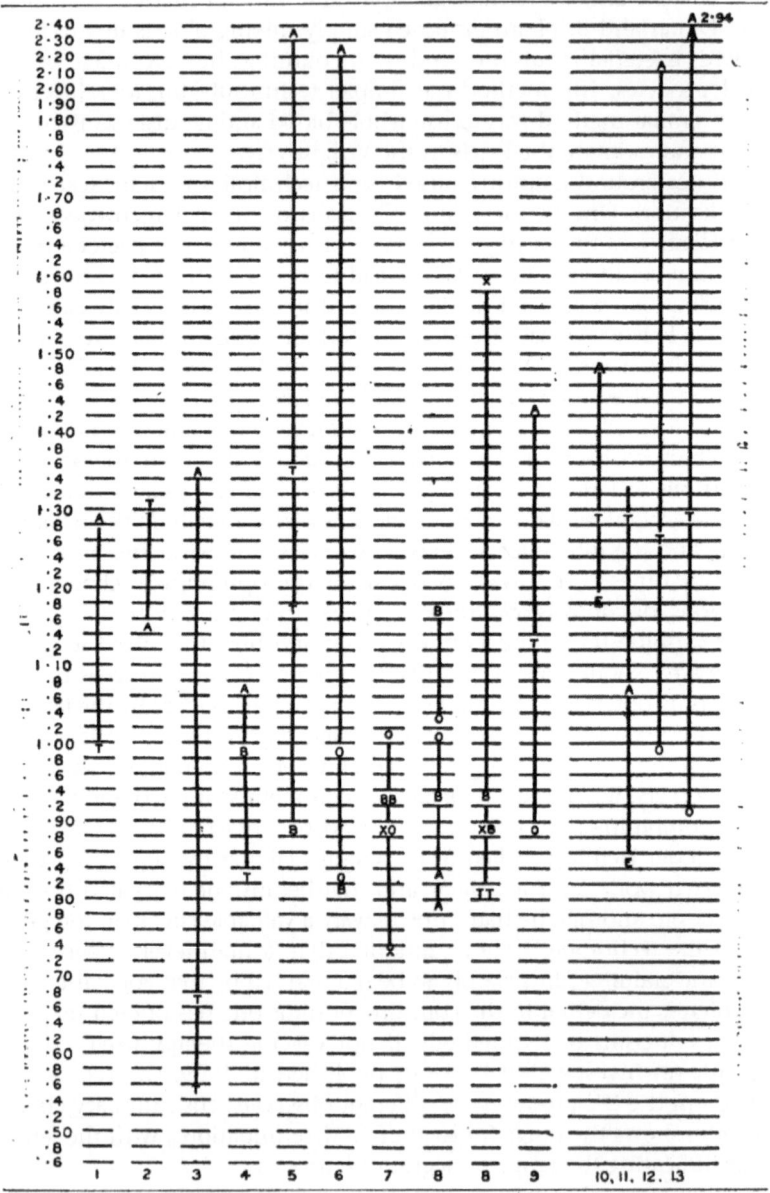

Figure 9.5 E. C. Hort, 'Can Opsonic Estimations be Relied on in Practice?'
Source: *British Medical Journal*, 1909, i: 400.

worked on blood and anaemia in the 1890s and would have been aware of the problems of counting raised by Gowers's haemocytometer, yet it was no doubt this background that shaped the vaccine therapist's use of similar techniques.

While most discussion focussed on the accuracy and need to standardize the measurement of OIs, the production of vaccine was not without its problems. Many of these were similar: for example, the purity of cultures, the making of emulsions and counting. Wright faced objections about his assumption that 'normal' healthy blood contained 5 million red blood cells per cc and that this could be used as a standard. Writing in 1913, Glynn, Powell, Rees and Cox, stated,

> We shall ... show that the average error in [Wright's] method of standardisation ... is about 15 per cent, with a maximum of 54 per cent, and, what is more important, that it *may under-estimate* the strength of the vaccine by at least 100 per cent.[46]

Critics pointed out that the normal range was between 4.5 and 6 million per cc, a 33% variation, and that as this factor was the basis for the determination of the strength of vaccines, and given Wright's warning about the 'negative phase', the assumption could be dangerous. One answer was, of course, to check independently the red cell count of the normal blood using a haemocytometer, which added time and expense. There were attempts to count the number of bacterial colonies or to dry vaccine emulsions so that weights could be used, as with Tuberculins. Some doctors even tried to standardize with measurements of the specific gravity of vaccines.[47]

Medical interest in, and discussion of, vaccine therapy in the 1900s concentrated on technique, standardization and results. However, there was an undercurrent from doctors and bacteriologists who doubted the theory; considering either that opsonins were one factor among many in serum immunity, Cheyne's point, or that they might be no factor at all! Although Wright had more enemies than friends in medicine, his authority meant that few challenged his ideas. Greenwood's critique of Wright's statistics carried the implication that there might be less to vaccine therapy than many supposed, but outright questioning largely came from Continental Europe where the therapy did not attract the same following as in Britain and the US.[48] In 1908, Wright's theory, technique and results came under fire in Berlin from Professor Jürgens.[49] He pointed out that Wright assumed that leucocytes only ingested bacteria in the presence of opsonins, so in saline solution they would be inactive. Jürgens showed that there was, what might be termed, 'spontaneous phagocytosis' by leucocytes in saline solution, indicating that factors besides opsonins were involved in ingestion, and indeed that there might not be a serum factor. Jürgens went on to highlight the 'subjective' factor in all of Wright's methods and to conclude that his results in terms of recoveries were 'unimpressive'.

The usual response of Wright and his men to criticisms on technique and standardization was to say OI measurements were reliable at St Mary's and that errors of 10 per cent were acceptable.[50] This was the main point of Alexander Fleming's response to Reyn and Kjer-Petersen in *The Practitioner* in 1908.[51] With OIs, Fleming argued that the important thing was to count enough phagocytes, but there were no rules to guide how many was enough; that depended on qualitative features of the slide – the density of phagocytes containing bacteria, the numbers of organisms per phagocyte, the ease of counting – the good vaccine therapist had a 'feel' for the opsonic materials and their instrumentation. Wright stated in 1909 that the 'OI cannot be computed with exactitude'.[52]

Conclusion

In 1912, John Eyre, an advocate and practitioner of vaccine therapy wrote that 'at the present time every pathologist follows a different method in preparing and standardizing his vaccines'. For doubters, whose numbers had been growing since 1908 and who had their day at the Royal Society of Medicine debate in 1910, the variation and lack of standardization meant that the whole vaccine therapy enterprise was compromised, if not fatally flawed. In accounts of the rise and fall of vaccine therapy, historians have previously emphasized the costs in time and money of the OI estimations and, ultimately, the poor results with patients, for its move from panacea to niche therapy between 1909 and 1914.[53] In this essay, I have argued that questions around the standardization of methods were as important, indeed, costs, efficacy and standardization were interlinked. As early as 1909, Hort had cautioned,

> Regarding the question of reliability, it is quite impossible to ignore the fact that countless observations have been reported in which estimations have appeared to be of the highest value, either in diagnosis or in directing some particular line of treatment. This is by no means, however, a universal experience, even when every effort has been made to secure the service of the high priests of the art. There appears, indeed, to be a growing conviction that except possibly in the hands of an extremely small band of experts, the method is not of the general utility with which it has been credited.[54]

Men trained at St Mary's were certainly understood by themselves, if not their peers, to be technically superior: the 'high priests of the art'. Yet, it was not simply better technique that enabled vaccine therapy to work in their hands, they also had superior morality, essentially that of the ideal scientist: disinterested observation, critical judgement, consistency, self-criticism.[55]

Wright maintained that his methods were practical and adaptable such that every physician could aspire to be an immunizator, if they became more of a 'scientist'. However, at the same time he argued that absolute standards were unnecessary as clinicians could work with controlled and knowable margins of error as they largely only required indicative or relative values. In this sense, the notion of 'opsonic art' was very appropriate.[56] What mattered was whether the opsonic index was low or high, not whether it was 0.60 rather than 0.59, and perhaps more critically how its level changed in response to the inoculations. Similarly with vaccines, the important thing was to have confidence in the approximate strength of the vaccine and to be able to give graduated doses; hence, it did not matter if the dosage regime was 4.5 to 9 to 45 million bacilli per cc, rather than 5 to 10 to 50, the crucial factor was to stimulate the immune system in a controlled manner. In short, having a standard *process* of testing and production was more important than having a standard *product*.

Notes

1. M. Worboys, 'Vaccine Therapy and Laboratory Medicine in Edwardian Britain', in J. V. Pickstone (ed.), *Medical Innovation in Historical Perspective*, Houndmills: Palgrave Macmillan, 1992, 84–103.
2. R. T. Hewlett, *Serum and Vaccine Therapy: Bacterial Therapeutics and Prophylaxis, Bacterial Diagnostic Agents*, London: J. & A. Churchill, 1910.
3. In 1906 Wright was writing off serum therapy as a failure. A. E. Wright, 'A Criticism of the Foundations of Serum Therapy', *Clinical Journal*, 16 May 1906.
4. G. B. Shaw, *The Doctor's Dilemma*, London, 1905; I. Lowy, 'Immunology and Literature: *Arrowsmith* and *The Doctor's Dilemma*', *Medical History*, 1988, 32: 314–32.
5. R. Hare, 'The Scientific Activities of Alexander Fleming, Other than the Discovery of Penicillin', *Medical History*, 1983, 27: 347–72.
6. A. E. Wright, 'Lecture on Therapeutic Inoculation of Bacterial Vaccines and Their Practical Application in the Treatment of Disease', *British Medical Journal (BMJ)*, 1903, i: 1069–74, 1096.
7. Ibid.
8. Ibid., 1071.
9. Unsurprisingly, vaccine therapy is omitted from most historical accounts of biological standardisation, see: D. R. Bangham, *A History of Biological Standardization*, London: Society of Endocrinology, 1999, 20–32.
10. E. C. Hort, 'Can Opsonic Estimations be Relied on in Practice?', *BMJ*, 1909, i: 400.
11. L. Colebrook, *Almroth Wright: Provocative Doctor and Thinker*, London: Heinemann, 1954; Sir Z. Cope, *Almroth Wright: Founder of Modern Vaccine-Therapy*, London: Nelson, 1966, M. Dunnill, *The Plato of Praed Street; The Life and Times of Almroth Wright*, London: RSM Press, 2000.
12. M. Worboys, 'Almroth Wright and the Army Medical School, Netley, 1892–1902', in R. Cooter, M. Harrison and S. Sturdy (eds), *Medicine and Modern Warfare*, Amsterdam: Rodopi, 1999, 77–98.
13. A. E. Wright, *A Short Treatise on Anti-Typhoid Vaccination*, London: Constable, 1904.
14. A. Hardy, '"Straight Back to Barbarism": Antityphoid Inoculation and the Great War, 1914–1918', *Bulletin of the History of Medicine*, 2000, 74: 265–90.

15. G. Macfarlane, *Alexander Fleming: The Man and the Myth*, Oxford: Oxford University Press, 1985.
16. M. Wainwright, *Miracle Cure: The Story of Penicillin and the Golden Age of Antibiotics*, London: Blackwells, 1990.
17. Fleming wrote in 1939 that, 'At the present moment vaccine therapy is [...] having something of a setback now that medical practitioners have become chemically minded [...]. However, it is certain that before long the limits of the new chemotherapy will be better understood, and I hope later to produce evidence indicating that the best results in the treatment of certain bacterial infections will be obtained by a combination of vaccine therapy with the new chemotherapy'. *BMJ*, 1939, ii: 99.
18. Dunnill, *The Plato*; W. C. Noble, *Coli: Great Healer of Men*, London: Heinemann Medical, 1974.
19. J. Rosser Matthews, *Quantification and the Quest for Medical Certainty*, Princeton, Princeton University Press, 1995, 85–114; J. Rosser Matthews, 'Major Greenwood versus Almroth Wright: Contrasting visions of 'scientific' medicine in Edwardian Britain' *Bulletin of the history of medicine*, 1995, 69: 30–43; J. Rosser Matthews, 'Almroth Wright, Vaccine Therapy, and British Biometricians: Disciplinary Expertise versus Statistical Objectivity', *Clio Medica*, 67, 2002, 125–47.
20. A. E. Wright, 'Vaccine Therapy: Its Administration, Value and Limitations', *Journal of the Royal Society of Medicine*, 1910, 3: 4.
21. In 1909 Wright bemoaned the lack of enthusiasm for medical research in Britain saying that he 'thought it a pity that whereas sportsmen could be found who would institute an anti-grouse league for the investigation of grouse disease, yet it seemed impossible for a like number of person afflicted with diseases such as bronchitis, asthma, or the like to form an anti-bronchitic or anti-asthmatic clubs for the investigation of that specific complaint'. A. E. Wright, 'Retrospect and Prospect in Vaccine Therapy', *Lancet*, 1909, ii: 396.
22. A. E. Wright, 'On the Value of Apparatus and Technique in Medical Research', *Lancet*, 1921, ii: 642.
23. W. Leishmann, 'Note on a Method of Quantitatively Estimating the Phagocytic Power of the Leucocytes in the Blood', *BMJ*, 1902, ii: 73–5.
24. A. E. Wright and G. Lamb, 'Observations Bearing on the Question of the Influence which is Exerted by the Agglutinins in the Infected Organism', *Lancet*, 1899, ii: 1727–9.
25. St Mary's Hospital Gazette, 1907.
26. A. E. Wright, *Handbook of the technique of the teat and capillary glass tube and its application in medicine and bacteriology*, London, Constable, 1912; A. E. Wright, *Principles of microscopy: being a handbook to the microscope*, London, Constable, 1906.
27. A. Norman Leeming, 'Vaccines and the Opsonic Index', *Guy's Hospital Gazette*, 23, 1909, 159.
28. A. E. Wright, *Technique of the Teat and Capillary Glass Tube*, 239.
29. A. E. Wright, *Technique of the Teat and Capillary Glass Tube*, 243.
30. R. W. Allen, *The Opsonic Method of Treatment*, London, H. K. Lewis, 1907; R. W. Allen, *Vaccine Therapy and the Opsonic Method of Treatment*, London, H. K. Lewis, 1908; R. W. Allen, *Vaccine therapy, its theory and practice*, London, H. K. Lewis, 1910; R. W. Allen, *Vaccine therapy, its theory and practice*, London, H. K. Lewis, 1913; R. W. Allen, *The bacterial diseases of respiration, and vaccines in their treatment*, London, H. K. Lewis, 1913.

31. P. Fitzgerald, R. I. Whiteman and T. S. P. Strangeways, *Bulletin of the Committee for the Study of Special Diseases*, 1907, 1 (8).
32. F. W. Andrewes and T. J. Horder, 'A Study of the Streptococci Pathogenic for Man', *Lancet*, 1906, ii: 852–5.
33. C. Gradmann, 'A Harmony of Illusions: Clinical and Experimental Testing of Robert Koch's Tuberculin 1890–1900,' *Studies in the History and Philosophy of Biological and Biomedical Sciences*, 2004, 35: 465–81; M. Worboys, *Spreading Germs: Disease Theories and Medical Practice in Britain, 1865–1900*, Cambridge: Cambridge University Press 2000, 224–8.
34. A. E. Wright, 'A Lecture on Therapeutic Inoculations of Bacterial Vaccines and their Practical Exploitation in the Treatment of Disease', *BMJ*, 1903, i: 1069–74.
35. W. W. Cheyne, 'Professor A. E. Wright's Method of Treating Tuberculosis', *Lancet*, 1906, i: 78–82, 78. Cheyne was generally sceptical of Wright's theories of immunity, questioning whether cellular immunity was the most important factor.
36. Ibid., 78. Also see: Editorial, 'The Value of the Tuberculo-opsonic Index in the Diagnosis and Treatment of Tuberculous Infections', *Lancet*, 1906, i: 237–8.
37. W. D. Anderson, 'Letter', *Lancet*, 1908, i: 1106.
38. F. Rufenacht Walters, 'The Opsonic Test', *Lancet*, 1909, ii: 6–8.
39. E. C. Hort, 'Can Opsonic Estimations be Relied on in Practice', *BMJ*, 1909, i: 400–1.
40. Cheyne, 'Professor A. E. Wright's', 78.
41. Allen, *Vaccine Therapy*, 1908, 41.
42. A. Reyn and R. Kjer-Petersen, 'Observations on the Opsonins, with Special Regard to Lupus Vulgaris', *Lancet*, 1908, i: 919–23.
43. L. Noon and A. Fleming, 'The Accuracy of Opsonic Estimations', *Lancet*, 1908, i: 1203–4.
44. Ibid., 1204.
45. Hort, 'Can Opsonic', 400–1.
46. E. Glynn, M. Powell, A. A. Armstrong Rees and G. Lissant Cox, 'Observations upon the Standardisation of Bacterial Vaccines by the Wright, the Haemocytometer and Plate Culture Method', *Journal of Pathology and Bacteriology*, 1913–14, 18: 379–400.
47. On counting bacteria, see Glynn, Powell, Rees and Cox, 'Observations upon the Standardisation of Bacterial Vaccines', 395–6.
48. Robert Koch's Tuberculin remedy was similarly subject to less rigorously scrutiny in Germany than it was abroad. See: Christoph Gradmann, *Krankheit im Labor: Robert Koch und die medizinische Bakteriologie*, Göttingen: Wallstein Verlag, 2005.
49. See report in *BMJ*, 1908, i: 947–8; and 1908, ii: 48–9.
50. A. Fleming, 'Observations on the Opsonic Index', *Practitioner*, 1908, 80: 607–34. This was in a special issue of the journal in May 1908.
51. A. Fleming, 'The Accuracy of Opsonic Estimations', *Lancet*, 1908, i: 1203–4.
52. A. E. Wright, *Guy's Hospital Gazette*, 1909.
53. P. Keating, 'Vaccine Therapy and the Problem of Opsonins', *Journal of the History of Medicine and Allied Sciences*, 1988, 43: 275–96; Worboys, 'Vaccine Therapy', 84–103.
54. Hort, 'Can Opsonic', 401.
55. R. K. Merton, *The Sociology of Science: Theoretical and Empirical Investigations*, Chicago: University of Chicago Press, 1973.
56. C. J. Lawrence, 'Incommunicable Knowledge: Science, Technology and the Clinical Art in Britain 1850–1914', *Journal of Contemporary History*, 1985, 20: 502–20.

10
The Visible Industrialist: Standards and the Manufacture of Sex Hormones

Jean-Paul Gaudillière

Introduction

In 1938, *Medizinische Mitteilungen*, the medical journal published by the German pharmaceutical company Schering, included a special article on the firm's biological laboratory. Written by Walter Hohlweg, Schering's head of physiological research, this article, entitled 'Biological Work in the Chemical Laboratory', focused on the history of a special section of Schering's research infrastructure, the so-called Animal Research Section (*Tierversuche Abteilung*). The infrastructure described by Hohlweg was the setting where three of Schering's most successful products – the preparations of estrogens (Progynon), progesterone (Proluton), and testosterone (Testoviron) – were evaluated.

> When chemists become involved in hormone research, they are confronted with new tasks and even new laboratory equipment. It is important to understand the difference between, on one side, working with pipettes, normal solutions, and dye indicators, and, on the other, carrying out a quantitative determination for a substance, for which one needs to use rats, mice, rabbits, guinea pigs and other test animals as research object, with the aim of determining the specific effects of the inoculation of a hormone, and of quantifying these effects. (...) For this type of work new laboratories were necessary. One had to build and organize not only spaces and cages for the animals, but also operation rooms and histological laboratories. (...) This infrastructure was – at Schering – first established for testing purposes. The existence of these new possibilities has however led to the development of biological investigations, which initially had no immediate sense for testing.[1]

When Hohlweg wrote this article in 1938, animal testing already had a long history at Schering. First associated with experiments on the toxicity and side effects of drugs, it gained importance and was scaled up in the 1920s. By the time it attained the status of a fully-fledged activity requiring specific personnel and space, animal research served aims as diverse as quality control, potency measurement, analysis of new products, improvement of production protocols, as well as the writing of patent applications.[2]

This was not an isolated development. Testing laboratories have a central place in the history of pharmacy, and more particularly in the industrialization of drug making, a process that gained momentum at the end of the nineteenth century. As with the chemical and electrical industries, testing laboratories became sites where experimental and productive tasks were combined to foster innovation and standardize products. Although the role of pharmaceutical firms has often been acknowledged in histories of medicine and biology, neither the testing-laboratory nor the issue of drug standardization has found more than a marginal place. Publications on the history of drugs and biographies of pharmaceutical firms usually mention the importance of industrial research and the collaboration between internal laboratories and academic researchers. They sometimes even allude to the part standardization played in these collaborations and in the making of putatively more stable, reliable and effective therapeutic agents, but they have rarely if ever looked in detail at the process itself; the conflicts it generated, its social and cultural meanings, and so on.[3]

This is all the more unusual since standardization – and more generally questions of metrology – have become an important topic in studies of science and technology.[4] Ten years ago, writing the introduction to a book in part devoted to this issue, and called the *Invisible Industrialist*, Ilana Löwy and I remarked that standardization has both direct and metaphorical meanings.[5] Its most direct and restricted meaning refers to a formal agreement to use the same instrument, the same substance or the same procedure in order to enforce the comparability and the replication of measurements and – more generally – of technical operations. But a standard is also a moral value and a social technology. It is a certain understanding of quality, excellence or worth, which provides the ground for agreed judgement and nurtures consensus.

Keeping in mind these diverse meanings, questions about drug standards should therefore not simply address the organization and details of work within testing laboratories. The standardization of pharmaceuticals is linked to practices as diverse as the calibration of instruments, the definition of product composition, the choice of measurement units, the design of work protocols, the control of preparation, the evaluation of potency, the selection of dosage and route of administration. Drug standardization is equally concerned with what happens in the production plant or with events taking place within the clinical ward where the uses of drugs are – in practice – decided. Being a critical component of contemporary techno-science, problems of standardization do not only bear on technical choices but also on the economic and political issues grounding the norms that are mobilized in the regulation and management of therapeutic agents.

These issues of standardization apply to many industrial products, such as food products, whose historiography has often paid more attention to questions of normalization than has been the case with drugs. Standardization in medicine has its own specific dimensions, originating on the one hand in the normative status of care, and on the other hand in the variability of the living. During the last three decades, discussions about medical standardization have not necessarily been focused on industrial practices. When linked with 'evidence-based medicine', standardization has been used to denote the need to 'rationalize' medical practice, to reduce its variability, to align care with procedures that have allegedly been demonstrated to be effective in properly conducted clinical trials.[6] From this perspective, standards refer simultaneously to: 1) the 'standard' criteria for proof provided by the statistical methodology of the controlled clinical trial; 2) the guidelines and recommendations associated with the treatment of a particular pathology that emanate from consensus conferences, committees of experts, management organizations, etc.; 3) the general values and norms of medical action that are widely accepted by (professional) evaluation bodies. Thus the booming industry around EBM (Evidence-Based Medicine) reminds us that 'clinical standards' are autonomous and essential elements in the standardization of drugs, not least because they focus on what is the main target as well as the main source of tensions in the construction of drug markets, namely the difficulties in normalizing clinical patterns of action as they affect physicians and patients.

Another specificity of drug standards is rooted in the fact that, up until WWII, many therapeutic agents were not purified chemicals, but preparations originating in biological entities, particularly plants. During the interwar period traditional plant extracts found in the various pharmacopoeias were complemented by another class of biological drugs associated with the development of physiological and biochemical research, namely hormones, vitamins, enzymes, and therapeutic vaccines. The preparation of all 'biologicals' must take into account – and somehow circumvent – the variability of raw materials made from the collected parts of living organisms.[7] Standardization served as a material and social technology for controlling not only the quality of such materials, but also the unstable outcome of the preparation. In most instances, the composition of extracts was not known. Moreover, it was not necessarily thought to be a critical element in medical progress.[8] The boundary between effects and side effects, between cure and poison had therefore to be addressed on a different basis than the chemical ideal of purification, weight measurement, and a linear relationship between dose and effect. As a reaction to such variable and uncertain patterns, biological assays, based on the quantification of physiological properties, were the most frequently employed means to measure potency and define dosage. Their multiplication in the various sites associated with the making of a given therapeutic agent was often viewed as an additional source of discrepancies and chaos that one could only overcome with stronger standardization that could cut across laboratories, firms, and national lines. Nevertheless, this diversity of assays had important positive functions that undermined the desire for homogeneity, since they became almost indispensable for assessing the diversity of functions – and therefore the diversity of indications – that potent physiological agents could display.

To discuss these issues more precisely, the following paper concentrates on the trajectory of one type of biological drug, namely the hormones (steroids as well as pituitary factors) involved in reproductive physiology that came to enrich the pharmacist's and the doctor's arsenal in the 1920s and 1930s. Nelly Oudshoorn first analyzed the archaeology of the sex hormones through the lens of the Dutch case.[9] In this work, she emphasized four features associated with the development of biological measurement during the 1930s: a) these assays provided an operational definition of what would be counted as 'male' or 'female' factors, while providing numbers that could be employed to characterize the 'potency' of a given preparation; b) scientists then developed highly heterogeneous testing systems, which have their roots in different disciplinary cultures or disciplinary styles; for instance gynaecologists selected a uterus assay to objectify female hormones while comparative physiologists preferred measuring the length of feathers in hens; c) standardization of testing was to a large extent achieved in international comparisons of practices, with the League of Nations and the National Institute for Medical Research in London as leading players; d) this move was grounded in the dominant role biochemists and their molecules played in the study of hormones during the second half of the1930s. By then, the sex hormones had become sex steroids; they had been crystallized and purified, their metabolism was being successfully investigated and partial synthesis was under way. The industrialist is quite visible in the Dutch trajectory of the sex hormones. As N. Oudshoorn recounts, the firm Organon was the main provider of glands and research material for academic researchers, and it also supplied the drugs for clinical use, becoming a critical actor in the techno-scientific construction of gender identities. The main focus of *Beyond the Natural Body* is, however, on forms of biological standardization that took place within the physiological laboratory. Here, processes revolved around the control of experimental systems, with particular attention being paid to the various factors and parameters that might affect biological functions. The main aim was to stabilize local results by facilitating reproducibility, thus 'black-boxing' various biological entities that participated in the physiology of reproduction.

Hohlweg's quote, cited at the beginning, reminds us that the standardization of the sex hormones was not in the first instance an academic problem but an industrial issue, a process pursued in various ways by the large pharmaceutical companies responsible for the production of as well as some research into hormone preparations. In comparison with laboratory standardization, its industrial variant focused on testing rather than experimental systems, that is, on technical arrangements that were simple and robust enough to on the one hand provide a model for production processes and monitor their functioning, and on the other provide the means for the large-scale control of raw materials, as well as ensuring product quality with homogeneous composition.

Industrial standardization did not operate in isolation, within production sites alone. It was linked to practices of laboratory standardization that could be pursued either in-house or in collaborative academic settings. It was also closely related to standardization in the clinic, since the manufacturer's definition of quality or the measurement of potency aimed at the standardization of dosage, route of administration and indications in a global strategy for disciplining adverse reactions, toxic events, lack of tolerance and variable efficacy. Even if animal substitutes were occasionally handled and assessed in hospitals, clinical standardization relied in the first place on the comparison of human cases.

Using this simple but convenient analytical distinction between experimental, industrial and clinical work, we now want to go on to explore the multiple dimensions of biological drug standardization by comparing developments in two German firms, IG Farben and Schering. Although both companies produced hormonal substances in the1930s, they developed significantly different regimes of standardization.

The testing laboratory and its context

Both IG Farben and Schering created 'physiological laboratories' in the late 1920s, at a time when the preparation of powdered extracts of ovaries marketed as 'female' hormones to be used in 'glandular therapy' were being replaced by purification protocols using the urine of pregnant women or pregnant mares as raw material. Schering's commercial product was called Progynon, IG Farben's, Unden, with the two firms adopting different structures and, more importantly, different ways of situating biological testing in the company's research infrastructure.

As recounted by J. Lesch, Bayer (later incorporated into IG Farben) expanded its drug research infrastructure in the 1920s, turning the identification of chemicals with putative therapeutic properties into a systematic enterprise, thereby instituting an industrial organization of drug research.[10] Bayer's screening system that would become a reference after the development of the sulfonamides in the1930s was based on the mobilization of an army of organic chemists responsible for the synthesis of entire families of molecules related by their core structure. In parallel, the company developed bacteriological and pharmacological laboratories where these compounds could be tested either on pure cultures of bacteria, or inoculated into animals infected with model diseases. Focusing on the chemotherapy of infections, the system nonetheless opened niches for developments in other therapeutic domains, beginning with the synthesis of anaesthetics, barbiturates, or gland extracts. In vivo and in vitro testing was in all instances pursued within the pharmacological department under the direction of Gerhard Domagk.

Rooted in the culture of bacteriology, the local understanding of standardization focused on the homogeneity of tools, the control of gestures, and purity. This combination did not imply that all preparations of pharmaceutical interest were pure chemicals. Unden, for example, was not such a chemical. Nevertheless the local culture promoted the ideal that research and development would ultimately lead to all such preparations eventually becoming purified substances, and so they should be handled in the closest possible way to pure chemicals, until their molecular characterization could open up the way to an appropriate synthesis. Relationships of dose to effect were therefore among the most important targets for research, representing basic tools for achieving efficacy without causing the severe toxic symptoms that the dye derivatives once explored by Bayer and Hoechst so often did. In the case of hormones, this pharmacological approach was plagued by an additional difficulty in addition to the general question of how laboratory models would match clinical applications, namely the absence of any means to directly measure the quantities of the drug being used. Biological tests were introduced to fill this gap. Their use was, however, perceived as an inferior substitute for chemical determinations of purity, making it a practice that was to be minimized and replaced by more direct measurements as soon as they became available

Schering's testing laboratory was introduced into a different, more biological culture of innovation.[11] In 1938, when Hohlweg wrote the article cited above, the pharmacology section of Schering's main research laboratory had already existed for 15 years, during which time it had evolved into two distinct research groups working in collaboration with the production plants and the medical department. The first group, under the leadership of Hohlweg himself, was responsible for physiological research and testing. In this group, investigations were conducted according to the division of labour outlined above. The second group, under the leadership of Karl Junkmann, concentrated on the chemical properties and pharmacological effects of drugs. It was this second group that would be radically transformed by the 'in-house' linkage between biochemical and physiological investigations.

Junkmann started his career as a company researcher in 1925, when he was hired as a pharmacologist. During the 1920s he was involved in studies of a diverse range of compounds using mammalian model organisms. The list included X-ray contrast media, sulfa drugs, arsenic derivatives, adrenaline, strychnine, theophyline, thymol, acetophenon, and guanidine. He also worked on some hormone preparations, namely a putative cardiac hormone involved in the control of blood pressure, and pituitary extracts.[12]

Most of these projects focused on issues raised by the pharmacists and doctors using Schering's products, particularly questions of toxic effects or contamination. By 1930, the growth of Schering's research facilities made it possible to transform the pharmacological service into a fully-fledged laboratory, opening up the possibility of pursuing projects that were less dependent on the demands of the manufacturing plants. Junkmann and his colleagues then embarked on the development of three types of drugs; sulfonamides, a group of molecules having hypoglycaemic properties to be used in the treatment of diabetes, and hormones. Most of the work on sex steroids, however, remained the province of Hohlweg's physiological section. Junkmann concentrated on pituitary hormones, which dominated the research agenda until the outbreak of WWII.

Schering's biotechnological infrastructure was not, however, limited to these in-house laboratories. A dense network of relations was forming between the pharmaceutical company and a small number of academic partners, of which the *Kaiser Wilhelm Institut für Biochemie* (KWIB) – headed by Adolf Butenandt – was the most important. Between 1930 and the end of WWII, the development of sex steroids into a widely available means of controlling reproduction, ageing, and some pathological conditions came to be at the very centre of this collaboration. What did Butenandt mean for Schering? How was he bound into the company's research and production? In the leaflets printed by the firm, he is depicted as a supplier of biochemical knowledge in the form of chemical structures, protocols for preparing materials, and biosynthetic pathways. Through the files of the patent department, he comes across as the co-inventor of many processes, and the benefactor of contracts that doubled the budget of the KWIB. Schering also contributed to the Institute's research in other ways, supplying biological and chemical materials that ranged from common reagents to molecules specifically synthesized for the KWIB biochemists. Finally, Schering was a major source of scientific information and know-how, passing on rare journals, patent applications made by competitors, information about production processes and the drafts of articles written by its own scientists.

One outcome of this collaboration was to reinforce Schering's biological culture of innovation. In contrast to the dominant approach at IG Farben, Schering's strategy was not to achieve the total synthesis of biological drugs like the sex hormones, but consisted in the limited use of chemical reactions to mimic the biochemical processes taking place in the cell in order to transform the naturally occurring steroids into more effective or cheaper compounds. In other words, partial synthesis was the watchword of the1930s, with the hope that it would provide the means for the economical production of steroid derivatives. Within this paradigm, the critical contribution of the KWIB was less the deciphering of molecular structure than the description of these biological reaction pathways. This research was carried out exclusively by the KWIB and not by the in-house laboratories, even though the resulting reaction pathways were used as a model or at least a source of inspiration for production processes in the firm's own laboratories and pilot plants.

Biological assays as industrial tools

Both companies believed the definition of consensual biological assays for hormone preparations to be of vital importance. Animal rather than human testing was considered indispensable for evaluating the potency of hormone fractions (known as *charges* in the production setting), as they were of variable and complex composition, and little was known about their chemical composition. Until the late 1930s, in both firms, the precise, direct measurement of one or several physiological effects remained the only way to identify the presence of an active substance within these charges and to quantify it.

The industrial standardization of hormone preparations was, therefore, in the first instance a problem of *Wertbestimmung*. Just as in the case of the production of sera starting in 1894, biological assays played a critical role in measuring the potency of industrially prepared pharmaceutical material. The assessment of each batch of hormone with respect to the standard of a 'male' or 'female' biological unit became a mandatory step in the definition of the pharmacological dosages to be employed by doctors. As a consequence, the adjustment of protocols by production engineers (most of them chemists) as well as the determination of clinicians' prescription practices depended on these bioassays.

The measurement of potency also had legal dimensions. The legal definition of the pharmacist's professional responsibility made him liable if the composition of the products did not correspond with what was claimed or if their use at the normal dosage under normal circumstances resulted in harm to the patient.[13] This general regulation of the drug market encouraged the measurement of physiological effects as a form of certification. Other papers in this volume have shown how the determination of the potency of sera was both required by law and carried out by a centralized state institution in Germany.[14] While there was no such equivalent regulation for hormone preparations, the practice was nevertheless adopted by all the major producers. As with sera, defining the potency on the basis of precise quantitative biological measures could be used for marketing purposes as well (Figure 10.1). Standardized preparations were generally considered to be better and more reliable drugs, with the precisely determined figures echoing physicians' optimistic declarations about the prospects of scientific medicine in the twentieth century.

For Bayer or Schering, the value of biological assays was not, however, restricted to *Wertbestimmung*. More precisely, the practices of *Wertbestimmung* in the testing laboratory were not exclusively for the determination of

Zur peroralen
Hormontherapie
PROGYNON
bei ovarieller Insuffizienz
und kausal-genetisch verwandten Erkrankungen:

endokrin bedingte
Polyarthritis,
klimakterische
Erregungszustände,
Fälle von Migräne
und Epilepsie,
verschiedene
Hautaffektionen.

**PROGYNON B
OLEOSUM**

*Originalpackungen mit 15, 30 und 60 Dragees zu je 150 M. E.
6 und 12 Progynon-Ampullen zu 1 ccm mit je 100 M. E.
Progynon B oleosum: in Ampullen zu 1 ccm mit je 10 000 M. E.
50 000 M. E. und 100 000 M. E.*

SCHERING-KAHLBAUM A. G. BERLIN

Figure 10.1   Progynon advertisement featuring the potency of the doses

biological and – putative – clinical efficacy. Quantifying the concentration of active substance in a given batch was central to the surveillance and control of production. Testing was not only implemented at the end of the production process, but also in its very preliminary stages to test the content of raw materials. From the late 1920s onward, the preparation of the female hormone started with the treatment of concentrated urine from pregnant women using various solvents. Initially, Schering collected the liquid in hospital services but rapidly shifted to urine collection from more readily accessible 'pregnant' mares kept in farms in the Berlin region. The content of incoming urine was regularly tested in Junkmann's laboratory to verify that it was worth processing (Figure 10.2). When production shifted to the use of horses, potency testing also became a way of controlling the work done by the farmers who benefited from a contract with Schering. Another industrial function of the biological assay was to compare the effects of changes in the preparation process (modifying the solvents, changing concentrations or varying temperature and pressure). Thus, Schering could test the value of any new procedures developed by in-house scientists and engineers at the pilot stage before implementing them in large-scale production. Biological assays of batches produced under different conditions thus formed part of the industrial search for increased productivity.

When examined in operational or practical terms, testing in the plant and testing in the academic laboratory had much in common. The majority of the assays for estrogens, progesterone and testosterone introduced in the 1920s actually emerged from physiological and/or medical research. This was the case of the Allen-Doisy mouse test for follicular hormones, which was used in both Schering and Bayer's testing laboratories. The assay consisted in the microscopic examination of vaginal smears obtained from castrated female mice (and later, rats) after the animals had been inoculated with known quantities of the material under test (Figure 10.3). A positive result was associated with the disappearance of white blood cells and the appearance of epithelial cells. One 'mouse' or 'rat' unit was the minimum dose that induced a full change in histology.[15] In other words, the unit for the follicle-hormone, later estrogens, was defined as the volume of preparation capable of inducing the differentiation of the vaginal epithelium in rodents.

Standardization is often described as a means for reducing variability in order to achieve a reasonable homogeneity of products and the replication of scientific or medical results. Within the context of biological laboratories, this kind of total control was a distant goal whose realization required vast amounts of work and the mobilization of huge material resources. Incidentally, this movement would also result in the disappearance of an important source of experimental innovation, rendering production processes less imaginative. Laboratory standardization is therefore usually limited and highly local. In the context of early hormone production, biological tests were the means of a more thorough, although less ambitious handling of variability. The problematic sources of variation targeted in the physiological literature were numerous: the animals (rats did not react like mice; strains could be sensitive or resistant); the modes of administration (a single inoculation did not produce the same effect as fractionated

1. und 2. Quartal 1935

Kontrollauswertungen der Stutenharnstationen.

Es wurden die Lieferungen aller Stationen und zwar nur auf 50 000 M.E. pro kg Harn untersucht. Sind 50 000 M.E. pro kg Harn enthalten, so wird der Harn als positiv, andernfalls als negativ bezeichnet.

Ort:	Zahl d. Kannen:	Zahl d. Ausw.:	Zahl d. Tiere:	Ergebnis:
Aderhausen	8	2	10	negativ
	4	1	5	positiv
Altenbruch	9	3	15	positiv
	15	3	15	negativ
Altenau	12	1	5	negativ
Arnstein	105	1	5	positiv
	140	1	5	negativ
Barenthin	79	2	10	positiv
	44	3	15	negativ
Bisbeck	5	1	5	positiv
Beierberg	137	5	25	positiv
Belgard	4	2	10	positiv
Blumental	41	2	10	positiv
Breese	2	1	5	negativ
Britzfleth	36	4	20	positiv
	13	3	15	negativ
Buschheide	149	2	10	positiv
Buxtehude	122	2	10	positiv
Cuschärfe	1	1	5	positiv
Daersdorf	97	1	5	positiv
Dannenberg	70	13	65	positiv
	7	2	10	negativ
Diose	11	2	10	positiv
Dollberg	42	2	10	positiv
Dömitz	2	1	5	positiv
Duffing	16	1	5	negativ
Eche	4	1	5	positiv
Eppmühle	101	3	15	positiv
Eylau	50 10 F.	17	85	positiv
Essleben	502	4	20	positiv
Freiburg, Elbe	99	7	35	positiv
	22	3	15	negativ
Gaukönigshofen	15	2	10	positiv
Glöwen	197	3	15	positiv
	40	1	5	negativ
Grossblankenbach	98	3	15	positiv
	39	1	5	negativ
Grosshardorf	157	2	10	positiv
Grosslangheim	2	2	10	positiv
	3	1	5	negativ
Hammak	21	5	25	positiv
	6	1	5	negativ
Henriettenhof	91 3 F.	4	20	positiv
	8 F.	2	10	negativ
Heugraben	77	1	5	positiv
Himmelpforten	44	6	30	positiv
Hofheim	254 10 F.	4	20	positiv
	73	1	5	negativ
Höftgrube	14	3	15	positiv
	8	2	10	negativ

Figure 10.2   Register of urine testing at Schering

inoculations, and subcutaneous injections did not produce the same results as intravenous ones); the timing of the assay (mice did not react well in dry, cold weather); the number of animals used, etc. Thus, the physiologists focused on three aspects to achieve standardization: a) they employed

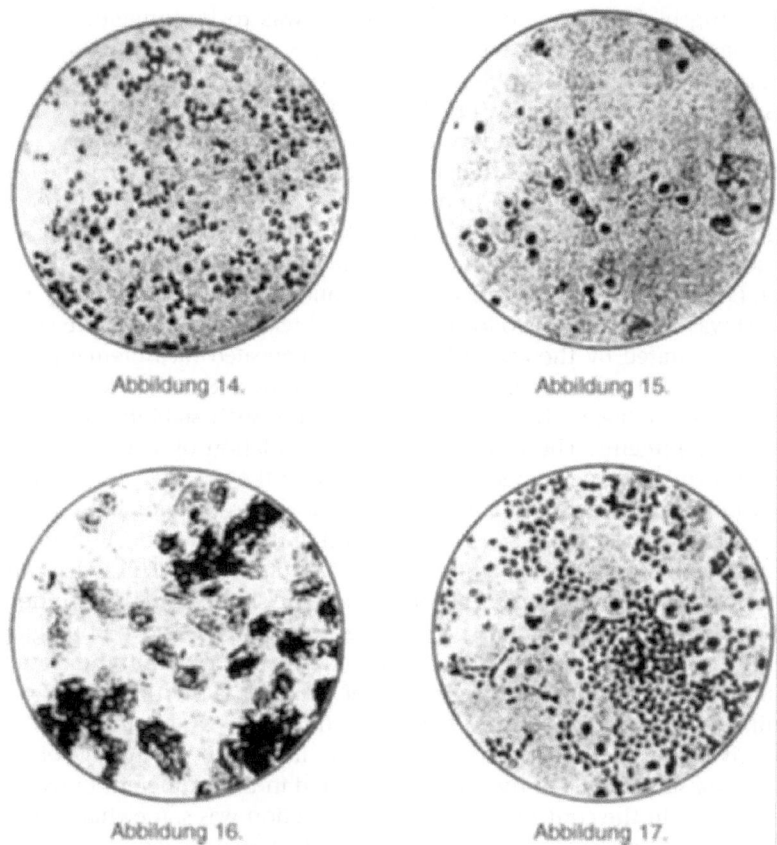

Abbildung 14.   Abbildung 15.

Abbildung 16.   Abbildung 17.

Figure 10.3   Mouse assay for estrogens ca. 1925

'normal' bodies (selecting strains of animals that behaved normally and could be bred in controlled ways); b) uniformity of practices (e.g. defining the number of smears to be taken and the number of inoculations); c) establishing the statistical significance of the results by using large numbers (e.g. establishing reference curves using hundreds of animals). A good assay was therefore less a matter of purity and composition, than a matter of stability, specificity and – most importantly – biological significance.

In the research stage, industrial standardization could follow the same routes, but when it came to issues of development, quality control and surveillance of production 'standard experimental systems' were replaced by 'standard test systems'. One obvious difference between the two derived from the difference in scale and productivity. An industrial assay needed to place the emphasis on being rapid, easy to learn, and cost-effective rather than being biologically meaningful since it was to be implemented on a much larger scale than a research procedure, which, while it had to be repeatable, could afford to be maximally comprehensive and informative. These constraints can be illustrated by the fate of the chicken crest assay for the male hormone that was developed by Schering's test laboratory.

Schering's procedure used the growth of the crest of a castrated chicken to measure the potency of male hormones and was developed in collaboration with the biochemists working at the KWIB mentioned above. The standard protocol proposed by Butenandt and his colleagues emphasized a form of precision based on the strict control of gestures and materials.[16] In the early 1930s, the test was defined in the following terms: 'Chickens from the white Leghorn race are castrated by the age of 8–10 weeks. Repeated measurements of the surface area of the crest will certify that in the next 8–10 weeks this sexual trait remains stable. Only the animals displaying such stability may be used for the experiments. The first phase is the inoculation of a control solution with known hormone content in order to check the animal's responsiveness. The animals not reacting according to the norm or showing a crest, which does not return to its initial size after the end of the test will be discarded. (...) Measurement of hormone potency is the growth of the crest. This is determined in the following way. Photographs of the crest are shot before the test inoculation, and after three and four days. In order to obtain these two photographs, one must place a piece of sensitive photographic paper behind the crest. The scene is briefly illuminated. The surface of the crest is then measured with a planimeter. The chicken unit is defined as the dosage which, in a solution of $1cm^3$ given in two successive days, induces an increase by an average of 20 per cent of the surface area of the crest as measured in three different birds'.[17] The version used in the control of industrial production was somewhat simplified (Figure 10.4). For each batch or preparation to be tested, Schering's technicians selected three dosages. Two birds were inoculated with each of these quantities, two photographs were taken one day later, leading to two measurements of the crest's surface area, and a qualitative binary assessment.

Industrial test development was, however, more complex than a simple 'Taylorian' quest for simplicity and productivity. These biological assays were manufacturing tools mobilized on a routine basis for the control of raw materials, routine production processes, commercialized drugs, and competing products (Schering analyzed the sex hormones made by Bayer, Roussel in Paris, and Ciba-Geigy in Switzerland). Specific goals as well as sites of implementation determined many characteristics of the reference protocols. This can be illustrated by the biological assay for progesterone, which was developed around 1930 at Schering in the context of a collaboration with Carl Clauberg's gynaecological clinic in Königsberg.[18]

The assay was based on the semi-quantitative examination of the rabbit uterus. One rabbit unit was the quantity of progesterone preparation necessary to induce the complete transformation of an infantile rabbit uterus

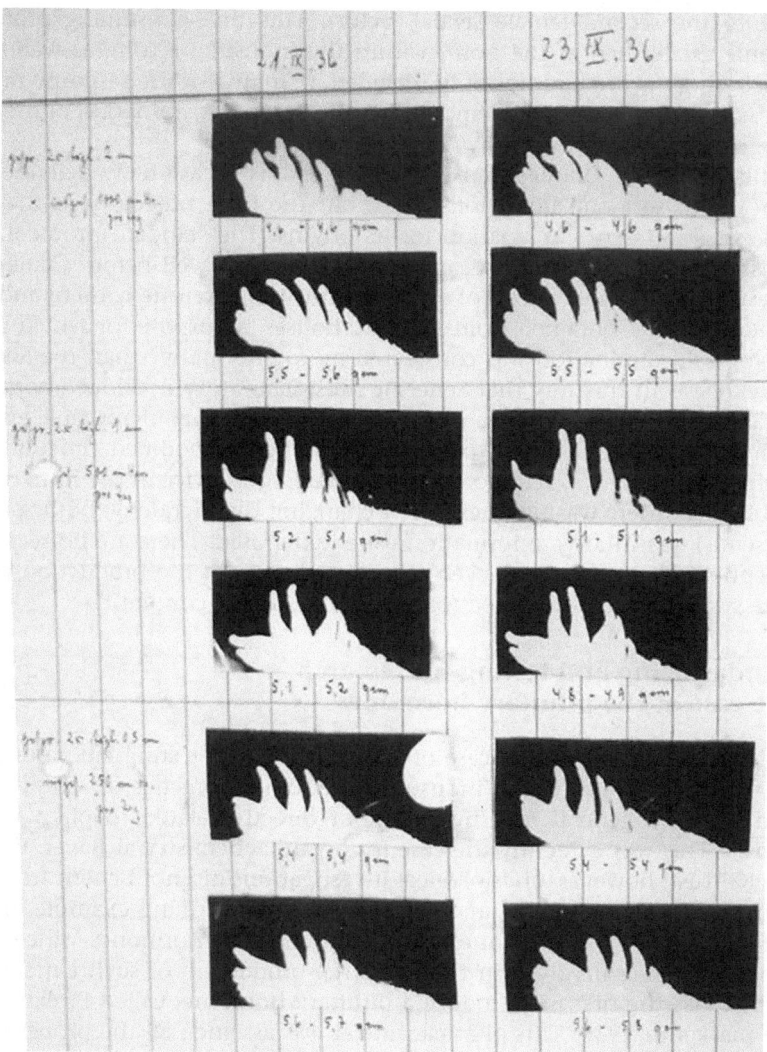

Figure 10.4 Series of photographs for determining the surface area of the crest of castrated chickens used by Schering to measure the potency of male hormones

into fully differentiated glandular tissue. In order to obtain any results, the animals had to be prepared with a dose of estrogens. The problem was that the test gave ambiguous results since both female hormones, estrogen as well as progesterone, presented similar effects on the uterus. Thus, the rabbit assay was complemented by microscopic observation in Clauberg's laboratory, which would distinguish growth (due to the estrogens) and differentiation (due to the second female sexual factor). This time-consuming protocol became established as the norm within the industrial setting as well, and so Schering's procedure not only included a ranking of the response of the uterus (+/++/+++), but also demanded a microscopic examination of the histology of the rabbit's ovaries.

This time- and labour-intensive operation was occasionally defended as a means to sort out ambiguous cases, but the local trajectory of the test suggests a different explanation for its origins. The standard protocol was a historical product that had appeared in the firm well before Clauberg's investigations in the context of evaluating the progesterone-content and the value of ovaries obtained from slaughterhouses, a purpose for which morphological examination was considered more informative than the animal assay. It was in this way that Schering's test laboratory developed both the infrastructure (histological rooms) and the skills for the microscopic examination of slides and organ preparations. When a modified progesterone testing procedure was reimported from Clauberg's service in the form of the rabbit assay there was no question of changing the hierarchy: both sources of (semi-) quantitative information were juxtaposed. There are no recorded exceptions to this pattern, even when batches from the production plant were systematically examined for their progesterone content.[19]

Standardizing and testing for research: The case of the pituitary hormones

As has been argued in the case of the expansion of testing laboratories in the chemical and electrical industries, the routine practice of assays could lead to more general investigations not directly related to plant operations.[20] This was evidently the case for organic chemistry at Bayer. Within the IG trust, however, physiological investigations did not benefit from the same favour they enjoyed at Schering. The most striking example of this difference is provided by the fate of pituitary sex hormones, since both companies were involved in the industrial production of such extracts. IG Farben was the first to introduce a pituitary hormone, called Prolan, onto the market in 1930. This pharmaceutical was assigned all the properties of the female sex hormone, which made the difference between the products rather unclear since the biological function of the new factor was to control the activity of the ovaries. Prolan was isolated from human urine following a process (involving differential precipitation using ammonium salts and alcohol) adapted from the work of Carl Zondek, a Berlin physiologist and gynaecologist, which IG Farben bought and patented in 1927.[21] The molecular studies conducted within the firm focused on the characterization of the active compound. In the records of the pharmacological section, Prolan was described as a protein-like material that did not tolerate heating but could be digested by enzymes or dialyzed and contained amino acids like tyrosine.[22] Most of the research on the effects of the hormone was conducted by outside collaborators like Zondek who pursued their own projects, only occasionally exchanging information with Bayer. The limited in-house 'biological' evaluations were oriented towards two goals: first, the elaboration of purification protocols, and second, evaluating any potential toxic effects of the drug. In 1929, for instance, local representatives of IG reported that clinicians who had been given Prolan preparations for clinical trials complained about the unexpected occurrence of fever following the inoculation. The adverse reaction was eliminated after the pharmacological laboratory proposed an additional precipitation step using barium salts.[23]

Perceived as mere production tools, the bioassays for Prolan remained simple. IG's version was a rat test focusing on the quantity of extract necessary to induce oestrus without any further qualification. The results of this industrial evaluation often disagreed with those obtained from the multiple assays employed in Zondek's service to evaluate the biological value of pituitary extracts, thus triggering complaints from both sides. While Zondek commented on the impure nature of the factory products, IG's scientists regretted the lack of academic interest in standardization.[24] An additional source of tension was Zondek's increasing emphasis on the existence of two different types of Prolan, which he named Prolan A and Prolan B. IG's biochemists could not separate them and came to regard the distinction as simply an invention of their outside collaborator.[25] Thus, although Bayer had a real interest in biological research it did not have the culture or the means to push it very far in practice in terms of its own in-house research.

In contrast, Schering's development of its capacity for physiological research and the practice of standardization led to more thorough and basic studies of the various hormonal factors. In the early 1930s, Schering's prestigious physiological laboratory was involved in two types of research directly related to testing. On one hand, the collaboration with the KWIB biochemists and Butenandt led to an analysis of the male or female activities of various analogues and derivatives of the natural sex hormones, which might serve as intermediates in the biological synthesis of the natural steroids and eventually become substrates or points of departure for industrial production. On the other hand, the testing laboratory handled animal brains, blood and urine samples from a few gynaecology clinics, as well as humans placentas on which they conducted studies of pituitary factors, including potency measurements. The initial goal was to develop a product for the thyroid hormone market, which led Junkmann to focus on the purification of a pituitary factor that could stimulate the secretion of thyroxine. With an augmentation in local know-how and a growing collection of biological material, there was a diversification of research interests and, by the mid-1930s, the pharmacological section was working on protocols for purifying extracts with effects on a whole range of organs beyond the thyroid, including the gonads.

This 'internalization' of biological expertise resulted in detailed analysis of the functional relations between the brain, the pituitary gland, and the ovaries at Schering. According to a report that appeared in the in-house scientific journal in 1930, the results obtained from combining the competencies developed within the pharmacological and the physiological sections of the research division were quite interesting. Junkmann's pituitary studies confirmed the idea that castrated animals presented a specific identifiable histology when injected with the extract: pituitary cells were enlarged, as if swollen by an increased production of the secreted material. As evidence of the hormonal origins of the phenomenon, the researchers showed that the so-called castration cells disappeared when the animals were supplied with estrogens in the form of commercial Progynon, at least when the effect was evaluated on the basis of microscopic evidence. Hohlweg reached the following conclusion:

> The changes in the pituitary following the ablation of the ovaries should not be considered to be the consequence of inactivity, but rather as the effect of a missing retro-inhibition (*Rückwirkung*) on the pituitary. The gland is put at rest by the ovary's follicle hormone (itself secreted under the influence of the pituitary). The artificial supply of Progynon is therefore able to inhibit the changes due to castration. The fact that the pituitary, which plays such an important role in the endocrine system, can only function in a normal way if the body is supplied with enough Progynon will be of major significance in the clinic.[26]

Over the next five years, this concept of retro-inhibition or *Rückwirkung* would itself take on major significance for Schering's researchers.[27] Investigating the mechanism of this feedback relationship between ovaries and pituitary they returned to morphological investigations. Besides simple transplantations, the main experiments consisted in attempts to isolate the pituitary from its neural environment. Ablation of the gland was followed by implantation in the body of a second animal, establishing vascular connections but no neural ones. Following castration, the endogenous pituitary exhibited castration cells while the implanted organ remained normal. A similar transfer was conducted using a 'castrated pituitary' in a previously castrated mouse and followed by a supply of Progynon. Normalization took place only in the endogenous gland. As Hohlweg and Junkmann commented:

> The structure of the castration pituitary depends on the increased secretion of gonadotropins. The lack of follicle hormone in the blood pushes the brain's sexual centre into overcoming this deficiency. A pituitary devoid of normal neural connections cannot react to this impulse and has no possibility to communicate the production of important amounts of the follicle factor. This is the case of the grafted pituitary, which lacks

the relevant neural apparatus. A normal pituitary also remains normal when it is implanted in a castrated animal, while the pituitary of a castrated animal is brought back to normality after being grafted into a castrated animal.[28]

Hohlweg and Junkmann summarized this series of experiments by formulating the notion that production of female hormone was regulated by means of brain *Rückwirkung*. The ovary was not influencing the pituitary directly through the concentration of estrogen in the blood but through the mediation of a neural centre.

Since IG had already patented a preparation process based on alcohol precipitation, Schering's strategy for the commercial development of this kind of gonadotropin was twofold: a) to look for an alternative process for purifying the pituitary factors out of the same material (here, they attempted precipitation using various salts); b) to find unexploited biological fluids as had been the case with the female hormone, and here their attention was largely turned to pregnant mares. By the mid-1930s, Junkmann had accumulated enough results using horses in Berlin to convince Schering's management that they should establish a horse farm in Argentina from which they could organize regular shipment of dried serum.[29]

The research into purification failed to provide interesting innovations for most of the 1930s; while Junkmann successfully reproduced existing (and patented) protocols, the design of a new one provided highly variable results. An important part of Junkmann's time was devoted to the stabilization of an industrially useable version of the mouse biological assay for gonadotropins. A typical 'research' assay for the gonadotropin content of horse or human urine included a combined evaluation of changes in ovarian weight, yellow body formation, appearance of vaginal cells, and uterus weight (Figure 10.5).

In contrast to the progesterone case, Junkmann and his colleagues ruled out the large-scale use of microscopic observations in assessing the specificity of pituitary factors, which would have involved the preparation of tissue slides to document the simultaneous maturation of the ovary and the differentiation of the uterus epithelium. Testing for the production plant would rely on morphological criteria that were thought to be simpler and easier to assess. Thus, the pharmacological laboratory embarked on a statistical analysis of ovarian weight in order to evaluate the range of natural variation, define the minimal number of animals to be examined to ensure statistical significance and select the dosages for inoculation.[30] They did not, however, completely abandon the histological definition of oestrus, and designed in parallel a semi-quantitative scale to evaluate the morphological changes not in epithelial slides, but in vaginal smears.

Standardization was finally achieved in the rat with the definition of a 'rat unit' for gonadotropin as the quantity sufficient to induce oestrus in at least half of the implanted animals within three days after the inoculation of the

Junkmann Juli — Dezember 1936.　　　　　　　　　　　　6.

Zur Auswertung wurden jeweils 2 ccm teils Serum teils Gesamtblut
auf 6 Injektionen verteilt in der üblichen Weise injiziert und
nach 100 Stunden das Resultat abgelesen.

Besitzer	Pferd	gedeckt am	Scheiden-Abstr.	Öffn.	Ute-rus	Folli-kel	Co.lu.
Seidenschur	Imposante	15.3.	+++	+++	+++	+	++
Roehr	Guste	1.4.	++	+++	+++	++	++
Pein	Junge Braune	11.4.	+++	+++	+++	+++	++
Herm	Namenlos	15.4.	+++	+++✗	+++	+	+++
Roesecke	Cletine	8.5.	+++	+++	+++	+++	+++
Roesecke	Laura	13.5.	+++	+++	+++	???	+++
Roesecke	Irma	13.5.	+++	+++	+++	+	++
Seidenschur	Friedel	5.	+++	+++	+++	+	+
Burg	Eva	6.	+++	+++✗	+++	+++	+++
Roesecke	Junger Fuchs	5.	+	+++	++	++	+
Erdmann	Junge Braune	4.	++	+++	+++	+	–
Roesecke	Brigade	13.5.	+	+	+	–	+
Ploigs	Sensa	7.3.	+	+	+	–	–
Seidenschur	Alte br.Liese	6.	+	+	++	+✗	?
Seidenschur	Junge Braune	6.	+	+	+	?	?
Baetge	Endetal		+	+	+	+	–
Pein	Fuchs	1.3.	–	++	–	–	–
Burg	Feodora	3.	–	–	++	–	–
Pein	Alte Braune	3.	–	+	–	–	–
Ploigs	Flotte	4.	–	–	+	–	–
Roehr	Nelly	20.4.	–	+	–	–	–
Erdmann	Alte Braune	4.	–	–	–	–	–
Pein	Florentine	5.	–	+	+++	–	–

Die vorstehenden Sera sind ausgewertet am 17.19.36., das folgende
am 30.11.36.

| Borg | Eva | 2.6. | – | – | – | – | – |

Die nun folgenden Seren wurden am 10.12.36.ausgewertet.

Legde	Dame	18.4.	–	–			
Legde	Amanda	3.4.	–	–			
Sandberg	Dora	10.4.	–	+			
Sandberg	Elsa	10.4.	++	–			
Riebe	Licente	5.5.	++	++			
Wolf	Geduld	8.4.	+	–			
Wolf	Nahore	15.4.	–	–			
Viebig	Lotte	3.	–	–			
Viebig	Liese	3.	–	–			
Viebig	Hanni	3.	–	+			

Pferdesera am 19.12.36.ausgewertet:

Heydmann	Liselotte	5.	–	+			
Ebel	Intos	5.	–	+			
Lüdtke	Liese	30.4.	–	+			
Jahnke	Sherry	30.4.	–	+			
Unbekannt			–	–			
Plättke	Rappel	1.5.	–	+			
Plättke	Fuchs	1.5.	–	–			
Stavenow	Ansägerin	4.4.	–	+			
Vettin	Loreley	25.5.	–	–			
Ammann	Braune	21.4.	–	+			

Figure 10.5　Measurement of gonadotropins in sera from horses, Junkmann 1936

test material. In addition to the global and qualitative 'induction of oestrus',
Schering's workers adopted two other indicators to measure potency. The
first one – following the previously mentioned statistical assessment of natural variability – was simply the growth of the ovary (one unit corresponded to

a 500 per cent growth in half the animals). The second indicator was a measure of the number of cells typical of the second phase of the reproductive cycle found in vaginal smears. Systematic testing of the serum of pregnant mares sent from Argentina based on this procedure started in 1938. Finally, Junkmann's standardization of the pituitary factors also integrated the construction of a system of equivalence between Schering's local units and those used in more official circles, like Evans' physiological laboratory in the US.

These various ways of investigating and defining the pituitary factors point to different cultures of standardization. At IG Farben, standardization was substance-driven and product-oriented, and focused on issues of toxicity and dosage. It can best be described as a pharmacological standardization, in contrast to Schering's model, which was more physiological. Schering linked the quest for industrial norms with investigations into functional effects, so that bioassays were not only used as a reliable means to produce potency figures but also as complex *in vivo* systems for identifying biological/biochemical facts of technical interest. This form of 'research standardization' was rooted in the peculiar structure of Schering's collaboration network that included not only the KWIB biochemists, but also long-term medical partners who used their products in a clinical setting.

Industrial standards and the construction of clinical practices

Industrial sources rarely mention standardization in relation to hospitals, physicians and clinical work. When they do, standardization is approached from the restricted perspective of *Wertbestimmung*, that is, the measurement of potency to determine the choice of dosage. The relationship between industrial standards and clinical use is, however, far more complex. Following the trajectory of insulin, Christiane Sinding has convincingly shown how clinical work focusing on the close monitoring of patient's individual reactions to the various lots Eli Lilly supplied, led to the definition of clinical standards distinct from the industrial unit of insulin.[31] She also demonstrated that the 'standardization' of practices originating in the clinical interactions between diabetes' specialists and 'their' patients was another major source of information for the company. Eli Lilly used the letters and reports of clinical experiences to assess and eventually modify its preparation processes. The case of the sex steroids reveals a similar two-way communication between the hospital and the production plant. A good example is the *Abstricheinheit* used in the determination of pituitary factors that Junkmann adapted from the clinical examination of vaginal smears. The development of commercial sex hormones also reveals transfers in the opposite direction, with industrial standards radically modifying the definition of 'normal' medical usage.

While both IG Farben and Schering set up medical divisions, which organized the contacts and collaboration with hospitals and reputed specialists, once again the relationships with clinicians were not organized in the same way in these two firms. IG Farben inherited a network of local groups or '*Buros*' that Bayer and the other firms in the trust had developed starting around 1900. Each local *Buro* included not only administrative staff and a pharmacist, but also one or two medical employees responsible for the organization of clinical testing, who regularly visited hospitals and medical services. The *Buros* collaborated with the division of pharmacology in choosing the clinicians who would be allowed to use samples of a new drug. They were responsible for collecting information about patients' reactions, ideally in the form of written reports, but more generally via telephone calls and on-site communication. This was a system of short-term collaboration for the assessment of toxicity and dosage that played an essential role in the final phases of drug development. A specific network of 'testers' was established for each major drug, in which IG Farben's medical partners were left free to select appropriate patients and indications. The first goal in each case was the documentation of dosage and side effects. Thus, for example, following physicians' complaints that its pills of the female hormone Unden, which contained 10 mouse units each, were not concentrated enough, IG shifted to packages containing vials of oil-based solutions at 100 mouse units. Subsequently, reports about pain following the injections similarly resulted in changes, this time the addition of a second precipitation step to the preparation protocol for Unden.

The industrial standardization of clinical practice also worked through more subtle channels, as when it was associated with the growth of 'scientific marketing'. This practice, which evolved starting in the 1930s involved the firms disseminating recommendations and selected results of clinical investigations. In the beginning, the main tools employed by IG representatives remained simple, consisting mostly in flyers written by the division of pharmacology and containing directions for use.

In contrast, Schering's medical department worked on a long-term basis with a few gynaecologists in charge of hospital services. Thus, there were no more than a dozen regularly collaborating clinicians who benefited from privileged access to new preparations and information from the company and participated in the early phases of drug development. A consequence of the closeness of this collaboration was that the industrialist's impact on clinical practice was more profound than it was within Bayer's network of pharmacological testers.

Two clinics played a particularly significant role in Schering's trials of estrogens and progesterone: the service of C. Clauberg in Königsberg mentioned above and C. Kaufmann's *Frauenklinik der Charité* in Berlin.[32] As we have already seen, the former focused on studies of progesterone and – at the clinical level – on the management of sterility, while the latter worked on the uses of estrogens and the treatment of amenorrhea. Kaufmann's clinical work was marked early on by his collaboration with Schering. His first important article was a paper published in 1933 in which he reported the treatment of a 'sterile' woman using a combination of Progynon and Proluton to mimic the changing hormonal concentrations of a woman's ovarian cycle (Figure 10.6). This regimen was a success in the sense that he managed to induce the menses.[33] Developed over the treatment of a dozen cases, Kaufmann's regimen was adopted by Schering's medical department, circulated and promoted all over Germany. The regimen was a standard in two different respects. First, it defined a normal reproductive cycle on the basis of curves showing changing concentrations of the sex steroids that were gradually accepted as references by gynaecologists and that provided the 'standard' against which dosage and pill content were defined. Second, the Schering-Kaufmann regimen gradually found its way into medical textbooks and its monthly design became the norm for the first-line treatment of amenorrhea.[34]

Kaufmann continued to test Schering's products throughout the 1930s, searching for optimal doses and wider indications. The most significant transformation was the 'molecularization' of menopause, as the management of its adverse side effects increasingly relied on substitution therapy. Kaufmann's list of symptoms ranged from circulation problems, bleeding, headaches and rheumatism, to nervous hypersensitivity, all of them attributed to an insufficient secretion of steroids. By the late 1930s, having grown accustomed to the manipulation of artificial hormones, he considered the administration of estrogens to represent the simplest and most effective way of handling physiological troubles exhibited by ageing women (Kaufmann, 1938).[35]

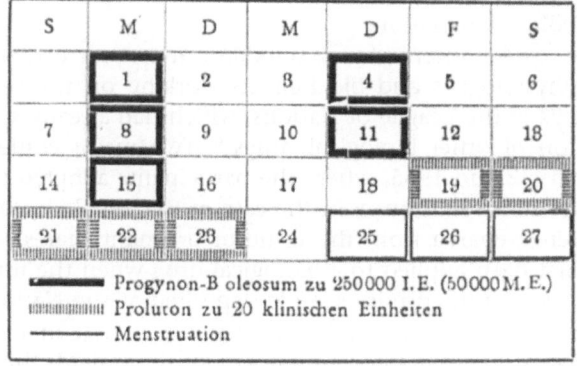

Dosierungsschema nach K a u f m a n n für die Therapie der Amenorrhoe

Figure 10.6 Kaufmann's regimen for normalizing menses

The principal products of this industrially sponsored definition of clinical norms were the brochures and booklets that described the various indications for Schering's sex hormones, provided advice about diagnostics and case selection, defined dosage and recommended treatment protocols. A widely circulated table from 1934, for example, classified the uses of Progynon into three categories: first, the genital domain, that is, the regularization of menstruation or the treatment of various menstrual disorders; second, the question of hormonal regulation, focused on problems associated with pituitary secretion or the side effects of thyroid disorders; third, more general indications including skin and articulation disorders, sleeplessness, menopause, and depression.[36]

The international standardization of the sex hormones

One final symptom of the difference between Schering's physiologically oriented standardization and the practices prevailing at IG Farben resides in the different status the two companies assigned to international standards for the sex hormones. Although the biological assays from the 1930s were made into standards in the sense that they were controlled, calibrated and served to build equivalences between products, comparability and robustness remained local achievements. The mouse, rat or rabbit units they served to produce circulated within the firms and within limited outside spaces. These networks in which the units were used linked each company and its representatives with the local pharmacists and physicians who sold or prescribed a given specialty. Within Germany, inter-laboratory or more precisely inter-company comparison neither emerged as a demand of the state or federal authorities as had been the case for the serum against diphtheria, nor as a demand of elite physicians in teaching hospitals as was the case in the US following the lead of the pharmaceutical council of the American Medical Association.[37]

The campaign for international standardization was actually launched by leading physiologists and biochemists working on the sex hormones under the aegis of the League of Nations, which had already sponsored the standardization of other biological drugs.[38] Two major conferences were organized in 1932 and 1935, where the participants adopted the chemical model of a reference substance; in the case of the female hormone, a pure lot of estrogen prepared from the urine of pregnant mares. This kind of model had first been applied to a biological drug when the unit of insulin was defined using a dried pure sample deposited at the National Institute for Medical Research in London.[39] This official reference material was used in comparative assays to define a number of local standards that operated at the institutional level. Following this experience, Henry Dale explained during the 1935 conference on standardization that the aim of such a gathering was to get rid of all the messiness, uncertainty, and lack of homogeneity associated with the various physiological tests currently in use. Adequate standardization would be one that linked the chemical characterization of the substance with weight determination and physical measurements.[40] As a consequence of the process, the unit of estrogen would no longer be an animal unit, but a weight unit, even if the relationship between weight and potency remained based on biological assays.

Schering and IG Farben reacted in different ways to this 1935 conference. Building on the local chemical culture, IG Farben rapidly adopted the weight units.[41] Moreover, in the case of estrogen production, it reorganized its quality control procedures around the colorimetric assays for steroids then under development.[42] These chemical tests had been discussed during the international conference without being turned into a reference method. Schering also adopted the international unit, but on the basis of an internal equivalence between a locally unchanged measurement of rat and rabbit units and the formal international standard. In other words, biological assays were not replaced by chemical assays and concentration measurements.[43]

Conclusion

Standardization in medicine is often presented as a problem of health professionals' autonomy put under threat by a powerful third party, like managed-care institutions or pharmaceutical firms. Standards are certainly material and social technologies that contribute to disciplining conduct and imposing patterns of action. The complex relationship between the standardization movement led by the companies producing biological drugs and the related developments in laboratories or in the clinic discussed in this paper show, however, that the successful pursuit of alignment and control through biological test systems also played mediating roles, created equivalences, and opened up spaces both for the circulation and for the confrontation of experiences. Beyond this highly general comment, the practices of industrial *Wertbestimmung* associated with the development of the sex steroids deserve a few specific remarks.

First, *Wertbestimmung* was a core element in the development of 'biologicals' across the twentieth century. In contrast to plant and organ extracts, which, while increasingly industrialized, developed out of the pharmacist's traditional culture of preparation, the 'biologicals' were new therapeutic agents isolated from living organisms that were purified – if not molecularized – entities mass-produced in industrial plants. Systems of *Wertbestimmung* developed in association with the making of sera, therapeutic vaccines, vitamins, or hormones thereby share important features beyond the final goal of quantification. First, all these products depended on the use of animal bodies as reagent vessels objectifying the complex biological properties targeted by the agent to be tested. This double meaning of biological assays created tensions between the 'exploratory' and the 'confirmatory' uses of such systems. While the case of Bayer illustrates the control function of biological testing, the multiple dimensions of *Wertbestimmung* at Schering show that the boundaries between research and control were often blurred, and that this porosity could become an industrial asset.

Second, *Wertbestimmung* has a wider significance than being simply an element in the trajectory of biological drugs. It is one manifestation among many of the culture of industrial standardization that nurtured the transformation of *materia medica* into ready-made and mass-marketed pharmaceutical specialties. Industrial standardization focused on technical arrangements that were simple and robust enough to permit: 1) the routine control of raw materials, the homogeneity of products, and therapeutic value or potency; 2) the design and surveillance of new production processes; and 3) the construction of markets and the framing of prescriptions. This industrial standardization did not however operate in isolation. Its relationship to laboratory-based standardization pursued in academic settings has been documented in many ways. It was also closely related to the standardization taking place in the clinic; not least because the manufacturer's definition of quality and measurement of potency aimed at the disciplining of adverse reactions, toxicity, tolerance and efficacy by means of recommended dosages, routes of administration and indications. As our exploration suggests, the expansion of industrial testing not only benefited from laboratory and clinical standardization, but also nurtured these same practices.

A third and final remark concerns the diversity of test systems grounding *Wertbestimmung*. This heterogeneity was certainly rooted in the diversity of products and relevant experimental practices. It was also reinforced by the scientific, organizational and commercial trajectory of pharmaceutical firms. It might be convenient to distinguish the chemical form of *Wertbestimmung* Ehrlich invented for the evaluating the sera against diphtheria, and the physiological form of standardization associated with the measurement of the potency of the sex hormones. It is, however, equally important to distinguish the chemical or, more precisely, pharmacological form of standardization that dominated biological testing at Bayer and the physiological form of standardization that prevailed at Schering. The contrast may be seen as a by-product of each company's early history (Bayer was a chemical firm, Schering a pharmacist's workshop), but it is also a consequence of the specific – and unpredictable – ways in which each firm developed in-house research in the 1920s and 1930s.[44] By taking into account the relationships these firms maintained with collaborating biologists and clinicians, this paper suggests that such pharmacological and physiological forms of standardization were also two different ways of constructing therapeutic value and medical markets. I have summarized these models in Table 10.1 below.

Table 10.1   Two forms of industrial standardization of biological drugs

	Pharmacological standard (Bayer)	Physiological standard (Schering)
Main sites	Chemical laboratory + testing service	Biological research and testing unit
Test system and target	Bioassay, molecular content, dose-effect curve	Animal model, functional pattern
Industrial issue	Composition, dosage, process development	Potency, dosage, process development
Aims and nature of clinical links	Toxicity and side effects, ad-hoc tester network	Toxicity and multiple indications, long-term association with a core-set of clinical researchers
Tools for market construction	Local representatives, *exposés*, articles in the company journal, ads in the medical press	Elite trend-setters, articles in the company journal and in the medical press

Notes

1. Hohlweg, 'Biologisches Arbeiten im Chemischen Laboratorium', *Medizinische Mitteilungen*, 1938, 10 (1): 4–9.
2. J.-P. Gaudillière 'Professional or Industrial Order? Patents, Biological Drugs, and Pharmaceutical Capitalism in Early Twentieth Century Germany', *History and Technology*, 2008, 24: 107–33.
3. J. Parascandola, 'Industrial Research Comes of Age: The American Pharmaceutical Industry, 1920–1940', *Pharmacy in History*, 1985, 27: 19–21; J. Swann *Academic Scientists and the Pharmaceutical Industry: Cooperative Research in the Twentieth Century*, Baltimore: Johns Hopkins University Press, 1988; J. Liebenau, *Medical Science and Medical Industry. The Formation of the American Pharmaceutical Industry*, Baltimore: Johns Hopkins University Press, 1987.
4. See, for example P. Keating and A. Cambrosio, 'Interlaboratory Life: Regulating Flow Cytometry', in Gaudillière and Löwy (eds), *The Invisible Industrialists. Manufactures and the Production of Scientific Knowledge*, London: Macmillan, 1998, pp. 250–95.
5. Gaudillière and Löwy (eds), *The Invisible Industrialists*, 1998.
6. S. Timmermann and M. Berg, *The Gold Standard. The Challenge of Evidence-Based Medicine and Standardization in Health Care*, Philadelphia: Temple University Press, 2003.
7. J.-P. Gaudillière, 'Hormones, régimes d'innovation et stratégies d'entreprise : les exemples de Schering et Bayer', *Entreprise et Histoire*, 2004, 36: 84–102.
8. J.-P. Gaudillière, 'Professional or Industrial Order?', 2008, note 2, p. 109.
9. N. Oudshoorn, *Beyond the Natural Body: An Archeology of the Sex Hormones*, London: Routledge, 1994.
10. J. Lesch, *The First Miracle Drugs: How the Sulfa Drugs Transformed Medicine*, Oxford: Oxford University Press, 2007.

11. J.-P. Gaudillière, 'Better Prepared than Synthesized: A. Butenandt, Schering AG and the Transformation of Sex Steroids into Drugs', *Studies in History and Philosophy of the Biological and the Biomedical Sciences*, 2005, 36: 612–44.
12. J.-P. Gaudillière, 'The Normal and the Artificial: Rückwirkung and Its Bio-industrial Origins', in M. Laublicher and H.-J. Rheinberger (eds), *The Concept of Regulation and the Origins of Theoretical Biology*, in press.
13. W. Wimmer, *Wir haben fast immer etwas Neues. Gesundheitwesen und Innovationen der Pharma-Industrie in Deutschland*, Berlin: Duncker and Humblot, 1994; J.-P. Gaudillière and V. Hess (eds), *Ways of Regulating: Therapeutic Agents between Plants, Hospitals and Consultation Rooms*, Berlin: Max Planck Institute für Wissenchaftsgeschichte, Preprint Series no. 363, 2008.
14. See the contributions to the present volume by Volker Hess, Axel Hüntelmann, and Jonathan Simon.
15. E. Allen and E. A. Doisy, *Physiological Reviews*, 1927, 7: 600.
16. W. Schoeller and W. Gehrke, 'Zur standardiesierung des männlichen Sexualhormons', *Wiener Archiv für innere Medizin*, 1931, 21: 329–36.
17. A. Butenandt and K. Tscherning, 'Über Androsteron, ein krystallisiertes männliches Sexualhormon', *Hoppe-Seyler's Zeitschrift für Physiologische Chemie*, 1934, 229: 167–89.
18. C. Clauberg, 'Der biologische Test für das Corpus Luteum-Hormon', *Klinische Wochenschrift*, 1930, 9: 2004–5. C. Clauberg, 'Die exakten Testierung des spezifischer Hormons des Corpus Luteum', *Klinische Wochenschrift*, 1931, 10: 1949–52.
19. Schering Archiv, Bericht für 1936, Wissenschaftlichen Arbeiten und Untersuchungen, B5/155.
20. L. S. Reich, *The Making of American Industrial Research: Science and Business at GE and Bell*, Cambridge: Cambridge University Press, 1985; D. Hounshell and J. K. Smith, *Science and Corporate Strategy: Dupont R&D*, Cambridge: Cambridge University Press, 1988.
21. Reichspatentamt, Patentschrift 688850, Verfahren zur Gewinnung des Hypophysenvorderlappenhormons aus menschlichen oder tierischen Körperflüssigkeiten unter gleichzeitiger Trennung von vorhandenem Ovarialhormon.
22. Bayer Archiv. Bestand 103/14. Physiologisches Labor. Berichte Weyland, 1928, 1931.
23. Bayer Archiv. Bestand 103/14. Physiologisches Labor. Berichte Weyland, 1929. On the difference between IG Farben and Schering's clinical networks, see J.-P. Gaudillière, 'Hormones, régimes d'innovation et stratégies', 2004, pp. 96–9.
24. Bayer Archiv. Bestand 166/8. Prolan. Pharmazeutsiche Wissenschaftliche Abteilung an IG Farben Pharmazeustiche Buro Berlin, 11. Februar 1929.
25. Bayer Archiv. Bestand 103/14. Physiologisches Labor. Berichte Weyland, 1931.
26. W. Hohlweg, 'Über allgemein-endokrine Wirkung des Progynon', *Medizinische Mitteilungen*, 1930, 5: 124–7.
27. J.-P. Gaudillière, 'The Normal and the Artificial', in press.
28. W. Hohlweg and K. Junkmann, 'Die Hormonal-Nervöse Regulierung der Funktion des Hypophysenvorderlappens', *Klinische Wochenschrift*, 11: 321–3.
29. K. Junkmann Berichte, 1936, pp. 5–7. Schering Archiv, Bestand 05, Ordner 30.
30. K. Junkmann Berichte 1938. Schering Archiv, Bestand 05, Ordner 30.
31. C. Sinding, 'Making the Unit of Insulin: Standards, Clinical Work and Industry, 1920–5', *Bulletin of the History of Medicine*, 2002, 76: 231–70.
32. J.-P. Gaudillière, 'Better Prepared than Synthesized', 2005, p. 632.
33. C. Kaufmann, 'Echte Menstruation bei einer kastrierten Frau durch Zufuhr von Ovarialhormonen', *Zentralblatt für Gynäkologie*, 1933, 57: 42–6.

34. C. Kaufmann, 'Die Behandlung der Amenorrhoe mit hohen Dosen der Ovarialhormone', *Klinische Wochenschrift*, 1933, 12: 1557–62; *Progynon and Prolution Reine Ovarial Preparate*, Schering AG, ca. 1940.
35. C. Kaufmann, 'Die Behandlung der Eierstocksinsuffizienz durch Keimdrüsenhormone', *Archiv der Gynäkologie*, 1938, 166: 113–31.
36. *Die ovarielle Hormontherapie mit Progynon, Progynon B oleosum, und Proluton*, Schering-Kahlbaum AG, ca. 1934.
37. Council on Pharmacy and Chemistry, 'Reports of the Council. Estrogenic Substances: Theelin', *JAMA*, 1933, 100, 1331–8.
38. See the contribution by Pauline Mazumdar to the present volume.
39. C. Sinding, 'Making the Unit of Insulin', 2002, p. 245.
40. H. Dale, 'Note préliminaire', in *Conférence sur la standardization des hormones sexuelles*, Genève: Organisation d'Hygiène de la Société des Nations, 1935, 120–2.
41. Bayer Archiv. Bestand 166/8. Unden. Exposé ca. 1937.
42. G. Pincus, G. Wheeler, G. Young and P. Zahl, 'The Colorimetric Determination of Urinary Estrin', *Journal of Biological Chemistry*, 1936, 116: 253–66.
43. K. Junkmann Berichte 1938. Schering Archiv, Bestand 05, Ordner 30.
44. J.-P. Gaudillière, 'Hormones, régimes d'innovation et stratégies', 2004, pp. 91–2 and 96–7.

11

'We need for digitalis preparations what the state has established for serumtherapy ...' From Collecting Plants to International Standardization: The Case of Strophanthin, 1900–38

Christian Bonah

Introduction

In an address to the Medical Society of Basel on 7 November 1935, the German-Jewish physician Albert Fraenkel (1864–1938)[1] (who had been dismissed from his official position as honorary professor of pharmacology at the University of Heidelberg and director of the medical tuberculosis clinic 'Speyererhof' by the Nazi regime in 1933) proclaimed that 'as has become self-evident for serumtherapy introduced widely in 1893, so should it be for drug therapies in general: before the introduction of a new medication, its action should be evaluated not only on normal, healthy animals or organs but on artificially sickened animals or pathological organs'.[2] By making reference to the evaluation strategies developed for serumtherapy and the 'Ehrlich model' of *Wertbestimmung*, Fraenkel was not offering a superficial analogy or simply paying lip-service to laboratory-based therapeutics. Instead, this orientation was the result of a longstanding commitment to what he himself termed 'the path to rational therapy'.[3]

Starting from Fraenkel's assessment of the situation in the 1930s, and using the example of digitalis therapy for acute or chronic heart failure during the first four decades of the twentieth century, the following contribution treats the following issue: how were the principle and procedure of *Wertbestimmung* initiated by Paul Ehrlich[4] in the context of biological therapeutic agents gradually extended from serumtherapy and vaccines to other fields of therapeutics in the early 1900s?[5] At the centre of the analysis lies a case study of a therapeutic agent – Strophanthin – that cannot be considered either as a 'magic bullet' or as a wonder-drug like penicillin.[6] In other words, the development of digitalis and assimilated substances such as Strophanthin, as discussed in this chapter, does not constitute a landmark in the annals of pharmaceutical invention. These kinds of drugs should instead be considered as valuable yet 'normal' pharmaceutical substances in the history of Western pharmacy. They belong to the classical group of substances first empirically employed as plant extracts in the form of alcohol-based tinctures or water-based infusions. Under the auspices of the rising field of pharmacology that was establishing itself as a basic medical science,[7] they became chemically purified and identified at the end of the nineteenth century, although chemical synthesis remained impossible until after WWII. In the meantime evaluation, standardization and quality control of increasingly industrially produced preparations required procedures for monitoring product quality. In what follows, we will examine the reasons for this application and generalization of *Wertbestimmung* as well as the routes taken to achieve it. Accordingly, the presentation is not a comprehensive micro-history of the development and production of Strophanthin by the physician Albert Fraenkel and the C. F. Boehringer company[8] between 1897 and 1938, but rather draws on significant episodes of the drug's history to illustrate what Fraenkel and Boehringer referred to as the 'physiological *Wertbestimmung'* of the ready-to-use Strophanthin solutions commercialized for the first time in 1906.

Before going any further, however, it seems necessary to specify the uses and implications of the notion of *Wertbestimmung*, here, generally translated with the double sense of evaluation and standardization. Moving from Paul Ehrlich's initial technological notion of *Wertbestimmung*, referring to a set of physiological techniques aimed at testing the effects of ill-defined antitoxins, the concept of standardization is extended to cover a group of pharmaceutical and economic meanings concerning quality control for therapeutic agents produced from plants by a variety of industrial entrepreneurs and used by clinicians for general treatment. Following demands for product uniformity, two approaches were adopted to maintain levels of compatibility and commonality, with, on one hand, stringent control of plant collection[9] and, on the other, control of the end product. Furthermore, standardization should be understood as a never-ending networking process where progressive transformations of the preparations in their production and medical use periodically recreate the necessity of standardization. The developments analyzed here are thus understood as a stabilization process connecting laboratory settings, products, and practices with medical theories and practices, and technical, bureaucratic, and organizational systems.[10] A central feature of the process is the fact that the pragmatic final therapeutic judgement – Strophanthin therapy is safe and works – was only as stable and resistant to criticism as the weakest of the links established in the overall Strophanthin-network implemented to guarantee its truth. In practice, as will be shown in what follows, this implies that Strophanthin therapy could be destabilized not only by fundamental physiological and pharmacological criticisms but also by industrial shortcomings, human failure, the apprehensions of physicians, or the existence of commercial competition. Accordingly, we develop the argument that therapeutic principles and products necessarily possess their inherent medical and scientific properties or qualities, but that these are not sufficient in themselves to establish permanent therapeutic use and success. Rather, they need to be maintained and adapted over time and space within a network of actors and material constraints in order to guarantee the enduring successful use of the product.

This contribution draws on the micro-history of Strophanthin, a 'normal' everyday therapeutic agent of the early twentieth century, to analyze in detail the practical significance of *Wertbestimmung* in the field of general drug therapy. The contribution is based on the analysis of the scientific publications of Albert Fraenkel and his close collaborator Georges Schwartz between 1897 and 1938, the industrial archives of the pharmaceutical company C. F. Boehringer and sons producing what their interwar advertisements present as the 'first injectable Strophanthin in the world' (Figure 11.1),[11] and the private correspondence between A. Fraenkel and G. Schwartz.[12]

In the four sections that follow, this contribution will explore the multiple meanings of *Wertbestimmung* and its place in the process of invention that turned the substance Strophanthin into a fully-fledged system of Strophanthin therapy, transforming an exotic arrow poison into a Western therapeutic agent. By 'Strophanthin therapy' in this context, we mean the rational use of k-Strophanthin as a pharmacologically grounded, industrially produced, laboratory tested and clinically adopted method of intravenous administration of standardized Strophanthin solutions for the treatment of chronic and acute heart failure in humans. The first section will present the initial transfer of *Wertbestimmung* from its original context of serum control to a pharmacological application of the same principles.

In a second section we analyze the relationships between *Wertbestimmung* and the clinical invention of Strophanthin therapy by Albert Fraenkel, a process that takes us up to 1906, the year of the official presentation of injectable Strophanthin to the German medical community at the Twenty-third Congress for Internal Medicine held in Munich from 23 to 26 April. In presenting this history, we will argue that the determination of the substance's pharmacological action combined with the globalization of local economies for digitalis and Strophanthin collection, preparation and distribution created a new context favourable for the centralized evaluation and standardization of the end product as an alternative to the local control of plant raw material. Fraenkel's response, in turn accepted by the industrial producers, was the establishment of an appropriate biological unit of evaluation, the so-called frog-unit.

In a third section, we will argue that the industrialization of Strophanthin production at Boehringer, following the Munich congress of 1906, created new and different requirements for evaluation and standardization. While Boehringer executives were confident that their production process for the

Figure 11.1 Boehringer advertisement from the 1930s for Kombetin and Myokombin

solutions was reliable, considering them to be both safe and uniform, they nevertheless requested Fraenkel's physiological testing to provide supplementary reassurance in the case of a high-potency drug being prepared for intravenous administration.[13]

In the fourth and final section, we will analyze the role of standardization in the marketing and promotion of the new therapeutic agent. The analysis will focus on the constant efforts on the part of both the researcher, Fraenkel, and the commercial producer, Boehringer, to maintain and reinforce the bonds of their Strophanthin therapy network during the period when the therapy and the product became established as a part of routine medical practice in Germany. Despite the acceptance and diffusion of the therapy, they continually confronted disaggregating forces that threatened the place they had established for the product and its use. The story will end with the final attempts by Fraenkel, supported by Boehringer and persecuted by Nazi Germany, to engage in the international promotion of Strophanthin therapy in Great Britain and the US.

Overall, our argument is that the standardization of digitalis preparations into Strophanthin therapy involved various practices of *Wertbestimmung*, mobilizing divergent and separate social worlds of laboratory scientists, clinical physicians and industrial entrepreneurs. Collaborations, practices and interactions did not extend in any linear form from the laboratory to industry and on to the clinic; they were multifocal and multidirectional, thus challenging a classic linear model of innovation.

From serological to pharmacological *Wertbestimmung*: Evaluating digitalis preparation strength, 1897–1902

Ever since William Withering (1785), treatment of heart weakness (cardiac failure) has relied heavily upon the use of extracts of the relatively common plant foxglove (or digitalis). His influential treatise *An Account of the Foxglove and Some of Its Medical Uses*[14] was based on empirical clinical studies and established not only dosage regimens of the drug but also a clinical parameter, pulse rate, as an indicator for assessing its efficacy.

Around 1800, the first seeds of a hitherto unknown African plant named Strophanthus were introduced to Europe by the French botanist Pyramus de Candolle. Sixty years later, the Livingstone expedition to the Zambezi described *Strophanthus kombi* (or k-Strophanthin), which was used as an arrow poison, and, in 1865, William Sharpey identified the action of the toxin on the heart. Five years later, Thomas R. Fraser isolated a pure chemical substance, a glycoside, as the active principle of the arrow poison, and in the 1880s Oswald Schmiedeberg (1838–1921) established the pharmacological family of digitalis and digitalis-like glycosides, including common digitalis and the Strophanthin plant preparations. During the 1870s, Strophanthin became a standard preparation for studying the pharmacological and physiological action of the digitalis group on the hearts of frogs and higher mammals. Finally, an essential addition to these physiological studies of heart function was the integration of systolic, diastolic and vagal effects in the 1920s. This was followed by Starling's numerical characterization of the circulation of the blood in terms of stroke volume and cardiac output per minute, known as minute-volume.

Albert Fraenkel's medical education during the 1880s coincided with the hey-day of physiological and pharmacological investigations of the cardiac system. Born in 1864 in Mussbach (the Palatinate, administratively attached to Bavaria) to a German Jewish family of wine merchants, he entered the medical faculty in Munich at the age of 19, where he studied basic medical sciences (*Physikum*). After six months of military service in Würzburg, he registered for his clinical medical education at the then German university of Strasbourg, a period that fell within the golden decade of the Strasbourg medical faculty. Leading German scientists like Wilhelm von Waldeyer in Anatomy, Friedrich von Recklinghausen in pathology, Friedrich Goltz in physiology and especially Felix Hoppe-Seyler in physiological chemistry and Oswald Schmiedeberg in pharmacology had just received brand new imperial institutes and were forging the university's reputation as a leading European research centre (*Arbeitsuniversität*).[15] In 1888, Fraenkel took his qualifying exam (Staatsexamen) and left Strasbourg for his second military service in Bayreuth. After contracting tuberculosis in 1889 and subsequent air and sun cures he registered for complementary medical courses in Zurich. In 1890 he moved to Berlin following Robert Koch's announcement of his 'revolutionary' treatment of tuberculosis with Tuberculin.[16] Disappointed by Tuberculin's ineffectiveness, Fraenkel left Berlin to set himself up as a general practitioner in Badenweiler, where he also became the medical director of two local sanatoria.[17] Starting in 1893, Fraenkel undertook experimental work on digitalis at the pharmacological institute of the University of Heidelberg, first under Woldemar Schröder and later under Rudolf Gottlieb. From the early 1890s, Fraenkel's summer months were spent doing clinical practice at the sanatoria, while his winters were dedicated to his work at the pharmacological laboratory in Heidelberg.

Albert Fraenkel first became acquainted with Strophanthin in the pharmacological laboratory of his teacher, Oswald Schmiedeberg, in Strasbourg. Following Frazer's isolation of its active glycoside in 1870, Strophanthin had become an indispensable tool for laboratory experiments on isolated heart function, and after learning how to use the substance from his mentor Carl Ludwig, Schmiedeberg adopted it himself for physiological tests and teaching. The physical and chemical qualities of the substance, particularly its solubility in water, made it practical to administer it intravenously, which allowed the precise quantification of amounts delivered to test animals and guaranteed a minimum delay between administration and the onset of the poison's action. For two generations, Strophanthin served as a privileged standard preparation for cardiac investigation by physiologists. It gave rise to a characteristic digitalis effect featuring shorter periods for the systole (heart muscle contraction) and diastole (distension), increased systolic force of the heart contraction, and a deepening of diastolic distension.

Upon his return to the pharmacological laboratory in Heidelberg in 1893, Fraenkel profited from his own situation between the laboratory and the clinic to try and combine pharmacology as a science, clinical investigation and treatment. In this quest to develop 'rational therapy' Fraenkel approached the question of digitalis therapy from two sides, the pharmacological laboratory and the therapeutics of circulatory failure. His published articles started on the laboratory side. An initial 'tonographic' investigation of digitalis action published in 1897[18] analyzed the modifications that physiological poisons made to blood pressure. In this context 'tonographic recordings' were images produced by recently improved sphygmographs (*Blutwellenzeichnern*), a physiological apparatus that registered and inscribed curves denoting the 'pressure changes in the arterial tubes'. Comparison of curves obtained in animal experiments indicated on the one hand that different animal species displayed specific curves and on the other hand that it appeared possible to distinguish the action of individual poisons on the shape and duration of curve segments. Recorded pressure changes represented the combination of two factors, central heart activity and peripheral blood vessel tonus. In his investigations, Fraenkel established that 'tonograms' could be of significant value for pharmacological studies since they allowed the researcher to differentiate between central and peripheral causes of changes in blood pressure. In the case of digitalis testing he ascribed the elevation of blood pressure to the poison's central action on the frog's heart. Overall, this first publication indicates an initial line of inquiry in which Fraenkel aimed to extend physiological experimental recording techniques to clinical observation, thereby allowing the quantification of clinical evaluation and judgement.

In 1902, Fraenkel narrowed the gap between experiment and treatment by approaching the dichotomy from the opposite end. Clinically, heart failure was a rather common condition whose symptoms were respiratory difficulties combined with a sensation of suffocation, oedema and loss of strength. Treatment relied on plant preparations of the common foxglove or the exotic Strophanthus, which displayed similar cardiac activity. Oral administration used water-based infusions or alcohol-based tinctures. Adopting the perspective of practicing physicians, Fraenkel oriented his second investigation towards the question of the 'physiological dosing of digitalis preparations'.[19] Indeed, he started his publication on this research with the remark that practitioners knew that traditional plant-based preparations possessed quite varied levels of efficacy depending on their origins. Furthermore, the changing content of the active principles in the plant raw materials used to make the products also contributed to variations of potency in the final product. Even for the common foxglove, the 'year, time of year and localization of the plant collected as well as its age and its preparation are factors that determine the quantity of active principle contained in a given infusion or tincture'.[20] Apothecaries with a reputation for particularly high-quality preparations paid specific attention to these factors when purchasing their raw materials, as chemical analysis of the proportion of digitoxin, one of the digitalis glycosides, showed that it varied from 0.26 to 0.62 per cent for digitalis leaves collected in the Black Forest.[21]

The central argument presented by Fraenkel in his paper was that neither screening and control of the conditions of plant collection nor 'chemical analysis alone are capable of responding to the issue of the real pharmacological activity and value of galenical digitalis preparations, since they are usually composed of several different glycosides, that, although they can be separated chemically cannot be determined quantitatively in parallel'. Accordingly, Fraenkel proposed that 'a better and much easier measure for therapeutic efficacy of digitalis preparations is a physiological assay. The same thing is true for officinal Strophanthus preparations'.[22] Echoing an assessment made by his mentor from the Heidelberg pharmacological institute, R. Gottlieb, at the 1901 Congress for Internal medicine, Fraenkel considered that 'Digitalis preparations distributed by pharmacies should be controlled for their pharmacological action as is done for serum'.[23]

Adapting serum control methods (*Wertbestimmung*) to poisonous plants required the identification of a constant and characteristic symptom that occurs regularly for a given dosage within a well defined time frame. Systolic cardiac arrest of the frog heart seemed to be the appropriate measure for evaluating digitalis. For this process, the frog as a biological experimental system needed to be defined as strictly as possible, meaning the specification of the age, sex, and variety of frog to be used. The size of the frogs was also standardized by referring drug activity to 100 g of frog. Furthermore, Fraenkel took into consideration the issue of whether toxic activity on the frog heart allows extrapolation for therapeutic activity in humans. Variation of animal test objects ranging from guinea pigs to cats showed strict analogies in the drug's activity leading to the conclusion that 'systolic cardiac arrest of the frog heart is as reliable a test object for digitalis, as the onset or absence of toxicity in a guinea pig for the evaluation of toxins and antitoxins'.[24]

Fraenkel carried out several experimental test series with digitoxin, a chemically pure and particularly valuable digitalis product, at least according to physicians. For these tests, Merck provided Fraenkel with an initial stable pharmacological assay against which various other preparations could be compared. At the same time, the experimental series confirmed his conviction that 'the frog heart provides for the evaluation (*Werthigkeitsbestimmung*) of digitalis preparations everything that can be desired of a pharmacological test object'.[25] Applying his method, Fraenkel detected considerable variations in strength between the various preparations, as he had predicted. Digitalis infusions gathered from all over the district of Baden varied from 100 to 275 per cent, digitalis tinctures from 100 to 400 per cent and for the locally produced but originally exotic Strophanthus preparations the variation was from 100 to 6000.

The practical conclusion from these experiments was that there was considerable uncertainty in therapy as one could never really know how much of the active principle was entering the patient. Low-strength preparations would often provide insufficient quantities of active principles diminishing the value of the treatment. The virtual absence of the active principle could also lead physicians to conclude that the treatment was ineffective, when the problem was really one of insufficient dose. High-strength preparations, on the other hand, particularly in the case of the more active Strophanthus preparations, could lead to overdosing, potentially threatening patients' lives. Or it meant overestimation of the risk of the therapeutic agent itself. The concluding lines of the paper promote the idea that

> The preceding experiments show that it would be rather easy to establish a procedure for testing the physiological activity first of conservable digitalis and Strophanthus tinctures, and then of freshly prepared infusions in order to standardize them. Possibly the recognition of these principles could incite the chemical industry to put galenical preparations with determined values of potency on the market. In a different direction it seems advisable that the state should also create an institution, like it has already endorsed an institute for the scientific evaluation of serum, giving physicians the possibility of testing digitalis preparations, employed in practice, experimentally on whole animals to establish their value and activity.[26]

This recommendation is complemented by the remark that sera only presented the risk of contamination or inefficacy and yet they were subject to regulation, while digitalis preparations carried the additional and considerable risk of overdosing, and yet could be marketed without any control. Thus, Fraenkel's 1902 investigation provided the justification for a campaign to define what would, in 1906, become the so-called frog unit, meaning the dosage of a given product that produced systolic cardiac arrest one hour after administration, per 100 g of frog weight. The frog unit was, therefore, a biological measure or a pharmaco-physiological assay that could serve as a standard for the comparative evaluation of various digitalis preparations and physiologically similar products in the contexts of both clinical use and pharmaceutical production.

Fraenkel considered pharmacodynamic standardization to be the appropriate option, not because he was fundamentally opposed to the analytical determination of the presence of chemical species as a means of standardization but because the preparations were rarely pure substances and because the presence of several digitalis glycosides made such chemical determination impossible at that time. The strong dependence of this physiological or pharmacological approach on the model of Ehrlich's *Wertbestimmung* for serum becomes strikingly clear through both the direct references made by the actors and through the similarity of the procedures. Indeed, it is worthwhile making a direct comparison. Standardization (*Wertbestimmung*), as a historical phenomenon that occurred at a particular moment in the history of drug development, was initially linked to a problem of underdetermination and to a widely recognized lack of knowledge about the substances concerned. In their 1916 handbook, *'Serums, Vaccines and Toxins in Treatment and Diagnosis'*, Cecil Bosanquet and John Eyre had this to say about the basis for the 'standardization of serums':

> Since it has hitherto been found impossible to isolate the actual toxins of bacteria, so that no process of weighing or measuring can be applied to them, it is necessary to devise some other way of calculating their strength. A physiological test of some sort is the only available means of measuring their effects. For this purpose it is necessary to find some animal which reacts in a constant manner to the poison, dying regularly within a certain time as the result of a given dose of toxin.[27]

For sera, the standard dose of a poison, or 'minimal lethal dose', was the quantity of a particular product that would invariably cause the death of a guinea pig of a certain weight on the fourth day. A 'unit' of antitoxin could then be fixed as the amount required to neutralize exactly 100 minimal lethal doses of a given poison. This method of standardization was the one inaugurated by Ehrlich.[28] Thus, a physiological test was necessary in the first instance to establish a toxic unit, and then to calculate the antitoxic unit based on the toxic one. Finally, one specific antitoxin would become 'a standard' for comparative evaluation of other preparations. Similarly, for digitalis, a high value digitoxin preparation served to establish the frog unit which then became a standard for the evaluation of other digitalis preparations.

It can be argued that it was the progressive extension and internationalization of the drug market that prompted the *Wertbestimmung* of digitalis and digitalis-like preparations. As plant collection evolved from local and individual practices by pharmacists within a given location to national schemes for the purchasing and pooling of raw materials from various origins and locations, control over the quality of collected plants became increasingly difficult. This was even more so in the case of the exotic Strophanthus plant seeds imported from overseas at the end of the nineteenth century. Chemical testing to check the quality of raw materials is an industrial response to the growing globalization of collection and the corresponding disappearance or at least dissipation of local quality control and traceability. Only a standardized evaluation of the increasing number of preparations could ensure that clinical evaluation and use responded to the demands of a 'rational therapy', which presupposed that identical products would cause identical effects. *Wertbestimmung*, or the use of a biological assay to characterize and evaluate a product was a response to the burgeoning and at the same time dissipation of the pharmaceutical market and the limits of purely chemical evaluation for policing this new market. In this atmosphere of 'rational therapy', *Wertbestimmung* was not only applied to the products themselves, it was also mobilized to bring a number of clinical imponderables under control, notably drug absorption and distribution and the possible variation of these factors within the diseased human body.

Wertbestimmung for clinical use: Inventing Strophanthin-therapy 1902–6

In the three years that followed, Fraenkel returned to his experimental studies, further characterizing the action of digitalis on blood pressure,[29] and investigating the cumulative effect when digitoxin was combined with Strophanthin (Figure 11.2a).[30] He became increasingly interested in Strophanthus, even though the commercially available preparations displayed the highest variability among the range of digitalis-like products, and were thus subject to the greatest uncertainties when used in the clinic.[31]

Working with this poison in the laboratory, Fraenkel demonstrated that it had a higher level of pharmacological activity than digitalin or digitoxin. Furthermore, the substance had a quicker yet long-lasting action (initially described as a cumulative effect, although this concept was reinterpreted differently in the 1920s. See Figure 11.2a). These experimental pharmacological explorations depended on an initial physiological assay of the products to determine their strength. The application of this method also served to convince Fraenkel and his new collaborator Georges Schwartz that k-Strophanthin, commercially available from the C. F. Boehringer company, was a particularly stable and pure preparation. In 1905, an unforeseen development finally determined the clinical choice of Boehringer's k-Strophanthin. Two years earlier, one of Bernhard Naunyn's assistants, Kurt Kottmann, dared for the first time to extend the experimental laboratory method of intravenous administration of Digalen to patients at the Strasbourg University clinic.[32] Fraenkel considered that the water soluble and chemically pure substance Strophanthin might be more appropriate for intravenous administration than the oily Digalen preparations and approached the new director of the Strasbourg University clinic, Ludolf von Krehl, asking for permission to test intravenous Strophanthin administration in his clinic. Having received von Krehl's authorization, Fraenkel conducted the first human experiments in the winter of 1905–6 on 25 patients.

  In this context, the mode of intravenous administration redefined the justification and the practice of *Wertbestimmung*. Strophanthin, characterized by its high potency, could only be used in very small dosages. Accordingly, it was the therapist's responsibility to ensure that the dosing of the administered substance fell within a therapeutic range guaranteeing optimal efficacy without inducing unwanted side effects, especially vomiting, a common result of digitalis overdosing. The therapeutic solution to this kind of problem was the administration of the pure active principle in precise quantities adjusted in proportion to the fraction actually absorbed and excreted. In practice, this implied first establishing with precision the strength of a given preparation, then determining an exact therapeutic range for patient administration, before quantifying how much of the administered substance reached its site of action (heart and the blood vessels) within the human body. This third step demanded the quantitative determination of absorption, concentration, accumulation and excretion of the substance in the human body.

In Fraenkel's view, intravenous administration allowed the clinician to overcome the problem of variable absorption. According to him, the oral administration of even a pure, accurately assayed preparation did not allow a precise determination of the administered dose since 'the way from the stomach to the heart is a long one, and the loss of active principle on this pathway and therefore the duration of its effects are incalculable'.[33] Thus, in 1905, intravenous administration was a clinical means to deliver precisely determined quantities of active principle to the drug's site of action, and not a response to issues of therapeutic urgency. Indeed, the advantages of intravenous delivery in terms of the rapid alleviation of symptoms would only later become a significant argument in favour of Strophanthin therapy. Intravenous injection thus can be considered as the technical solution to the biological problem of variability of absorption and excretion when a drug is given orally. From this perspective, intravenous injection completes the biological standardization of the product by standardizing the administration procedure itself. The meaning of *Wertbestimmung* is transformed and extended here to what can be designated as clinical standardization. In this transformation, laboratory techniques are directly transferred into the clinic, thereby exacerbating the characteristic of the high potency of even small dosages. Even very minor changes in the concentration or strength of a preparation could lead to unwanted side effects and overdosing when the drug was given intravenously. The clinical standardization of administration in this sense thereby supplied further impetus to efforts to standardize and control the product as much as possible.

Not only did intravenous administration heighten the need for appropriate physiological *Wertbestimmung*, it also prompted new *Wertbestimmung* parameters. Now solutions needed to be sterile, as contamination could result in attacks of fever or even septicaemia. Of course, possible contamination by microbes was also a parameter in the *Wertbestimmung* for serum preparations.

Last but not least, it was under these circumstances that Fraenkel and Schwartz realized that clinical parameters registered by graphic recordings (Figures 11.2b, 11.2c) could not be interpreted by reference to the drug's action on the isolated heart alone, but that they reflected the state of the whole circulatory system including complex compensatory mechanisms that integrated increased heart contractility with opposing vagal reactions, which could slow down the pulse rate and thereby keep the blood pressure more stable. It was this more comprehensive understanding of the complexities of the circulatory system that led Fraenkel to the recognition in the passage cited in the introduction that: 'as has become self-evident for serumtherapy (...) so should it be for drug therapies in general: before the introduction of a new medication, its action should be evaluated not only on normal, healthy animals or organs but on artificially sickened animals or pathological organs'.[34]

By this move, the initial signification of *Wertbestimmung* as a biological assay that could make up for the deficiency of control over plant collection and the impracticability of chemical analysis was translated into an autonomous concept: pharmacodynamic *Wertbestimmung*, taking into account the complexity of the animal and human body as integrated organ systems. Thus, on his path to 'rational drug therapy' Albert Fraenkel adapted and transformed the initial concept of serum *Wertbestimmung* to its application in drug therapy by not only drawing upon an experimentalization of the clinic but also by recognizing the body as a complex integrative organ system.

(a)

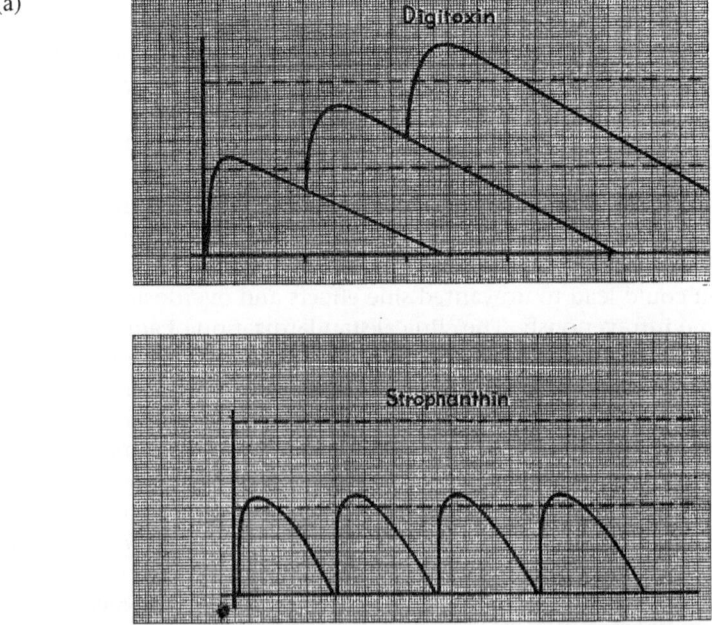

Figures 11.2a, 11.2b, 11.2c Clinical evaluation of Strophanthin

(b)

(c)

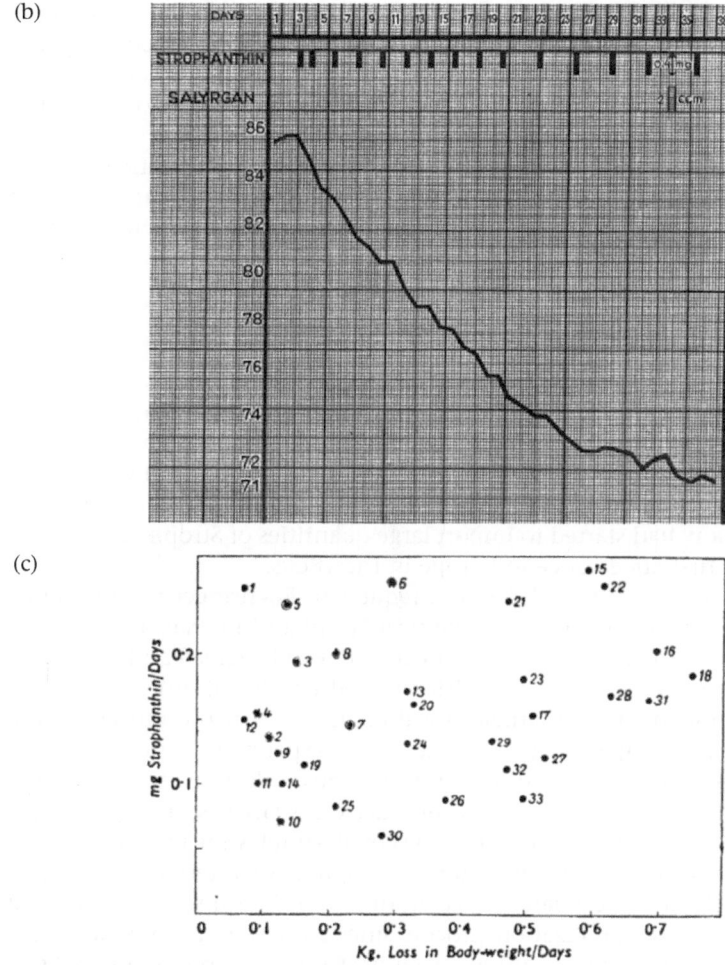

Figures 11.2a, 11.2b, 11.2c   (Continued)

In this process, 'rational therapy' demanded innovations in the laboratory approach to *Wertbestimmung*. While the methods relied on knowledge about the physiological and pharmacological action of the drug on whole animals (*Ganztiere*) as in the case of serotherapy, this needed to be complemented by specific information on the effects of the drug in the pathological context, which took into account the integrative function of the circulatory system and its effect on the drug's action. In the clinical context, the approach implied that variations in preparations needed to be measured using stand-ardized biological physiological units of activity – *Wirkungseinheiten* – and that intravenous administration of the drugs should be adopted in the clinic to reduce variations in absorption. Both of these initiatives – laboratory and clinical – contributed to the idea that doctors knew precisely what quantities of the active principle were arriving at its site of action. Thus, we can see that Strophanthin therapy was both based on and contributed to the invention of a pharmacodynamic evaluation, with the 'frog-unit' as its specific standard. Thus, Fraenkel succeeded in going beyond the initial sense of *Wertbestimmung* as a biological assay method for comparing products, adding specific forms of clinical and pharmacodynamic evaluation to the concept.

Wertbestimmung as an industrial principle for quality control, 1906–38

The initial intravenous solutions used for the first clinical trials were pro-duced locally by Fraenkel and Schwartz by diluting and sterilizing oral pharmaceutical preparations from C. F. Boehringer and Sons in Germany.[35] C. F. Boehringer had a great deal of experience with the exotic plant seeds, as the company had started to import large quantities of Strophanthus since the poison's first appearance in Europe in the 1860s.

On 11 March 1906, Fraenkel sent a request to Boehringer for a personal visit in order to present his recent clinical Strophanthin trials to the direc-tors of the company. The pharmaceutical firm readily agreed and three days later the physician paid his first visit to Friedrich Engelhorn, the director of the Mannheim-based pharmaceutical company.[36] In the process of the discussion, the company agreed to start the experimental production of phials of sterilized Strophanthin IV (intra-venous) solutions, and to test the feasibility of manufacturing such a product on a larger scale. Meanwhile, they would provide Fraenkel with provisional samples for his presentation of Strophanthin therapy at the Munich Congress for Internal Medicine to be held barely six weeks later (23–6 April 1906). The discussions focused particularly on the sterilization process since heat was the only available means and Strophanthin solutions were heat sensitive, presenting risks of alteration. Furthermore, normal glass phials released an alkali with which the Strophanthus preparations readily reacted. Accordingly, Fraenkel and Boehringer agreed at the end of the meeting that the company would pro-duce 30 phials of a 1/1000 dilution of regular oral Strophanthin solutions. This corresponded to the dosage Fraenkel had employed in the clinical tri-als. Ten phials were heated for one minute, ten others for three minutes, and the last ten for five minutes. Using half the phials, Fraenkel would then test for the physiological effect of the different samples, and Boehringer for the absence of germs. The other half were to be stored in order to re-evaluate solutions after some weeks of conservation.

On 26 March, barely two weeks after the initial meeting, Boehringer sent Fraenkel the three sets of phials (*Tuben*) containing heated Strophanthin solutions. The samples were accompanied by the following request 'We would like to ask you to test the content of our shipment [the tubes] for their physiological potency (*physiologische Wirksamkeit*). We will conduct bacteriological analysis of the three sets of tubes. Furthermore we will set aside a part of the solutions in order to examine them some time hence in order to assess whether the solutions can be conserved without chemical alteration. We would like to ask you to proceed in a similar fashion and to test the physiological effect once again some time from now'.[37]

Two days later, Fraenkel acknowledged receipt of the shipment, indicating that he would proceed with the 'physiological evaluation' (*physiologische Wertbestimmung*) immediately.[38] The next day, Fraenkel was prompt to report the results of his analysis, indicating that the tubes sterilized for three minutes showed significant loss of physiological effect whereas the heat sterilization for one minute apparently did not affect the 'physiological strength' of the product. Fraenkel triumphantly added: 'It might be of interest to you to discover through this example the value of *Wertbestimmung* for digitalis preparations and I will send you the test protocol for the evaluation'.[39]

Things could not have worked out better for Fraenkel; Boehringer was interested in manufacturing the IV product even though some scepticism remained concerning its economic viability. Using Fraenkel's pharmacological principles, industrial test products could be evaluated for purity, absence of bacteria and physiological effectiveness. Finally, the physiological evaluation had immediately proved its value by indicating that the sterilization process affected the physiological action for the product in the tubes that had been sterilized for three minutes, while the one minute sterilization had not changed its potency. Without *Wertbestimmung*, the three-minute tubes would have been distributed and found ineffective in clinical use. Ignorance of the product's degradation would have led to a negative assessment of Strophanthin-therapy as such, a setback avoided thanks to the initial evaluation of the product's potency using physiological *Wertbestimmung*. During the early phase of the production of injectable Strophanthin, Boehringer also needed to determine whether the sterilization of the solution affected potency, and whether time (conservation) could influence the quality of the product. Neither control of the plant raw material nor chemical analysis could provide what physiological *Wertbestimmung* could: a global evaluation of the efficacy of the end product. This aim was achieved through the prior standardization of a biological evaluation system, in the case of Strophanthin, one based on Fraenkel's frog-unit.

Once the physiological efficacy had been confirmed for the product in the one-minute tubes, Boehringer went on to confirm that heat sterilization for one minute was sufficient to guarantee the absence of any bacteria. When Fraenkel invited the company to make a presentation of packages of the phials at the Munich Congress, however, the company declined, arguing that the expense was out of proportion to their expected profits. Nevertheless the company did provide Fraenkel with 25 phials that he could show to his colleagues. Boehringer's predictions proved correct, as demand for the new product was slow in coming. In August 1906, the company needed to produce a new batch of Strophanthin phials and requested Fraenkel's services for performing the physiological tests. The tests were satisfactory, and Boehringer released the new solutions, taking the opportunity to propose a formal collaboration with Fraenkel for their regular physiological control. Thus, *Wertbestimmung* gained a third sense as a form of physiological quality control in a technological system of mass production. Boehringer's rationale behind the agreement appears clearly in a letter written to Fraenkel by its director Friedrich Engelhorn: 'Even if we think that we are sure that our production method always produces a chemically and physiologically identical product, it is nevertheless reassuring in the case of a medicament like Strophanthin to control its strength (*Wertigkeit*) from batch to batch'.[40] A financial agreement to remunerate Fraenkel's expertise opened an intense and close collaboration that lasted for 32 years.

By early 1907, the injectable Strophanthin was regularly available either from Boehringer or through its subsidiary Kade in Berlin, who filled and sealed the phials. In May, Professor Starck informed Boehringer that he was withdrawing his authorization for the company to reproduce his publication, which was favourable to Strophanthin, as marketing material for physicians. Starck's decision was motivated by the fact that recent injections of Strophanthin had consistently induced 'unpleasant side-effects'.[41] Boehringer requested samples of the purchased Strophanthin phials and sent them to Fraenkel for physiological testing, while they themselves analyzed the product for the presence of bacteria. Bacterial controls of the sterility of the solutions hinted at insufficient sterilization by their subsidiary Kade. Insufficient sterilization during the industrial production process could thus threaten or at least hamper the clinical acceptance of Strophanthin therapy as such. Fraenkel repeatedly emphasized that the final efficacy and safety of the product could only be assured if the whole production and distribution process respected strict standards of production and quality control. In the industrial context, therefore, *Wertbestimmung* becomes a permanent, everyday process involving constant surveillance aimed at eliminating all deviations from the established standards.

In July 1908, Fraenkel warned Boehringer that a new batch of solutions that he had tested showed a marked increase in pharmacological action, thus creating a risk of overdosing.[42] He also pressed Boehringer for information concerning the precise nature of the plant raw material they used to make their k-Strophanthin. As early as 1904 chemical analysis had identified different types of Strophanthin originating from at least three plant subspecies *Strophanthus kombé* (k-Strophanthin), *Strophanthus gratus* (g-Strophanthin or ouabain) and *Strophanthus hispidus* (h-Strophanthin). Fraenkel conjectured that one reason for the increased strength he had observed might be that the raw material for these batches included different plant seeds, or some other impurities. Thus Fraenkel asked whether Boehringer might want to use other raw materials like *Strophanthin gratus*, as was the case for Strophanthin Thoms, produced by a competitor.[43] The company was not interested in this proposition since it considered that the crystalline g-Strophanthin displayed quite different chemical and physical qualities from their k-Strophanthin. Boehringer was also reluctant to reveal industrial production secrets even to their close collaborator Fraenkel,[44] and they instead insisted on his repeating the physiological testing. The resulting tests confirmed the initial results, leading the company to change its strategy. They decided that controlling the raw material from Africa was an impossible task and that physiological *Wertbestimmung* could serve as an alternative means of control, providing a basis for adjusting the concentration of drug preparations to give the desired levels of activity.[45] This relocation of *Wertbestimmung* changes the sense of the biological assay once again. The idea was to integrate it into the production process as an 'upstream' quality control mechanism, thereby situating it as a response to the impossibility of 'disciplining' the tropics and the system of plant collection in place (Table 11.1).

The irony of this episode is that a month later, Fraenkel revealed that when the same solutions were assayed once again they had now become 'normal'. His explanation of this unexpected result was that the reported variation was probably due to the fact that 'the frogs used in July were particularly sensitive (*besonders empfindlich*)'. The initial conclusion of increased pharmacological activity was thereby ascribed to variation in Fraenkel's biological experimental system and its scale, and not to the product at all. This episode demonstrates that the circular reference of the 'biological scale' always includes the possibility that it is the assays and not the solution under test that vary.

Table 11.1 Steps of Strophanthin evaluation in the stabilization process of Strophanthin-therapy between 1906 and 1930

Raw material		Industrial preparation		Quality testing	
	==>		==>		
					==> Increased action
–Poor raw materials –Variability of raw material	<==	What is the basis of k-S (industrial secret)	<==	Repeat and test quality testing	↓
↓		↓		↓	↓
Control of collection impossible		Change plant No: similar problems		Limits of *Wertbestimmung*	Clinical disputes safety/efficacy

The outcome of this change in analysis was that all the former doubts and fears concerning the k-Strophanthin were dismissed in light of the normalization of the physiological tests, and sales continued as before. Nevertheless in 1908 and the years that followed, the company informed Fraenkel that the sales of roughly 1000 boxes of Strophanthin solution had yet to cover the company's marketing expenses, which included the reprinting and distribution of favourable medical journal articles as well as samples sent to physicians. This leads us to suppose that this new form of injectable Strophanthin was greeted with hesitation in the context of clinical use, and also offers a useful historical counter-example to the typical story of drug development, where drug innovation is seen as automatically being a highly profitable enterprise.

The final illustration of the practical functioning of standardization concerns the actual clinical use of Strophanthin solutions. In 1914, Fraenkel informed Boehringer about frequent complications encountered with the treatment due to overdosing by physicians. He proposed a pragmatic solution to remedy the all-too-human shortcomings of individual physicians who were employing the drug incorrectly. The idea was to rein in overenthusiastic practitioners by reducing the content of the phials by half, from 1mg to 0.5mg. The new presentation of the solutions would oblige physicians to use lower dosages of the drug for injection as they would hesitate before opening two phials at a time. Boehringer replied that the company considered the present format with maximal doses of 1mg in each phial as appropriate since it 'should be no problem for physicians to draw only half of the contents of the phial into a syringe'.[46] In opposition to Fraenkel, the company was afraid that the need for multiple manipulations might discourage physicians who wanted to employ higher doses from using the Mannheim product.

The conflict between this commercial logic and medical precaution was resolved by the addition of a package insert consisting of an information leaflet for physicians, underlining the recommended dosage regimens and proper injection technique (Figure 11.3).[47] In 1916, further complaints by other physicians convinced Boehringer to modify the presentation of the product but this time it was Fraenkel's turn to reject this approach, as he considered that this technical fix would probably not significantly modify physicians' behaviour. It was only in 1925 that Boehringer and Fraenkel finally agreed to reduce the content of phials to 0.5 mg of Strophanthin, hoping that this change would help to standardize human clinical behaviour. Thus, in the end, the packaging of the drug became a means to discipline human behaviour in clinical practice.

There is another important element in the transformation of the by now standardized product of k-Strophanthin into what has been designated above as Strophanthin therapy, and that is the clinical adoption of intravenous administration as a routine medical technique. During the first decade of the twentieth century, intravenous injection was still considered an experimental technique; well integrated into the laboratory, it was not yet part of routine clinical practice.[48] Physicians' reluctance to adopt the technique (which was not necessarily due to any principled opposition) was an important obstacle to putting the system of Strophanthin therapy into place. The reluctance to give intravenous injections could be due to global apprehensions concerning a technique considered to be experimental when used with human beings, or to concern about purely technical difficulties created by the rarity of the necessary technical devices such as needles, syringes and sterilization apparatus in routine medical settings. The principal source of anxiety was probably the physicians' lack of training and skill as they had never learned how to inject substances intravenously. These problems explain in part the reactions of the medical community, which Fraenkel and some of his contemporaries called 'injection incapacity' (*Spritzinsuffizienz*) or 'vein-horror' (*Venenhorror*). These terms point to a vicious circle, in which an untrained physician misses a vein in his attempt to inject the patient and then has to repeat the inoculation procedure on a suffering and unsatisfied patient. The result is that both patient and physician become psychologically resistant to the procedure and seek to avoid it in the future.[49]

Retrospectively, Fraenkel considered that despite the numerous publications following the initial presentation of the method in 1906, Strophanthin therapy faced enormous difficulties in becoming a routine medical practice. He considered that the diffusion of a given therapy was always the outcome of a complex mixture of 'objective and personal elements',[50] but concluded that the initial reluctance to adopt Strophanthin therapy was intimately linked with physicians' fears of the intravenous technique, a problem that only time would overcome. In order to avoid practical difficulties such as the poor positioning or insufficient clamping of the arm to be injected, improper syringes for obtaining the 'venous contact' (Venengefühl) or insufficiently sharpened canulae,[51] Boehringer and Fraenkel decided to add a second insert to packages presenting the proper materials for injection as well as repeating the correct procedures to adopt.

These largely theoretical explanations were complemented by consulting tours (*Konsultationsreisen*) where Fraenkel would travel to physicians in order to teach them how to inject properly. In 1927, these tours gave rise to the creation of training courses for Strophanthin therapy run by Fraenkel at the recently built Speyererhof sanatorium. These training courses and information leaflets formed the central elements in the ongoing marketing efforts of Fraenkel and Boehringer to maintain and reinforce the clinical bonds of their Strophanthin therapy network, with the aim of establishing the therapy and the product as routine medical practice in Germany. After changing the presentation of the product, the company tackled the training and attitudes of the physicians themselves. In this process, standardization and conditioning on the one hand, and marketing and promotion of the new therapeutic agent on the other were always bound up together.

Given the route for administration of the product, it was inevitable that injectable Strophanthin would be promoted as a physician-only drug. In

Kombetin (k-Strophanthin Boehringer)
nach Prof. Dr. Alb. Fraenkel-Heidelberg

Injektionstechnik.

Um eine Reizwirkung auf das Unterhautzellgewebe zu vermeiden, ist eine einwandfreie intravenöse Injektion erforderlich. Vorbedingung hierfür ist eine ideal scharfe Nadel und eine brauchbare Vene. Im allgemeinen wird die Injektion in die Vena medicna cubiti gemacht. Der linke Arm des Kranken wird von links auf eine Tischecke gelagert, sodaß der eigene rechte Arm, der die Spritze hält, eine feste Unterlage zum Aufstützen hat. Bei Bettlägerigen wählt man besser, um nicht behindert zu sein, den rechten

Arm. Nach Stauung der Vene (am besten mit Staugurt) wird mit dem Zeigefinger der linken Hand die Haut an einem Punkt angespannt, der 2 cm handwärts und 1 cm links von der gewählten Einstichstelle liegt. Der Spannungszug des Fingers geht mit dem Venenverlauf parallel. Dadurch wird die Vene fixiert. Beim Einstich ist die Spritze so in der rechten Hand zu halten, daß der Zeigefinger die Kanüle führt, während der kleine Finger dem Stempel anliegt (Abb. 1).

Nachdem man sich durch Ansaugen von Blut von der richtigen Lage der Nadel in der Vene überzeugt hat, wird die Stauung gelöst (Abb. 2) und der Spritzenstempel mit dem Daumen der linken Hand langsam vorgedrückt (Abb. 3). Zur Kontrolle der richtigen Lage der Kanüle während der Injektion wird wiederholt Blut bis an den Anfangsteil der Spritze gesaugt. Die hieraus sich ergebende längere Dauer der Injektion ist therapeutisch erwünscht. Die technisch einwandfrei ausgeführte Injektion dauert etwa 2 Minuten und ist vollkommen reizlos.

Wenn die intravenöse Injektion aus irgendwelchen Gründen nicht erwünscht oder indiziert ist, so kann man Strophanthin auch intramuskulär mit Myokombin, perlingual mit Strophoral und rektal mit Kombetin-Suppositorien verabreichen.

 C. F. Boehringer & Soehne G.m.b.H., Mannheim

E. 27. 510.XXX. I.

Figure 11.3   Boehringer insert for Strophanthin packages

this context, Fraenkel was opposed to pharmaceutical advertisement and Boehringer was of the same opinion. Thus, promotion of the product included presentations and exhibitions at medical congresses with offprints sent to physicians, direct mailings to spa-physicians – who often treated chronic heart disease – and presentation in pharmacies. Indeed, the company approached the national German commission for the pharmacopoeia (*Pharmakopoe Kommission*) of the Imperial Council of Health (*Reichsgesundheitsrat*, RGR), but with limited success, since one of its members, Professor Thoms, was a champion of g-Strophanthin preparations licensed under his own name as 'Strophanthin-Thoms'. Indeed, the crystalline g-Strophanthin preparation had already entered the German pharmacopoeia before WWI. Subsequently, the 1914 '*Kriegssanitätsordnung*' listed g-Strophanthin but did not include Boehringer's k-Strophanthin.

Despite the drug's absence from the official pharmacopoeia, Boehringer was nevertheless able to use *Wertbestimmung* as a commercial argument in favour of their product. Thus, in order to inspire confidence in their physician-customers they placed the following text on the packaging: 'Tested by Prof. Fraenkel at the Speyererhof'.

During the interwar period, as the production of the digitalis and Strophanthin preparations became more and more industrialized, the Standardization Commission working for the League of Nations' Health Organization established an international standard of assay for preparations following the biological evaluation methods (*Wertbestimmungsmethoden*) used by pharmacologists.[52] But even this international standard did not put an end to the question of the quality and strength of Strophanthin preparations. In 1935, Fraenkel complained that 'we have to admit that the multiplicity of the so called *Ersatzpräparate* that flood the market and invade physicians' offices around the entire world leads to the opposite of high quality rational digitalis therapy. The quality of treatment has diminished and confusion is everywhere. We witness underdosing here and overdosing there, and the one who pays for it is the patient with heart insufficiency'.[53] Despite the efforts made by individuals, businesses, as well as national and international organizations, drug standardization is never arrived at all at once. Instead, these various actors contribute to the ongoing but neverending efforts in the context of modern pharmaceutical production to stabilize therapeutic products and practices. While we have been looking in detail at the Strophanthin therapy network, it is clear that there are similar, equally complex stories of standardization associated with most if not all high-potency drugs.

The wider significance of *Wertbestimmung* as an industrial practice for controlling the quality of pharmaceutical products can be appreciated through the financial negotiations between Boehringer and Fraenkel. The company first proposed paying Fraenkel for his *Wertbestimmung* services in 1906. Although Fraenkel initially requested a portion of the sales revenues of the preparation, he eventually accepted a system of payment by honoraria, not least because of the minimal benefits the company realized on the drug for over a decade. These *Wertbestimmung* services were not covered by any official contract until after WWI, when, in 1923, following the German economic crisis, Fraenkel renewed his request for payment on the basis of a percentage of sales revenues. Boehringer proposed an official contract for the next 10 years with a fee of 10 Pfennigs for every 100 phials sold which amounted to 0.5 per cent of the sales price. In 1924, the fee was raised to 25 Pfennigs, and in 1929 to 50. Finally, in 1933, when Fraenkel was deprived of all his other sources of income by the Nazi regime, the company increased the fee to 60 Pfennigs per phial (and no longer per 100 phials as before) rising to 64 Pfennigs if more than 800 000 phials were sold. This contract continued to provide a revenue to Fraenkel's widow after his death in 1938, as the fees were not simply dependent on the *Wertbestimmung* service provided by the researcher. During this difficult time of humiliation, persecution and financial deprivation for Fraenkel, Boehringer also acknowledged his 'longstanding promotion and invention' of the now viable network of Strophanthin therapy. The value of the product depended not only on its evaluation and its presentation on the market but also on the training and preparation of the physicians who used it. This combination has remained of central importance for the promotion of pharmaceutical products ever since.

Conclusion

Wertbestimmung (evaluation) in Ehrlich's original sense of the term can be understood as the concept and practice of physiologically testing the effects of a product whose strength cannot be evaluated simply by weighing or counting. But beyond this purely technical understanding of product standardization as a means for measuring the effects of ill-defined antitoxins, this approach to standardization also grew into a state-run system of quality control for biological therapeutic agents produced by a variety of industrial entrepreneurs. The provision of supposedly identical substances by multiple private producers led to demands for a central agency to ensure their uniformity. Thus, from an economic point of view, standardization can be seen as the process of establishing a standard for 'therapeutic value' among competing actors in the pharmaceutical market.

In a broader sense, standardization can be defined as the development and implementation of concepts, doctrines, procedures, and laboratory practices and designs to achieve and maintain the required levels of compatibility and commonality in operational, procedural, material, technical, and administrative fields to prepare a substance for use and exchange. Exchange in the case of Strophanthin not only meant that the substances could be shipped and distributed to different countries; it also concerned mediation between the different worlds of the laboratory, industry and medical therapeutic practice. The overall standardization process analyzed here can thus be considered as a stabilization process that connected laboratory settings, products, and practices with medical theories and practices through the elaboration of scientific, technical, bureaucratic, and organizational systems. Thus, the safety and efficacy of Strophanthin were established along three lines of evidence: laboratory animal experiments, clinical observations and case reports, and finally, statistical compilations and analysis. In practice, such proof depended partially on the particular regimen that produced the evidence but also on the new preparation's purity and stability.

Once the relative safety and efficacy of Strophanthin had been established, the laboratory product could then be integrated into medical practice. The success or failure of the overall Strophanthin therapy network depended as much on the initial risk/benefit analysis of the substance itself as on its subsequent use. While doctors may be considered to some extent as independent of the product itself, they nevertheless codetermined the final evaluation of its therapeutic value. The appropriation of the laboratory product by medical practice in turn challenged scientific and laboratory standardization via an interrogatory feedback mechanism. Although the ultimate goal of this second phase of clinical standardization was similar to the first one, namely to establish the compatibility of results in terms of safety and efficacy, the immediate aspects to be standardized were substantially different.

What we have presented here is the extension of a practice initiated in the context of less stable 'biologicals' like vaccines and sera to chemical drugs (albeit extracted from plants). This generalization of the procedure implies the idea of biological evaluation independent from sera, which brings with it the question of the mode of institutionalization. Indeed, we see that, even when displaced, the Ehrlich Institute in Frankfurt remained a model for a centralized quality control agency, which reached its acme with the efforts to establish international standards and methods of evaluation. In this process, *Wertbestimmung* became the intellectual, technical and administrative basis for pharmacologists' quest for 'rational drug therapy' in general, an approach based on the idea that a uniform substance provokes a uniform pharmacological response. While this approach started out as a response to a lack of knowledge about the 'ingredients' of biological drugs, and became relevant in the case of plant extracts due to problems associated with controlling plant-collecting practices, *Wertbestimmung* eventually came to be the central principle of rational therapy as a whole. These practices were the cement that bound together the elements of the Strophanthin-therapy-network; cardiac patients, oedema, k-Strophanthin, phials, intravenous injection technique, physicians, and so on. All these elements formed an integral part of successful and 'rational' therapy and guaranteed the 'therapeutic value' of Strophanthin. The entire network, held together by Fraenkel's evaluation procedures, had to be kept in place in order to guarantee that the therapy worked. If this structure became weakened, because, for example, the cement no longer held, cracks would start to appear, and products, practices and people would be put at risk. The threat of disintegration was never very far away, and a single broken link in the network could compromise the carefully constructed central affirmation: Strophanthin therapy does work.

Notes

1. Peter Drings, Jörg Thierfelder, Bernd Weidmann, and Friedrich Willig, *Albert Fraenkel: Ein Arztleben in Licht und Schatten, 1864–1938*, Landsberg: Ecomed, 2004.
2. 'Dass man auch bei der medikamentösen Therapie, wie dies bei der 1893 eingebürgerten Serotherapie selbstverständlich ist, vor der Einführung eines Mittels seine Wirkung am künstlich krankgemachten Organ oder Ganztier und nicht nur an normalen Tieren und Organen studieren sollte'. Albert Fraenkel, 'Von der empirischen zur experimentellen Digitalistherapie', *Schweizer Medizinische Wochenschrift*, 1936, pp. 434–40, 435.
3. Albert Fraenkel (ed.), *Der Weg zur rationellen Therapie*, Leipzig: Georg Thieme, 1933.
4. For a presentation of the initial procedures, see the contributions by Cay-Rüdiger Prüll and Axel Hüntelmann in this volume. For French analogies, see the contributions by Jonathan Simon and Gabriel Gachelin.
5. For the extension of the principle of *Wertbestimmung* to vaccines, see the contribution by Mick Worboys and for other biologicals, the one by Jean-Paul Gaudillière.
6. For a general overview and bibliographies see, John Parascandola (ed.), *History of Antibiotics: A Symposium*, Madison: American Institute of the History of Pharmacy, 1980. Ralph Landau, Basil Achilladelis and Alexander Scriabine (eds), *Pharmaceutical Innovation: Revolutionizing Health*, Philadelphia: Chemical Heritage Press, 1999. Liebenau, Jonathan, Higby, Gregory and Stroud (eds), *Pill Peddlers: Essays on the History of the Pharmaceutical Industry*, Madison: American Institute of the History of Pharmacy, 1990. Miles Weatherall, *In Search of a Cure: A History of the Pharmaceutical Discovery*, Oxford: Oxford University Press, 1990. Walter Sneader, *Drug Discovery: The Invention of Modern Medicines*, Chichester, New York: John Wiley and sons, 1996. François Chast, *Histoire contemporaine des médicaments*, Paris: La Découverte, 1995. Sophie Chauveau, *L'invention pharmaceutique. La pharmacie française entre l'État et la société au XXᵉ siècle*, Paris: Les empêcheurs de penser en rond, 1999. Robert Bud, *Penicillin: Triumph and Tragedy*, Oxford: Oxford University Press, 2007. Andrea Tone (ed.), *Medicating Modern America: Prescription Drugs in History*, New York: New York University Press, 2007. John E. Lesch, *The First Miracle Drugs: How the Sulfa Drugs Transformed Medicine*, Oxford: Oxford University Press, 2007. Jeremy A. Greene, *Prescribing by Numbers: Drugs and the Definition of Disease*, Baltimore: Johns Hopkins University Press, 2007. Viviane Quirke, *Collaboration the Pharmaceutical Industry: Changing Relationships in Britain and France, 1935–1965*, New York: Routledge, 2008.
7. John Parascandola, *The Development of American Pharmacology: John J. Abel and the Shaping of a Discipline*, Baltimore: John Hopkins University Press, 1992. George Weisz, *Divide and Conquer: A Comparative History of Medical Specialization*, Oxford: Oxford University Press, 2006. Christian Bonah, *Instruire, guérir, servir: Formation, recherche et pratique médicales en France et en Allemagne pendant la deuxième moitié du XIXe siècle*, Strasbourg: Presses Universitaires de Strasbourg, 2000, pp. 109–19.
8. For a general history of the C. F.Boehringer and sons company in Mannheim see: Ernst Peter Fischer, *Wissenschaft für den Markt. Die Geschichte des forschenden Unternehmens Boehringer Mannheim*, Munich: Piper, 1991. *Denkschrift der C.F. Boehringer & Soehne GmbH Mannheim-Waldhof anlässlich ihres 75 jährigen Bestehens 1859–1934*, Mannheim: Boehringer, 1934.
9. Christophe Bonneuil, 'Crafting and Disciplining the Tropics: Plant Science in the French Colonies', in John Krige and Dominique Pestre, *Science in the Twentieth Century*, Amsterdam: Harwood Academic Publishers, 1997, pp. 77–96.

10. For comparison on this wider understanding of *Wertbestimmung*, see the contribution to the present volume by Axel Hüntelmann.
11. Boehringer advertisement for 'Kombetin-Myokombin' around 1930. Stadtarchiv Mannheim (hereafter StA MA), C. F. Boehringer and Söhne Archives, Fraenkel papers, 31/1995, package 9.
12. My attention was drawn to the archival resources for a study of Albert Fraenkel's work on Strophanthin by Michael Ehmann. A first factual account of the relations between Fraenkel and Boehringer has been published in 2004 by Egon Dietz. Egon Dietz, 'Albert Fraenkel, C. F. Boehringer und die intravenöse Strophanthintherapie', in Peter Drings, Jörg Thierfelder, Bernd Weidmann, and Friedrich Willig, *Albert Fraenkel: Ein Arztleben in Licht und Schatten, 1864–1938*, Landsberg: Ecomed, 2004, pp. 195–240. See as well: Christian Bonah, 'Albert Fraenkel, die Medizinische Fakultät Strassburg und die Entstehung der der Strophanthintherapie', in Peter Drings, Jörg Thierfelder, Bernd Weidmann and Friedrich Willig, *Albert Fraenkel: Ein Arztleben in Licht und Schatten, 1864–1938*, Landsberg: Ecomed, 2004, pp. 155–86.
13. StA MA, C. F. Boehringer and Söhne Archives, Fraenkel papers, 31/1995/2 (Correspondence 1906–10). Letter from Boehringer to Fraenkel, 26/03/1906.
14. Withering, William, *An Account of the Foxglove, and Some of Its Medical Uses*, London: Swinney, 1785.
15. For a detailed analysis of Fraenkel and his medical training in Strasbourg, see Bonah, 2004, pp. 155–86.
16. On Koch and Tuberculin, see Christoph Gradmann, *Krankheit im Labor: Robert Koch und die medizinische Bakteriologie*, Göttingen: Wallstein, 2005, especially Chapter 3.
17. Jörg Thierfelder, 'Albert Fraenkel. Eine biographische Skizze', in Peter Drings, Jörg Thierfelder, Bernd Weidmann and Friedrich Willig, *Albert Fraenkel. Ein Arztleben in Licht und Schatten, 1864–1938*, Landsberg: Ecomed, 2004, pp. 17–69.
18. Albert Fraenkel, 'Tonographische Untersuchungen über Digitaliswirkung', in *Archiv für experimentelle Pathologie und Pharmakologie*, 40, 1898, pp. 40–52.
19. Albert Fraenkel, 'Ueber die physiologische Dosirung von Digitalispräparaten', in *Die Therapie der Gegenwart*, 1902, pp. 106–111.
20. Fraenkel, 'Ueber die physiologische Dosirung', 1902, p. 106.
21. According to a 1897 study by Keller cited in Fraenkel, 'Ueber die physiologische Dosirung', 1902, p. 107.
22. Fraenkel, 'Ueber die physiologische Dosirung', 1902, p. 107.
23. Fraenkel, 'Ueber die physiologische Dosirung', 1902, p. 107.
24. Fraenkel, 'Ueber die physiologische Dosirung', 1902, p. 111.
25. Fraenkel, 'Ueber die physiologische Dosirung', 1902, p. 108.
26. Fraenkel, 'Ueber die physiologische Dosirung', 1902, p. 111.
27. Cecil Bosanquet and John Eyre, *Serums, Vaccines and Toxins in Treatment and Diagnosis*, London: Cassell and Company, 1916, p. 45.
28. See for more details the contributions by Cay-Rüdiger Prüll and Michael Worboys in this volume.
29. Albert Fraenkel, 'Ueber Digitaliswirkung am gesunden Menschen', in *Münchner Medizinische Wochenschrift*, 32, 1905, pp. 1537–9.
30. Albert Fraenkel, 'Vergleichende Untersuchungen über die kumulative Wirkung der Digitaliskörper', in *Archiv für experimentelle Pathologie und Pharmakologie*, 51, 1904, pp. 84–102.
31. For a more detailed description of this phase of the development, see Bonah, 2004, pp. 174–6.

32. Results were published in 1905: Kottmann, Kurt, 'Klinisches über Digitoxinum solubile Cloetta', in *Zeitschrift für klinische Medizin*, 56, 1905, pp. 128–66.
33. Fraenkel, *Der Weg zur rationellen Therapie*, p. 118.
34. Fraenkel, 'Von der empirischen zur experimentellen Digitalistherapie', p. 435.
35. For a company history, see Fischer, *Wissenschaft. Denkschrift.*
36. StA MA, C.F. Boehringer and Söhne Archives, Fraenkel papers, 31/1995/2 (Correspondance 1906–10). Letter from Boehringer to Fraenkel, 12/03/1906.
37. StA MA, C.F. Boehringer and Söhne Archives, Fraenkel papers, 31/1995/2 (Correspondance 1906–10). Letter from Boehringer to Fraenkel, 26/03/1906.
38. StA MA, C. F. Boehringer and Söhne Archives, Fraenkel papers, 31/1995/2 (Correspondance 1906–10). Letter from Fraenkel to Boehringer, 28/03/1906.
39. StA MA, C.F. Boehringer & Söhne Archives, Fraenkel papers, 31/1995/2 (Correspondance 1906–10). Letter from Fraenkel to Boehringer, 29/03/1906.
40. StA MA, C.F. Boehringer & Söhne Archives, Fraenkel papers, 31/1995/2 (Correspondance 1906–10). Letter from Boehringer to Fraenkel, 20/08/1906.
41. StA MA, C.F. Boehringer & Söhne Archives, Fraenkel papers, 31/1995/2 (Correspondance 1906–10). Letter from Boehringer to Fraenkel, 08/05/1907.
42. StA MA, C.F. Boehringer & Söhne Archives, Fraenkel papers, 31/1995/2 (Correspondance 1906–10). Letter from Fraenkel to Boehringer, 22/07/1908.
43. StA MA, C.F. Boehringer & Söhne Archives, Fraenkel papers, 31/1995/2 (Correspondance 1906–10). Letter from Fraenkel to Boehringer, 22/07/1908.
44. StA MA, C.F. Boehringer & Söhne Archives, Fraenkel papers, 31/1995/2 (Correspondance 1906–10). Letter from Boehringer to Fraenkel, 24/07/1908.
45. StA MA, C.F. Boehringer & Söhne Archives, Fraenkel papers, 31/1995/2 (Correspondance 1906–10). Letter from Boehringer to Fraenkel, 31/07/1908.
46. StA MA, C.F. Boehringer & Söhne Archives, Fraenkel papers, 31/1995/3 (Correspondance 1911–18). Letter from Boehringer to Fraenkel, 27/04/1914.
47. StA MA, C.F. Boehringer & Söhne Archives, Fraenkel papers, 31/1995/3 (Correspondance 1911–18). Letter from Fraenkel to Boehringer, 30/04/1914.
48. Liliane Pariente, *Naissance et évolution de quinze formes pharmaceutiques*, Paris: Louis Pariente, 1996. Anne Rasmussen, 'Les enjeux d'une histoire des formes pharmaceutiques : la galénique, l'officine et l'industrie (XIXᵉ-début XXᵉ siècle)', *Entreprises et histoire*, 'Industries du médicament et du vivant', 36, 2004, pp. 12–28. Rasmussen, Anne, 'Préparer, produire, présenter des agents thérapeutiques. Histoires de l'objet médicament', in Christian Bonah and Anne Rasmussen (eds) *Histoire et médicament aux XIXᵉ et XXᵉ siècles*, Paris: Editions Glyphe, 2005, p. 159–88.
49. Fraenkel, *Der Weg zur rationellen Therapie*, p. 65.
50. Fraenkel, *Der Weg zur rationellen Therapie*, p. 65.
51. Fraenkel, *Der Weg zur rationellen Therapie*, p. 65.
52. Iris Borowy, and Wolf D. Gruner, *Facing Illness in Troubled Times, Health in Europe in the Interwar Years, 1918–1939*, Berlin: Peter Lang, 2005. Iris Borowy, *Coming to Terms with World Healt: The League of Nations Health Organization, 1920–1946*, forthcoming. I would like to thank Iris Borowy for providing me with an early version of a chapter on the LNHO Standardization Commission. For more on this topic, also see the contribution by Pauline M. H. Mazumdar in this volume.
53. Fraenkel, 'Von der empirischen zur experimentellen Digitalistherapie', p. 437.

12
Changing Regulations and Risk Assessments: National Responses to the Introduction of Inactivated Polio Vaccine in the UK and West Germany

Ulrike Lindner

Introduction

A series of large-scale epidemics following WWII helped to confirm polio as a major health threat in the 1950s. The fact that its principal victims were children, coupled with horrific images of the crippling deformities associated with the disease ensured its place as an important issue in public health policy in Europe. In this period, vaccination was the only feasible preventive measure against the disease, with the US at the forefront of vaccine research. Thus, most countries took a keen interest in the new vaccines that were being developed in the US.

In this paper, I will discuss national responses to the introduction of Inactivated Polio Vaccine (IPV) – first developed in the US by Jonas Salk and therefore also known as the Salk vaccine – in West Germany and the UK. I will examine the introduction of IPV in both countries between 1955 and 1961, a period during which the risk assessment and regulation for IPV changed considerably. As IPV was a new vaccine, most countries that introduced it developed slightly different methods for risk assessment and also applied different safety standards for the production, testing and administration of IPV, which not only reflected the different national constellations and networks but also varied over time.[1]

Against the background of these changing regulations and perceptions of the vaccine, I will explain in detail the responses to the new vaccine in West Germany and the UK in the 1950s. In particular, I will analyze the factors that influenced the decisions of the two countries concerning how to produce their own vaccines and examine the issues that influenced the development of the regulations covering the safety and testing of IPV. The introduction of IPV was certainly not determined exclusively by the epidemiology and gravity of the disease in the two countries or even by the scientific consensus concerning the efficacy, potential risks and prospective costs of the vaccine. There were many other elements that have to be factored into the equation. One important consideration is the structural features of the two health systems concerned. In the UK, there was a single state health system with a centrally organized public health service, while in West Germany there was a weak, de-centralized public health service.[2] However, neither an exclusive focus on disease and vaccine characteristics nor on structural features of the relevant health systems is sufficient to explain the patterns of changing assessments around IPV regulation. Specific scientific traditions that exist in different countries, with their respective research and therapeutic cultures can also influence the introduction and use of new drugs, as has been demonstrated by Arthur Daemmrich concerning the introduction of new medicaments in Germany and the US.[3] In general terms, one can describe medical innovation as the process of introducing a new drug or medical technique against the backdrop of an evolving context. Furthermore, the technology itself is also subject to change during this process.[4] In recent years, more attention has been paid to the cultural and social background surrounding these processes.[5] Such approaches are particularly valuable when discussing the changing risk definitions, as in the case of IPV. In his work on AIDS, Peter Baldwin has convincingly argued that risk evaluations in particular should be understood as social and cultural issues.[6]

In the case of IPV, the process of vaccine introduction and regulation was clearly shaped by a variety of social and cultural influences.[7] These included national scientific cultures and traditions as well as public opinion, the influence of pressure groups, the relationships between vaccine manufacturers and public health authorities as well as international pressure.[8] I will seek to identify and analyze these factors and will try to show that the risk assessment and regulation that accompanied the introduction of the new IPV vaccine depended greatly on the cultural, political and economic environment in which they were put into place.[9]

Development of inactivated polio vaccine (IPV) and efforts at international regulation and standardization

Polio vaccine research was mainly carried out in the US. One reason for this concentration of research was that the US suffered from severe polio epidemics starting at the end of the nineteenth century, whereas in most European countries polio remained an endemic disease and grew to epidemic proportions only in the middle decades of the twentieth century.[10] Therefore, interest in the disease was momentous in the US, especially after a serious epidemic in 1916, whereas most European countries remained relatively indifferent.[11] Furthermore, the country had a famous polio-sufferer in the person of F. D. Roosevelt, who had contracted polio in 1921 at the age of 40. During the subsequent years, he was able to raise huge sums of money for polio patients, especially after being elected President of the US in 1935. Reports on his fight against polio transformed the disease, which had formerly been mainly viewed as a problem of poor and immigrant communities, into a health danger that could touch everyone.[12] In 1937, Roosevelt founded the National Foundation of Infantile Paralysis (NFIP, later also known as the 'March of Dimes') to separate his activities on behalf of polio sufferers from his fundraising for the Democratic Party. Under Basil O'Connor, a former partner in Roosevelt's law practice, the foundation developed into an extremely powerful pressure group able to raise huge amounts of money, around 20 million US dollars per year. Even if most of the money was used for the care of polio victims, a significant part went into funding research.[13]

Despite the public and scientific focus on polio and the financial opportunities such a development offered in the US, early attempts to develop polio vaccines in the 1930s remained unsuccessful, with the vaccines proving unsafe.[14] Furthermore, it was not yet known how many serotypes of polio there would turn out to be, and research in the 1930s also suggested that it would be very difficult to produce large quantities of vaccine.[15]

The most important step towards vaccine development was the discovery made by Enders, Weller and Robbins in 1949, when they succeeded in growing polio virus in non-nervous-tissue culture in their laboratory in Boston.[16] Indeed, they received the Nobel Prize in 1954 for this technique, which was used to produce vaccines against measles, mumps and other viral diseases.[17] Another important step towards a vaccine was the typing of the polio virus after 1948. In a huge collaborative effort that received financial support from the National Foundation, researchers from several universities in the US formed a typing committee that sought to resolve this problem. After two years of research, they showed that three distinct types of polio virus existed, under which the 196 analyzed strains could be classified.[18] It was as a member of the typing committee, along with other leading researchers in the polio field (including Bodian, Francis, and Sabin), that Jonas Salk first became interested in polio-vaccine research.[19] With the typing work successfully completed and following the recent discovery by Enders, Weller and Robbins, a polio vaccine seemed to be near at hand.

At that time, the most striking characteristic of polio vaccine development was the disagreement among experts regarding the qualities of 'inactivated' and 'attenuated' polio virus vaccine. Most virologists believed that an inactivated vaccine would be insufficiently active, providing only a few months of protection. Furthermore, an inactivated vaccine would have to be injected and vaccination would have to be reiterated to obtain an acceptable level of protection.[20] Despite these objections, Jonas Salk continued to work on obtaining an inactivated vaccine: his former work on killed vaccines during the war, his experience with inactivated influenza vaccines and his experience in polio typing led him to the conclusion that an inactivated polio vaccine might be successful. He produced a formalin-inactivated polio vaccine and started a series of tests. After the failures of the 1930s, he was among the first researchers who dared to extend their tests from laboratory animals to human subjects.[21] His successful immunization tests with children were published in 1953, and he also presented his results at a meeting of the NFIP in the same year. Eventually, Salk won the support of the powerful NFIP that decided to back his inactivated vaccine.[22] With the help of the NFIP, a huge field trial was prepared involving 1.8 million children in 44 States of the US. The trial included two control groups; one group of children injected with placebos and another group of observed children who received no injections. In April 1954, the US Public Health Service approved the trial, which was carried out by a large team of researchers directed by the epidemiologist Thomas Francis and the NFIP. Thomas Francis presented the results in April 1955 at the University of Michigan. He declared the Salk vaccine over 90 per cent effective against Types II and III and 60–70 per cent effective against Type I polio virus. The news of these successful polio-vaccine trials made the front pages of newspapers all over the world. Immediately, many countries started to debate the desirability of polio vaccination.[23] On the same day, the Salk vaccine was licensed for use in the US. The hurried decision of the American licensing committee would later be heavily criticized, but at the time the committee found itself under enormous pressure from an expectant public and the press.[24] Thus, a programme of mass vaccination could start immediately in the US, with five million American children being vaccinated in 1955. One has to add that – despite the triumph and the enormous media coverage – acceptance of the vaccine was not uniform all over the US, particularly amongst adults. By 1961, only one-third of the adult population had been vaccinated.[25]

Shortly after the start of vaccination in the US, the programme suffered a serious setback due to the 'Cutter incident'. Two hundred and four children became paralysed after being inoculated with a vaccine produced by the US manufacturer Cutter. Immediately, the American Surgeon-General took the Cutter vaccine off the market.[26] In May 1955, the whole US vaccination programme was suspended, since it was not immediately clear what had caused the accident. It turned out that the batch of vaccine in question was not completely inactivated and so still contained active virus that had caused the illness in the children. Virologists first thought the reason might have been the so-called Mahoney-strain, the specific virus strain Salk had used for his vaccine, as this strain was considered to be highly aggressive. After some weeks, it was discovered that the reason was not the virus-strain, but a technical problem within the filtration process. Thus, a better filtration method and changes to the production procedure were considered sufficient to guarantee the inactivity of the vaccine.[27] In the US, vaccination was resumed after some weeks, although now with tightened safety regulations. After the first year, the US-vaccination programme was acclaimed as a striking success. However, the Cutter incident led to a new assessment of the risk associated with the Salk vaccine previously considered to be completely safe. Many other countries reacted to the problems with more caution than the US and delayed their own vaccination programmes.

The Cutter incident also served to stimulate international efforts aimed at developing satisfactory safety standards for IPV. These safety issues were repeatedly discussed at international polio conferences, meetings of the European Association against Poliomyelitis and at international conferences for biological standardization.[28] Thus, the safety of IPV was discussed at length during the second international biological standardization conference in Rome in 1956. The strict safety regulations that had been implemented in the US after the Cutter incident in May 1955 were seen as sufficient by most delegates, since no further accidents had been reported in the US. However, there was still a debate concerning how effective any safety control tests could be and the question remained whether there might not be significant residual risks. The representatives of most countries agreed that there should be at least a double control to guarantee the inactivity of the vaccine, and all the delegates at the Rome conference stressed the need for control tests using live monkeys as well as tissue culture. After all, it was argued, it was only tests with live monkeys that had revealed the Cutter vaccine to be dangerous.[29]

Thus, there continued to be uncertainties about safety controls and regulation of the vaccine, and in addition, a standardized vaccine preparation had not yet been developed. Standardization was internationally monitored by the World Health Organisation (WHO) expert Committee on Biological Standardization, a committee that met annually in order to establish detailed recommendations and guidelines for the manufacturing, licensing, and control of vaccines, sera, and other medicaments.[30] Before WWII, biological standardization was coordinated by the League Standardisation Commission that was set up after the first meeting of the League of Nations Health Organisation in 1921.[31] The work of the commission was taken up again by the WHO in 1948. From 1956 on, the committee discussed the need for providing international reference preparations for inactivated poliomyelitis vaccine, but it was thought that preparations were still premature and the committee should wait for further developments.[32] Two years later, the committee observed that several countries had tried to obtain stable dried vaccine for reference preparation and asked the *Statens Seruminstitut* in Copenhagen to coordinate international efforts to obtain a stable standard vaccine.[33] For this undertaking, the institute in Copenhagen worked together with the Moscow Institute for Poliomyelitis Prophylactics and the National Institute of Health, Bethesda, US, although these efforts ultimately proved unsuccessful. Prior to 1965, there were several collaborative efforts that all failed to produce a sufficiently stable freeze-dried trivalent inactivated polio vaccine that could serve as a standard vaccine, with the principal problem being the loss of antigenic activity.[34] As a result, the different national health administrations and control agencies were confronted with a relatively open field of interpretation that they could mobilize for assessing the risks and benefits of the IPV.

The introduction of IPV 1954–6

Looking at West Germany and the UK, we can observe that the introduction of IPV was accompanied by a changing evaluation of its risks and benefits in both countries.

In Britain, virological research on polio and polio vaccines hardly existed in the 1940s. The following complaint from the Infantile Paralysis Fellowship (IPF), the main pressure group for Polio victims in the UK, dates from 1951:

> The Executive Committee of the Infantile Paralysis Fellowship representing as they do some 9000 of our members feel some concern that research into the causes, prevention and cure of Poliomyelitis is not more actively pursued in this country. They feel alarmed to think that for the main part Great Britain has to rely on the research work of other countries, chiefly the United States, and strongly urge that without further delay, large scale research be started in this country.[35]

As Anne Hardy has noted, polio remained under the dominance of neurologists until the 1940s in Britain, and only very few research projects addressed polio in the years before 1947.[36] The interest in polio grew following the epidemic of 1947, when polio affected over 7000 people, compared with the usual figure of below 1000 per year.[37]

Nevertheless after 1947 the newly founded National Health Service (NHS) had serious financial problems and could not fund extensive virological research on the scale of the activity undertaken in the US. The most important institutions for research in Britain were at that time the Medical Research Council (MRC), and the Public Health Laboratory Service, both state-funded organisms. However, only very few researchers within these institutions were working on virological issues, as William Bradley, an official in the Ministry of Health, wrote in 1950: 'laboratory facilities for virus work will continue to be very limited in this country'.[38] He also concluded that British research had to concentrate on descriptive epidemiology, but even this work seemed to be limited by a shortage of staff and money.

> I hope this winter to repeat our hospital enquiry and to look especially for evidence re tonsillectomy, inoculations, pregnancy. Cross-infection in hospital and wards would justify detailed study. We shall not be able to get to them. Although such studies would be little more than descriptive epidemiology they might clarify the picture as to how the disease spreads. If I had more workers I should go forward.[39]

However, in contrast with West Germany, British researchers were in closer contact with US developments. Some British researchers, like John Beale, went to work on vaccine development in North America before returning to the UK in the 1950s.[40]

Besides the problems with virological research, the IPF, the British pressure group for polio victims, which was founded in 1939, was tiny in comparison with the NFIP, its equivalent in the US. With an annual income of around 10,000 US dollars the IPF mainly tried to help with care for polio sufferers, supporting some sport and recreation facilities: they ran a seaside hotel and a home for polio sufferers and coordinated swimming activities, but they certainly could not fund any research.[41] Besides the IPF, there was no other foundation or committee to coordinate or fund polio research in the UK in the 1940s. This same problem of coordination became apparent at the European level, and the *European Association against Poliomyelitis* invited all European countries to join the European body, forming national associations that could oversee scientific and technical aspects of polio research and therapy in each country. In 1951, the Ministry of Health finally agreed to install a Committee on Poliomyelitis to co-ordinate British research.[42] Nevertheless this committee never disposed of its own financial resources with which it could directly support research and therefore never had a great deal of influence.

Thus, the British scientific community looked to the US for new virological research. UK health officials had high hopes for Salk's IPV vaccine and closely followed its development. The Ministry of Health sent Sir Weldon Dalrymple-Champneys to attend the huge field trials in 1954, and he returned full of enthusiasm. Accordingly, the Ministry of Health prepared its own vaccination campaigns for spring 1955.[43] The plan was to inoculate children from several cities in England before testing them for their serological reaction. Not everyone in the Ministry of Health was so enthusiastic about the new vaccine, however, with many in the health administration viewing the new vaccine with suspicion.[44] The Minister of Health, Iain Macleod, offered cautious comments concerning the polio vaccination; in April 1955 he made the following announcement to the House of Commons:

> The new vaccine involves inoculating our children at repeated intervals with a preparation derived from the kidneys of dead monkeys. The House and the country will surely agree that they must carry out intensive tests as to the exact effects so that they could eliminate any possible dangers from it.[45]

This caution has to be seen in connection with the long history of the anti-vaccination movement in Britain and a strong tradition of public distrust in vaccination. The National Anti-Vaccination League had been founded in the 1870s to campaign against compulsory smallpox vaccination in the UK, and the League was still active in the 1950s (even if it was now a fairly marginal and far less powerful organization) and continued to oppose smallpox vaccination campaigns.[46] Thus, most politicians assumed that new vaccination programmes were likely to meet at least some resistance on the part of the public.

Despite these fears, the Ministry of Health went ahead with the British trials, using the new Salk vaccine and adopting the same US safety regulations for UK production. The Ministry was confronted by production problems right from the start. In the UK, the health administration collaborated with private producers in order to have sufficient quantities of IPV. When the main producer of vaccine in Britain, the pharmaceutical company Glaxo, was unable to provide sufficient vaccine in time, the trials planned for spring 1955 had to be delayed.[47] Indeed, Glaxo seemed to have considerable difficulties with vaccine production. John Beale, who returned from Canada to join Glaxo in 1956, reported that he still found the situation to be desperate, with serious problems of residual active virus in inactivated vaccine as well as problems with the potency of the vaccine.[48] Thus, in April 1955, vaccine production was still insufficient, and when the Cutter incident was reported in May 1955, the whole project was stopped.[49] G. S. Wilson, Director of the British Public Health Laboratory Service, now characterized the Salk vaccine as generally unsafe and called for more research to find appropriate virus strains for producing a safer vaccine.[50]

The Medical Research Council as well as the newly established Joint Committee on Poliomyelitis Vaccine now firmly advised against the importation of American Salk vaccine from the US as well as the use of British-produced Salk vaccine.[51] Thus, vaccine made from the Mahoney-strain (including the US-manufactured Salk vaccine) was forbidden in the UK and new safety regulations were elaborated. These regulations were far stricter and required vaccines to be made from polio strains considered to be less aggressive. British vaccine manufacturers, which had experienced production problems from the start, were unable to produce vaccine made from different virus strains in sufficient quantities. Consequently, there were no more vaccinations in the UK in 1955.[52]

The development of a reliable safety test for the vaccine was no easy undertaking either, since live monkeys had to be used for the neurovirulence tests of the vaccine. D. R. Bangham describes the early years of polio vaccine testing in his personal account:[53]

> Completely new safety tests had to be developed, rapidly, under conditions of great tension. Certain tests for neurovirulence, for both types of vaccine, involved the import, assessment and housing of large numbers of wild rhesus and green monkeys from India, and Macacus Irsu from the Philippines. In one of the earliest tests developed, suitable doses of vaccine were injected into the spinal cord of monkeys previously treated with cortisone, some days later the monkeys were killed, perfused, the cord was excised and histological sections of it prepared.

These were examined for evidence of neural damage if live virus was present by Dr Janet Niven, the histologist at NIMR (National Institute for Medical Research, Mill Hill), and Dr T Beswick. When mass vaccination for poliomyelitis was implemented in Britain, some 1000 monkeys/year were imported by NIMR for this test alone. Leonard Ward and his brother Harry gave startling accounts of the hazards in the earliest days of catching and handling dangerous, unselected wild monkeys in a holding room containing 20–30 at a time.

In the end, a new vaccination programme was established in 1956. Only British IPV that was made from the so-called Brunhilde virus-strain (which was considered less problematic than the Mahoney-strain) should be used, even though tests in the US had established that the Mahoney-strain had not been the reason for the Cutter accident.[54] The problem was that the British manufacturers (Glaxo and Burroughs Wellcome, later joined by Pfizer UK) were only able to produce small amounts of vaccine. In 1956 1.6 million children under ten years of age had been registered for vaccination, however, only 10 per cent could be inoculated with two injections in May and June before the available vaccine was used up.[55]

It is clear that reservations concerning vaccination in Britain and reaction to the Cutter incident were only two factors in the reassessment of risks after May 1955. The introduction of IPV was also strongly shaped by national considerations, including the attitudes of the British health administration and the complex relationships that existed between the NHS and pharmaceutical companies.

Turning to West Germany, although the epidemiological situation was quite similar to the UK, there were some significant differences. As in the UK, large polio epidemics appeared only after WWII. Polio had not been a prominent topic in public health until the summer of 1947, when an epidemic of 2500 polio cases broke out in Berlin, with most of the patients being children.[56] The outbreak met with marked indifference on the part of German physicians and health officials who had to deal with overwhelming medical and social problems in war-ravaged Berlin. An American medical team sent to Berlin in 1947 to monitor the epidemic and to advise on the care of the acutely ill was appalled by the lack of facilities and the post-war fatalism of the Germans.[57] Furthermore, there had been almost no scientific exchange with western countries during the final years of the Third Reich, meaning that the Germans were unaware of US developments in the area of virology. Directly after the war, German interest in polio research was negligible and financial support and facilities for local research in virology practically non-existent. However, with severe polio epidemics occurring almost every year after 1947, West German interest in polio grew, although, as had been the case in the UK, the research focused mainly on the epidemiology of the disease.[58]

In 1954, the trials in the US were followed with great interest. As in the UK, vaccine in West Germany was produced mainly by one private company, the Behringwerke (part of Hoechst AG, located in Marburg).[59] When the trials took place in the US in 1954, the Behringwerke started producing their own IPV vaccine based on Salk's procedure, adding aluminium hydroxide to enhance the effect of the vaccine.[60] Thus, even despite the lack of interest in vaccine development in the 1940s, the beginning of vaccine production seemed to be quite auspicious in West Germany.

In order to speed up the vaccination process, no new testing requirements were proposed in 1954 by the German health officials at the *Paul-Ehrlich-Institut* in Frankfurt, which would later become the federal office for controlling sera and vaccines in West Germany. Instead, they adopted the US-American 'minimum requirements' for vaccine testing, which became effective in Germany in April 1955. Additionally, the West German Ministry of the Interior asked the *Bundesgesundheitsamt*, the federal agency for controlling and testing medical products, to develop general regulations covering polio vaccination. In November the first West German federal state to launch vaccination, the *Land* of Hesse (where the Behringwerke was located), gave its permission for large-scale production of the vaccine. Hesse already approved vaccination in certain cases and allowed for the vaccination of volunteers, but without implementing any broader vaccination programme.[61] One such programme was planned in Hesse for spring/summer 1955, pending the regulations imposed by the *Bundesgesundheitsamt*. However, problems emerged between the manufacturer of IPV in Germany and the public health authorities. In April 1955, members of the *Bundesgesundheitsamt* were not allowed to enter the company's production laboratory and view the laboratory journals. Later, the Behringwerke failed to report an incident with a batch of vaccine, in which a test had led to the death of two monkeys. While further testing by the Paul-Ehrlich-Institute revealed no problems, the German health administration grew suspicious. The Behringwerke apparently encountered the same difficulties producing sufficient quantities of reliable vaccine as the pharmaceutical companies had experienced in England. Together with reports of the Cutter incident in the US, the events in the Behringwerke laboratory accentuated the German health administration's concerns, and led to a new risk assessment of IPV. Vaccination was then completely stopped in West Germany and the question of the risks associated with the use of IPV had to be thoroughly investigated before starting any comprehensive vaccination programmes.[62]

The *Bundesgesundheitsamt* eventually published a report in January 1956, which clearly came out against any IPV vaccination campaigns and was only in favour of vaccination in exceptional individual cases. Thus, there were no more vaccinations in 1955 and 1956, and the importation of Salk vaccine made from Mahoney-strains was strictly prohibited.[63] What we observe in West Germany, then, is that despite the debut of vaccine production in quite favourable conditions, the conclusion of the risk assessment exercise that followed the Cutter incident led to a harsher anti-vaccine position than that adopted in the UK.

The reaction in West Germany has to be seen in the context of the attitudes to be found in the West German scientific community. Many German health officials and scientists had been mistrustful of US vaccine developments from the very beginning. As already mentioned, there had been a lack of exchange with other countries during WWII and under the influence of the Third Reich. Germany, which had been so prominent in bacteriology, serology and immunobiology in the decades preceding the war, no longer had researchers of international stature in these fields. It was, therefore, hard for many German scientists to admit that they had lost their leading role in virology, with some going as far as deploring that there had never been a second Robert Koch.[64] Another important issue was the representation of German scientists in an international context, particularly now that there were two countries. Thus, for example, the Germans were simply not present at the International Polio Conference in 1948, and it was only in 1958 that a West German health official reported on the polio situation in his country at one of the International Polio Conferences. Political disagreements between East and West Germany about representation at international meetings made this issue even more problematic, and, in 1958, both East and West Germany sent delegates.[65] Even when West German scientists or health officials did attend international conferences, their status remained a major concern for the members of the aggrieved West-German post-war scientific community.[66] Thus, just to give one example, when the West-German delegate at the European poliomyelitis conference in 1958 reported back to the Ministry of the Interior, he seemed to be mainly concerned about how unfairly the German delegates had been treated by the organizers.[67]

Given the general sentiment of paranoia, it is not surprising that members of the German health administration, such as Franz Redeker (director of the health department of the Ministry of the Interior) and Georg Henneberg, director the Robert-Koch-Institute in Berlin (the German state institute for serology and immunobiology) questioned the sense of IPV vaccination after the incidents in the summer of 1955 and even stated that it might not be necessary for Germany to implement an IPV vaccination programme at all.[68] Other German epidemiologists and physicians were very much in favour of a vaccination programme, but people like Henneberg and Redeker had considerable influence over the development of polio vaccination. Thus, while we can claim with some certainty that the favourable attitude towards vaccination in 1954 was motivated by the serious epidemics that hit Germany so unexpectedly after the war, this attitude quickly gave way to reveal an underlying scepticism once the first problems with IPV arose.

Another problem in West Germany was the lack of any powerful independent pressure groups. The *Deutsche Vereinigung zur Bekämpfung der Kinderlähmung*

(German Association for the Fight against Poliomyelitis, DVBK) was only founded in August 1954.[69] As in the UK, this happened mainly due to pressure from the European Association against Poliomyelitis, which requested every country to form a national association capable of joining the European body. Whereas the UK set up a committee in 1951, the West Germans waited until 1954, even though Germany suffered from polio epidemics of equal severity. The German association was constituted by a group of health officials working in the area of public health and vaccination policies. Almost all the leading German experts were members of the association, meaning that there was hardly any difference of opinion between the committees of the association and the state health administration. Furthermore, the association was entirely state-funded, with no private support, and never undertook any fundraising efforts.[70] A regional Bavarian organization for polio sufferers, called 'Pfennigparade' (after the American NFIP's March of Dimes) was founded in Munich in 1952, but it mostly concentrated on the care of children crippled by the disease. Unlike the American NFIP, however, it never reached the status of a powerful national pressure group and was unable to raise the amounts of money needed to sponsor research.[71] Thus, polio patients and their families were hardly represented at all in the West-German discussions about the disease.

Just as in the UK, the Cutter incident and the reactions it provoked constituted only one factor among many that influenced the evaluation of IPV in West Germany. We can see that the attitudes of the members of the scientific community played an even more important role in shaping the vaccination process in West Germany than they had in the UK.

Polio vaccination after 1957

Following the initial assessments and reassessments of the risk associated with the IPV vaccine, both the UK and West Germany tried to introduce a working vaccination programme after 1957.

In the UK, the British company Glaxo had promised to produce enough vaccine for the vaccination campaign of the Ministry of Health in spring 1957, but here again there were problems and delays.[72] Similarly, Burroughs Wellcome was unable to produce sufficient vaccine, at least according to the discussions held at the Ministry of Health in the autumn of 1957:

> The first three batches of Burroughs Wellcome polio vaccine have now passed all the necessary tests and the MRC agreed that they should be released. In total they amount only to 138 litres which is not worth distributing to all local health authorities in Great Britain.[73]

In the meantime, the Ministry of Health was heavily criticized in the press and by the public for the delayed vaccination campaigns. The success of the US vaccination programmes had been widely reported in British newspapers and the public was well aware of the problems with the British programme, in particular the delays in producing sufficient quantities. The health administration was not in the situation of dealing with resistance to the vaccination as they had assumed earlier, but rather with a public that demanded a comprehensive vaccination programme. Thus, the decision made to forbid the US-vaccine produced from Mahoney-strains in 1955 had been principally a reaction to the Cutter accident. In the meantime, the US had vaccinated millions of children with the Salk vaccine without any further incidents, placing the problem in a different light.[74]

In July 1957 the Medical Research Council made the decision to recommend the importation of US vaccine for 1958, but it advised that all imported vaccine should still undergo British tests. Until then, the attitude of the Council and the health administration had been extremely cautious, with members of the MRC refusing to recommend any imports of US Salk vaccine for medical reasons. Only a few weeks earlier, one of the members of the MRC had even stated that US Salk vaccine could only be imported 'over his dead body'.[75] Thus, the risk assessment concerning the American vaccine changed quite suddenly in July 1957, obviously influenced by public pressure but also by the success of the vaccination campaigns in other countries.

The UK government now made contracts with the Canadian Connaught Laboratories and the US-American company Pitman-Moore.[76] The Ministry of Health decided to vaccinate all children under 15 years of age in 1958, and for this, they would use the imported vaccine but only – following the MRC guidelines – after British testing.[77] Nevertheless there were more delays in spring 1958, although this time with the testing of the imported vaccine, and so the MRC changed its guidelines once again and allowed the imported vaccine to be administered without the British tests. The position of the MRC was now:

> that the use of imported vaccine untested in this country should be limited to that of a temporary supplement to meet the present shortage ... [and] ... that every endeavour should be made to obtain maximum supplies of British vaccine and imported vaccine which has satisfied both British tests and tests in the country of origin, so that, at the earliest possible moment, the need to use imported vaccine untested in this country will disappear.[78]

In the meantime, the changing attitudes of the MRC met with considerable public criticism. The new guidelines also produced quite bizarre regulations for the actual vaccination campaign: Parents were supposed to be able to choose between non-tested imported vaccine, imported British-tested vaccine, and finally vaccine produced and tested in Britain. The complex and bureaucratic mechanism certainly did not help the progress of vaccination in the UK.[79] Local authorities complained about the regulations, and the need for additional tests meant that certain batches of vaccine passed their expiry dates.[80] In spite of such problems, the Ministry of Health did not change its policy and continued to follow the MRC guidelines.

Clearly, these regulations were not only the outcome of new risk assessments and cost/benefit considerations, as political factors also played an important role. Canadian and US-American products were cheaper than the British ones (£ 107–117 per litre compared to £ 125–175 per litre for vaccine made by British companies) and from this perspective would have been preferable for the NHS with its chronic budget problems. Other criteria seemed to be more important than the question of cost, however, as the issue of vaccination and of importing vaccine was regarded as a highly important national question, with many in the Ministry of Health in favour of buying only British vaccines.[81] Thus, producing and administering a British vaccine seems to have been not only a matter of safety but also a matter of national pride.

The actual process of vaccination proceeded quite efficiently under the British NHS. Vaccination was centrally coordinated by the Ministry of Health and locally organized by the Health Authorities.[82] Medical Health Officers and other personnel of the Local Health Authorities took charge of the inoculation programmes. Sometimes, however, problems occurred with the allocation of vaccine, with some Local Health Authorities suffering shortages, while others had ample stocks. In a number of regions, General Practitioners had to help with the inoculations, but altogether vaccination progressed quite evenly. Registration of children and statistical analysis of the vaccination programme was also coordinated by the NHS. Like other preventive measures in Britain, polio vaccination remained free.

Despite the problems with changing regulation and testing, the import of Canadian and US-American vaccine enabled an extension of the British vaccination programme. Furthermore, from 1958 onwards, the British manufacturers Burroughs Wellcome and Glaxo were also able to produce vaccines in larger quantities that were used for domestic vaccination. In 1958, 6.4 million people had had two inoculations and the number of polio cases started to decline.[83] All in all, Britain did put a fairly successful vaccination programme in place using IPV, even though other European countries like Denmark, Sweden and the Netherlands had considerably higher vaccination rates. The British programme was extended during the following years, relying on both British-manufactured and imported vaccine.

If we look at West Germany, a different picture emerges. Following the recommendations of the *Bundesgesundheitsamt*, there was no vaccination in 1955 and 1956, while the British had started at least a small programme in 1956. The attitude towards vaccination continued to be very cautious within the West German health administration and the scientific community. Then, from 1957 onwards, German scientists and health officials started to report regularly at International and European Polio Conferences.[84] These meetings allowed them to see quite clearly the backwardness of the situation of West Germany with respect to the international fight against polio. At the same time, they observed that IPV had been introduced very successfully in other European countries.[85] Furthermore, East Germany had started its own IPV programme, which appeared to be working well, a development that had a considerable impact on the West Germans.[86] West Germany was always keen to prove that its medical care was superior to that in East Germany and so the news of East-German successes in this field was received with great concern in the higher ranks of the West-German health administration. These different factors helped to change the attitudes of the West-German scientists and politicians in favour of IPV vaccination.

In 1957, IPV vaccination was finally introduced into West Germany. The *Bundesgesundheitsamt* agreed to polio vaccination, but restricted vaccination to the use of German vaccine produced by the Behringwerke and made from the Brunhilde-strain. The German health administration then experienced similar problems to those encountered in the UK, with the Behringwerke unable to produce sufficient vaccine for the 1957 programme. Thus, as in Britain, the *Bundesgesundheitsamt* had to change its risk assessment and its regulations, permitting the importation of US-American and Belgian vaccine in order to facilitate vaccination.[87]

However, there were significant differences between the developments in Britain and in West Germany. First, public pressure for vaccination was marginal in West Germany and there was no group like the British IPF that was forcefully promoting vaccination. Second, national considerations concerning vaccine production seemed to be less important in Germany than in the British case. Indeed, the *Bundesgesundheitsamt* explicitly stated in 1958 that imported and German-produced vaccines were equally safe and could both be used by the German federal states.[88] All vaccine had to undergo testing but no choice was offered to parents that differentiated between German and imported vaccine. In West Germany, international pressure and East–West competition seemed to be the most important factors for the change in its assessment of the risk associated with the polio vaccine.

The huge problem that emerged following the start of vaccination in West Germany was the organization of the programme. Since vaccination fell under the direction of the German federal states (*Länder*), the programmes started in 1957 were very uneven and uncoordinated, reaching an insignificant portion of the population. This failure to introduce a widespread programme was due to the structure of the West-German public health system. After WWII, the public health system was de-centralized, being organized at the level of the German federal states.[89] On the national level, West Germany had a health department within the Ministry of the Interior, but a Ministry of Health was only created in 1961. Most public health issues were organized at the level of the individual federal states. Vaccination as a public health measure was regulated by the individual German federal states and their health ministries. Thus, the states ended up with responsibility for vaccine production, licensing, importation, and implementing the relevant laws and regulations. At the national level, there were some coordinating institutions like the *Bundesgesundheitsrat*, a committee of health ministers from the different *Länder*, the *Bundesgesundheitsamt*, the aforementioned federal health control office, and several committees of medical officers and ministers sent by the individual German federal states.[90] However, without a central administration at the national level it was very difficult to organize a coordinated vaccination programme.[91]

Furthermore some of the *Länder* charged fees for polio vaccination. Thus, for example, Bavaria charged a considerable amount (7.50 German Marks) for one inoculation, which must have been a serious obstacle for low-income families. While families could apply for a subsidy if in need, this kind of bureaucratic procedure certainly did not facilitate the vaccination process.[92] Other *Länder,* including North-Rhine-Westphalia, distributed free vaccine, meaning that this federal state initially vaccinated a relatively high percentage of children between two and five years of age. However, in 1959 the Ministry of Finance in North-Rhine-Westphalia implemented fees and the rate of vaccination fell once again. Also, some *Länder* organized vaccination via the public health service, while others used general practitioners to inoculate children. No coherent official programme regarding vaccination was developed in the years that followed and there was no national campaign designed to encourage it.[93]

By 1958, it was becoming clear that West Germany was falling behind other European countries in bringing polio under control, but the West German government did not introduce any new programmes to speed up IPV vaccination. As a consequence, IPV was never successfully implemented in West Germany and reached only 5 per cent of the population in 1960 (compared to around 50 per cent in the UK). There continued to be serious epidemics, like the one that broke out in Munich in 1959,[94] and the number of polio cases remained high until 1962, when oral polio vaccine was eventually introduced enabling a mass vaccination programme.[95]

The dominant view in the West German government was that the public simply did not accept the IPV vaccine and that there was nothing that the state or the administration could do about it.[96] Health officials spoke of a general mistrust of vaccination in Germany, making reference to the vaccination catastrophe in Lübeck in 1933 when 76 babies died due to contaminated BCG vaccine against tuberculosis. Following Lübeck, BCG vaccination became impossible in Germany for the next two decades.[97] However, there is hardly any evidence for the existence of these kinds of argument in connection with polio. No powerful anti-vaccination movements or groups existed in West Germany in the 1950s. Furthermore, polio was perceived as a much more devastating threat than tuberculosis, affecting children in particular and often leaving them crippled, making parents more likely to favour IVP vaccination programmes. Also, in neighbouring East Germany IPV vaccination was working well and the East-German public had had the same experiences with Lübeck as their West German counterparts. A free, well-organized vaccination programme might have been very successful in West Germany, as is illustrated by the example of the North-Rhine-Westphalia case.

Conclusion

The evaluation of the risk associated with IPV as well as the regulations covering vaccine production and vaccination changed drastically during the period under consideration in both the UK and West Germany. At the beginning, in 1954, neither the UK nor West Germany issued strict regulations concerning IPV, as the risks associated with the Salk vaccine were deemed to be acceptable. Shortly afterwards, both countries were confronted with production and safety problems occurring in the factories of their own national producers of IPV. At the same time, the Cutter incident in the US suddenly rendered the risks associated with the vaccine very public. Thus, both the MRC and the *Bundesgesundheitsamt* developed strict regulations and testing procedures in 1955, and the importation of IPV made using the Mahoney-strains was forbidden in both countries. In West Germany, vaccination was completely stopped for two years.

In the UK, where vaccination was postponed in 1955, a programme nevertheless started in 1956 under the same strict guidelines that had been issued a year earlier. Under increasing pressure from the public, regulations were again changed in 1957 and 1958 to allow the importation of vaccine and thereby to facilitate the vaccination process. In 1958, the UK had a polio vaccination programme that worked relatively well. In West Germany, vaccination started in 1957, but only on a very small scale. There was hardly any pressure from the German public, but instead international pressure played a far more important role than in the UK. In West Germany the conclusions drawn from risk assessment changed in 1958 in a similar way to the UK, allowing the use of more imported vaccine. Overall, however, the vaccination process remained uneven and ineffective in West Germany, where IPV never really succeeded.

Why did the evaluation of the risk associated with IPV change so often and why did its introduction proceed so differently in West Germany and the UK? After all, both countries were confronted with serious polio epidemics practically every year in this period.

National attitudes within the health administrations and the scientific communities were very important as factors influencing the evaluation of risk and shaping the development of regulations. This is quite clear for these two countries where epidemiology dominated the field of polio research, and so neither carried out much original research on polio viruses and vaccines. Both countries, therefore, had no choice but to look to the US for vaccine development. However, there were significant differences between the UK and West Germany; whereas the West German scientific community was cut off from international developments in many ways after WWII, the UK had relatively strong ties with US research, both institutionally and in terms of exchange between individual researchers. Nevertheless despite these strong connections and early enthusiasm about the Salk vaccine, the British health administration and the British scientific community lost confidence in the US vaccine after the Cutter incident, leading the administration to act cautiously. In the UK, national considerations seemed to have been important within the scientific community and had a considerable impact on decision-making, meaning that British testing remained paramount even after the introduction of 'foreign' vaccines. In West Germany, the attitude towards the new vaccine was even more suspicious in both the scientific community and the health administration, and this sceptical attitude prevailed much longer than in the UK. West-German doubts about US achievements in virology were also accentuated by a general lamentation over the loss of Germany's leading scientific position in the post-war world. In West Germany, mounting international pressure was the main reason for initiating a change in vaccination policy. Thus, distinctive national scientific cultures strongly influenced the process of risk evaluation and regulation of vaccination.

An additional problem both countries had to face was the problematic connection between the public health systems and the organization of vaccine production that was undertaken by private pharmaceutical companies. These commercial suppliers had difficulties in rapidly producing vaccine of an adequate quality and in the quantities needed, which again led to problems associated with the importation of foreign vaccines.

The example of the introduction of IPV vaccine in two European countries shows quite convincingly that risk assessment and the development of regulations are very much influenced by factors such as national scientific cultures, the attitudes of health officials, the national considerations of governments and international pressure. These factors, though difficult to weigh up and specify with precision, have proven to be very important in understanding the ways in which the new vaccines were assessed and introduced in different national contexts.

Notes

1. For the adoption of IPV and OPV (attenuated polio vaccine) in three countries, see Ulrike Lindner and Stuart Blume, 'Vaccine Innovation and Adoption: Polio Vaccines in the Netherlands, the UK and West Germany 1955–1965', in: *Medical History* 50 (2006), 425–46; see for the introduction of BCG in Germany and France, Christian Bonah, 'As Safe as Milk or Sugar Water: Perception of the

Risks and Benefits of BCG Vaccine in France and Germany', in: Thomas Schlich and Ulrich Tröhler (eds), *The Risks of Medical Innovation*, London, New York: Routledge, 2006, 71–92, 71–2.

2. For accounts of the different health systems, see Jens Alber and Brigitte Bernardi-Schenkluhn, *Westeuropäische Gesundheitssysteme im Vergleich*, Frankfurt a.M., New York: Campus, 1992; for the analysis of the adoption of medical technologies in different countries with a strong focus on the influence of different health systems, see J Rogers Hollingsworth, Jerald Hage and Robert A Hanneman, 'State Intervention in Medical Care: Consequences for Britain, France, Sweden and the United States, 1890–1970', Ithaca: Cornell University Press, 1990.

3. For the notion of different therapeutic cultures, see Arthur Daemmrich, 'Pharmacopolitics: Drug Regulation in the United States and Germany', Chapel Hill: University of North Carolina Press, 2004; see also Linda Bryder, 'We Shall Not Find Salvation in Inoculation: BCG Vaccination in Scandinavia, Britain and the USA 1921–1960', in: *Social Science and Medicine* 49 (1999), 1157–67.

4. Thomas Schlich, 'Risk and Medical Innovation: A Historical Perspective', in: Thomas Schlich and Ulrich Tröhler (eds), *The Risks of Medical Innovation*, London, New York: Routledge, 2006, 1–19, 2.

5. See Schlich and Tröhler, 2006; Daemmrich, 2004; Jennifer Stanton (ed), *Innovations in Health and Medicine*, London, New York: Routledge, 2002; see also Thomas Wieland: 'Innovationskultur: Theoretische und empirische Annäherungen an einen Begriff', working paper of the Munich Centre for History of Science and Technology, 19 November 2004.

6. Peter Baldwin, *Disease and Democracy. The Industrialized World Faces AIDS*, Berkeley: University of California Press, 2005, 257.

7. See for the many different aspects that can influence a regulation process, Jean-Paul Gaudillière and Volker Hess, 'Introduction', in: Jean-Paul Gaudillière and Volker Hess (ed.), *Ways of Regulating: Therapeutic Agents between Plants, Shops and Consulting Rooms*, Max Planck Institute for the History of Science, Preprint 363, 2008, 5–17.

8. Many of the factors seem to be less rational, as has been already shown by Jennifer Stanton in her analysis of the introduction of the hepatitis B vaccine, see Jennifer Stanton, 'What Shapes Vaccine Policy? The Case of hepatitis B in the UK', in *Social History of Medicine* 7 (1994), 427–46; see also Bonah, 2006.

9. See for example Anne-Marie Moulin who understood vaccination as a 'total social fact', Moulin, 'A Philosophy of Vaccinology', in S Plotkin and B Fantini (eds), *Vaccinia, Vaccination, Vaccinology. Jenner, Pasteur and Their Successors*, Amsterdam: Elsevier, 1996, 17–23; see for the introduction of a new medical technique into different countries, Thomas Schlich, 'Degrees of Control: The Spread of Operative Fracture Treatment with Metal Implants. A Comparative Perspective on Switzerland, East Germany and the US 1950s–1990s', in: Jennifer Stanton (ed), *Innovations in Health and Medicine*, London, New York: Routledge, 2002, 106–25.

10. John R. Paul, *A History of Poliomyelitis*, New Haven, London: Yale University Press 1971, 148–60.

11. Anne Hardy, 'Poliomyelitis and the Neurologists: The View from England', *Bulletin of the History of Medicine* 71 (1997), 249–72.

12. See for Roosevelt and his fight against polio Hugh G. Gallagher, *West Germany's Splendid Deception*, New York: Dodd, Mead &Comp, 1985, also Naomi Rogers, *Dirt and Disease. Polio before West Germany*, New Brunswick: Rutgers University Press, 1992, 165–9.

13. Jane A Smith, *Patenting the Sun: Polio and the Salk Vaccine*, New York: William Morrow, 1990, 52–78.
14. Paul, 1971, 252–62.
15. John Beale, 'The Development of IPV', in: Stanley A. Plotkin and Bernardino Fantini (eds), *Vaccinia, Vaccination, Vaccinology: Jenner, Pasteur and Their Successors*, Amsterdam: Elsevier, 1996, 221–7, 221.
16. John F. Enders, Thomas H. Weller and Frederick C. Robbins, 'Cultivation of the Lansing Strain of Poliomyelitis Virus in Cultures of Various Human Embryonic Tissues', in: *Science* 109 (1949), 85–7.
17. Samuel L. Katz, Catherine M. Wilfert and Frederick C. Robbins, 'The Role of Tissue Culture in Vaccine Development', in: Stanley A. Plotkin and Bernardino Fantini (eds) *Vaccinia, Vaccination, Vaccinology: Jenner, Pasteur and Their Successors*, Amsterdam: Elsevier, 1996, 213–16.
18. Smith, 1991, 124–7.
19. Paul, 1971, 235; David M. Oshinsky, *Polio. An American Story*, Oxford: Oxford University Press, 2005, 115–21.
20. Arthur Klein, *Trial by Fury. The Polio Vaccine Controversy*, New York: Charles Scribner, 1972; see also Kurt Link, *The Vaccine Controversy. The History, Use, and Safety of Vaccination*, Westport CT: Praeger, 2005, 89–93.
21. H. V. Wyatt, 'Poliovaccines: Lessons Learnt and Forgotten', in: *History and Philosophy of Life Sciences* 17 (1995), 91–112, 98–9.
22. Smith, 1991, 135–42; Oshinsky, 2005, 188–213.
23. Sarah Marie Lambert and Howard Markel, 'Making History. Thomas Francis Jr, MD, and the 1954 Salk Poliomyelitis Vaccine Field Trial', in: *Archive of Pediatrics and Adolescent Medicine* 154 (2000), 512–17; see also Smith, 1991, 266–74.
24. Paul, 1971, 432–6.
25. James Colgrove, *State of Immunity: The Politics of Vaccination in Twentieth-Century America*, Berkeley: University of California Press, 2006, 113.
26. For a general account of the Cutter incident, see Paul A. Offit, *The Cutter Incident*, New Haven, London: Yale University Press, 2005, 83–131.
27. Oshinsky, 2005, 236–43; Wyatt, 1995, 101–2; Paul, 1971, 437–8.
28. See for example: 'Third Symposium of European Association against Poliomyelitis', in: *Revue d'hygiene et de medicine sociale* 4 (1) (1956), 87–93; or Poliomyelitis. Papers and discussions presented at the Fourth International Poliomyelitis Conference, Philadelphia 1958.
29. See the report on this conference in Heymann, 'Standardisierung von Sera und Impfstoffen', in: *Deutsche Medizinische Wochenschrift* 82 (1957), 356–9.
30. Cf. WHO Expert Committee on Biological Standardization, yearly reports (1947–)
31. See for the League Standardization Commission the article of Pauline M. H. Mazumdar in this volume on 'The State, the Serums Institute and the League of Nations'.
32. WHO Expert Committee on Biological Standardization, 9th report (1956), 10.
33. WHO Expert Committee on Biological Standardization, 11th report (1958), 13–14. See, for the history of the *Statens Seruminstitut*, Anne Hardy in this volume.
34. WHO Expert Committee on Biological Standardization, 14th report (1961), 13–14; 17th report (1965), 15.
35. The National Archives, Public Record Office (TNA, PRO), MH 55/1769, Infantile Paralysis Fellowship to Minister of Health, 27 September 1951.
36. Anne Hardy, 'Poliomyelitis and the Neurologists: The View from England', in: *Bulletin of the History of Medicine* 71 (1997), 249–72, 265–6.

37. Ulrike Lindner, *Gesundheitspolitik in der Nachkriegszeit: Großbritannien und die Bundesrepublik Deutschland im Vergleich*, München: Oldenbourg, 2004, 258.
38. TNA, PRO, MH 55/1775, Bradley to Green, MRC, 31 August 1950.
39. TNA, PRO, MH 55/1775, Bradley to Green, MRC, 31 August 1950, see also William Bradley, 'Poliomyelitis', in: *Journal of the Royal Sanitary Institute* 7 (1954), 519–41.
40. Beale, 1996.
41. Tony Gould, *A Summer Plague*, New Haven, London: Yale University Press, 1995, 165–7.
42. TNA, PRO, MH 55/1775, Bradley to Miss Newham, 9 October 1952.
43. Wellcome Library, Contemporary Medical Archive Centre, GC 139, F 22.
44. John R. Wilson, *Margins of Safety. The Story of Poliomyelitis Vaccine*, London: Collins, 1963, 122–5.
45. Anonymous 'The Salk Vaccine in Britain', in: *British Medical Journal* (1955), 1099.
46. Dorothy Porter and Roy Porter, 'The Politics of Prevention: Anti-Vaccinationism and Public Health in Nineteenth-Century England', in: *Medical History* 32 (1988), 231–52, 235. See also Nadja Durbach, *Bodily Matters: The Anti-Vaccination Movement in England 1853–1907*, Durham: Duke University Press, NC 2005.
47. British Parliamentary Papers and Reports, Session 1956/1957, Vol. XIII, Ministry of Health, On the State of Public Health 1955, HMSO 1956, 77.
48. Beale, 1996, 223.
49. Anonymous 'Poliomyelitis Vaccine Trials deferred', in: *British Medical Journal* (1959), 1539.
50. Wilson, 1963, 123.
51. TNA, PRO, MH 133/467, Joint Committee on Poliomyelitis Vaccine to Senior Administrative Medical Officers, 16 August 1955.
52. TNA, PRO, MH 133/467, Joint Committee on Poliomyelitis Vaccine, Meeting on 13 July 1955.
53. D. R. Bangham, *A History of Biological Standardization: The Characterization and Measurement of Complex Molecules Important in Clinical and Research Medicine. A Personal account by D.R. Bangham*, London: Society for Endocrinology, 1999, 159.
54. TNA, PRO, MH 133/467, Joint Committee on Poliomyelitis Vaccine, Meeting on 10 October 1955.
55. Lindner and Blume, 2006, 436.
56. Werner Anders, 'Epidemiologische Studien über die Poliomyelitis 1947/48 in Groß-Berlin', in: Albrecht Tietze and Paul Kühne, *Die Poliomyelitis. Bearbeitet nach den Erfahrungen bei den Berliner Epidemien 1947/49*, Berlin: de Gruyter, 1949, 12–40, 13.
57. Gould, 1995, 161.
58. Ulrike Lindner, *Gesundheitspolitik in der Nachkriegszeit: Großbritannien und die Bundesrepublik Deutschland im Vergleich*, München: Oldenbourg, 2004, 240.
59. Stengel von Rutkowski, 'Die Schutzimpfung gegen Poliomyelitis in Amerika und in Deutschland', in: Behringwerk-Mitteilungen no. 31, *Die Schutzimpfung gegen Poliomyelitis*, Marburg a. d. Lahn, 1956, 35–6; see also for the history of the Behringwerke, W. Bartmann, *Zwischen Tradition und Fortschritt. Aus der Geschichte der Pharmabereiche von Bayer, Höchst und Schering 1935–75*, Stuttgart: Franz Steiner, 2003.
60. R. Haas, V.Keller and R.Sauthoff, 'Die Wirkung des Aluminiumhydroxyds auf die immunisatorischen Eigenschaften inaktivierter Poliomyelitis-Vaccine', in: Behringwerk-Mitteilungen no. 31, *Die Schutzimpfung gegen Poliomyelitis*, Marburg a.d. Lahn 1956, 37.

61. Bundesarchiv Koblenz (BAK) B 142/22, Protokoll über die Besprechung der leitenden Medizinalbeamten der Länder, 10 February 1956.
62. BAK B 142/22, Vermerk, 30 September 1955; see also: Behringwerk-Mitteilungen no. 31, *Die Schutzimpfung gegen Poliomylieits*, Marbug a.d. Lahn 1956. See for a discussion of the issue of private vaccine producers versus public sector vaccine institutes, Stuart S. Blume, 'Lock In, the State and Vaccine Development: Lessons from the History of the Polio Vaccines', in: *Research Policy* 34 (2005), 159–73.
63. BAK B 142/22, Protokoll über die Besprechung der leitenden Medizinalbeamten der Länder, 10 February 1956.
64. See Georg Henneberg, 'Zum Problem der Poliomyelitisschutzimpfung in Deutschland', in: *Der öffentliche Gesundheitsdienst* 18 (1956/57), 181–7.
65. Habernal [Habernoll], Germany. Bundesrepublik Deutschland (West), in: Reports of the Official Delegates. Poliomyelitis. Papers and Discussions Presented at the Fourth International Poliomyelitis Conference, Philadelphia u.a., 1958, 36–8.
66. See for similar problems after the WWI, the article by Pauline M. H. Mazumdar in this volume. In the 1920s, German researchers were not admitted to the scientific committees of the League of Nations. Only after long negotiations a solution was found.
67. BAK, B 142/23, Bericht über das Poliomyelitis-Symposium in Madrid für das Innenministerium, 1958.
68. Henneberg 1956/57, 181–7; see also Lindner and Blume 2006, S. 431.
69. Vorarchiv Arbeitsministerium Nordrhein-Westfalen 1200, Sitzung 7 April 1960, Beilage Denkschrift: Fünf Jahre Tätigkeit der Deutschen Vereinigung zur Bekämpfung der Kinderlähmung; see also Günther Maass, 'Von der Vereinigung zur Bekämpfung der Kinderlähmung zur DVV', in: *Immunologie & Impfen* 2 (1999), 40–2.
70. BAK, B 142/23, Sechstes europäisches Symposion über Poliomyelitis in München, 6–9 September 1959; see also Maass 1999, 40–2.
71. Stiftung Pfennigparade, 40 Jahre Pfennigparade, München 1992.
72. British Parliamentary Papers and Reports, Session 1958/1959, Vol. XV, Ministry of Health, On the State of Public Health 1957, HMSO 1958, 85.
73. TNA, PRO, MH 55/2462, Emery to Dodds, 28 October 1957.
74. Ministry of Health, On the State of Public Health 1957, 86.
75. TNA, PRO, MH 55/2467, Deputy Secretary to Woodlock, 3 June 1958; TNA, PRO, MH 55/2463, Note to Dodds, 30 May 1958.
76. Ministry of Health, 'On the State of Public Health 1957', 85–86.
77. TNA, PRO, MH 55/2462, County Councils Association to Enid Russel-Smith, Ministry of Health, September 1957.
78. TNA, PRO, MH 55/2463, Memorandum, 12 May 1958.
79. TNA, PRO, MH 55/2462, Russel-Smith to Dacey, County Councils Association, 27 September 1957.
80. TNA, PRO MH 55/2464, Judd to Dodds, 16 October 1958.
81. Lindner and Blume 2006, 437.
82. TNA, PRO, MH 55/2469, Memorandum on Poliomyelitis Vaccine, 9 May 1960.
83. British Parliamentary Papers and Reports, Session 1959/1960, Vol. XVI, Ministry of Health, On the State of Public Health 1958, HMSO 1959, 82.
84. See report on the Poliomyelitis-Symposion in Madrid in 1958, in: BAK B 142/23.
85. See reports on poliomyelitis vaccination in England and France in the files of the West German health department: BAK B 142/22, Poliomyelitis in England und Wales, Poliomyelitis in Frankreich, June 1956.
86. BAK B 142/23, Zusammenstellung der Polio-Prophylaxe in den europäischen Ländern, 6 December 1961.

87. BAK B 142/23, Vermerk über das Sechste Europäische Symposion über Poliomyelitis in München, 5 September 1959.
88. Zweites Gutachten über den Stand der Schutzimpfung gegen die spinale Kinderlähmung. Erstattet vom Bundesgesundheitsamt nach dem Stand vom 31. März 1958, Berlin (West) 1959; see also Dieter Hess, 'Poliomyelitis von 1956 bis 1960 in den kreisfreien Stadten und Landkreisen', med. Diss., Erlangen, 1963.
89. Lindner, 2004, 62–6, see also Ludwig von Manger-Koenig, 'Der öffentliche Gesundheitsdienst zwischen Gestern und Morgen', in: *Das Öffentliche Gesundheitswesen* (1975) 37, 433–48.
90. Lindner, 2004, 39–41.
91. BAK B 142/22, Protokoll über die Besprechung der leitenden Medizinalbeamten der Länder, 10 February 1956; see also Lindner and Blume, 2006, 434.
92. Polio-drohende Gefahr trotz fortschreitender Schutzimpfung, in: *Bayerische Staatszeitung*, 12 August 1960.
93. Lindner, 2004, 249–50.
94. Hess, 1963.
95. See Statistik der Bundesrepublik Deutschland, Gesundheitswesen, different volumes. For information on the introduction of OPV in West Germany, see Lindner and Blume, 2006.
96. See a statement of the Minister of the newly founded Ministry of Health, Dr Elisabeth Schwarzhaupt, who refers to the failure of IPV during the 1950s. Verhandlungen des Deutschen Bundestages, 4. Wahlperiode, 26. Meeting of 12 April 1962, p. 1069.
97. Lindner, 2004, 248–50; see for the Lübeck catastrophy, S. Hahn, 'Der Lübecker Totentanz. Zur rechtlichen und ethischen Problematik der Katastrophe bei der Erprobung der Tuberkuloseimpfung 1930 in Deutschland', in: *Medizinhistorisches Journal* 30 (1995), 61–95.

13
Standardization Before Biomedicine: On Early Forms of Regulatory Objectivity

Alberto Cambrosio

Some time ago, I was invited to comment on a set of papers, now collected in this volume, to be presented at a Heidelberg meeting on the history of the standardization of therapeutic agents from 1890 to 1950. I must confess that I was somewhat puzzled and surprised by this invitation for, in spite of some forays into the visual history of early twentieth-century immunology, most of my research centers on the post-1950s period and has a sociological flavor. As some of my historian colleagues would certainly argue, I therefore was (and still am) doubly unqualified to discuss historical accounts of *Wertbestimmung*, first because of my association with sociology and, second, because of my contemporary focus. But this, of course, also provided me with a greater degree of speculative latitude: an outsider's freedom, so to speak.

In this postface, I will take advantage of this freedom to re-specify the topic of the present volume. Its authors and editors will certainly stand by its title and claim that the book is indeed about therapeutic agents. While this is certainly the case, an admittedly selective and tendentious reading allows me to reframe the volume's content as one about the beginnings of biomedicine.[1] Several processes discussed in this book do indeed foreshadow key aspects of the "interactive history of invention and convention" (to quote Krislov's elegant formulation)[2] that came to define biomedicine during the second half of the twentieth century. At this point, a clarification is in order. Peter Keating and I have argued elsewhere[3] that the evolution of Western medicine since WWII resulted in the development of new practices grounded in the direct interaction of biology and medicine: we call "biomedicine" this new configuration of work that generates spaces and modes of representation and intervention that are irreducible to either the normal or the pathological. We have since extended our analysis to include the claim that the post-war realignment of biology and medicine has been accompanied by the emergence of a new type of objectivity, *regulatory objectivity*.[4] Based on the systematic recourse to the collective production of evidence, regulatory objectivity consistently results in the production of conventions, sometimes tacit and unintentional, but most often arrived at through concerted programs of action. The conventions produced by regulatory objectivity create the conditions for a clinical objectivity that relies on the existence of entities (including therapeutic agents) and protocols that, while framing the intimate encounter between doctors and patients, are produced and maintained far outside doctor–patient relationships. These entities and protocols qualify as instances of *regulation* in two distinct, albeit related ways: they are the outcome of regulatory practices and they act, in turn, as regulatory devices for biomedical activities.

The relation of this set of claims to the present book is to be sought in the fact that insofar as *Wertbestimmung* refers to quality control and standardization, it is a form of regulation. In turn, as shown for instance by Christian Bonah's chapter on digitalis and Strophantin, quality control and standardization are obligatory passage points for translating substances into legitimate therapeutic entities and thus, also, for regulating practices that rely on the use of chemical and biological compounds. While the term regulation is often used in a narrow sense to refer to the activities of national regulatory agencies, I am using it here more broadly to denote a wide array of more or less formal initiatives sponsored *inter alia* by professional organizations or, in more recent times, by national and international consortia and networks. This broader definition does not merely imply a quantitative extension of the term, but also a qualitative shift of its meaning: while the more narrow description is consistent with an understanding of regulation as an external constraint imposed upon preexisting medical practices and substances, the broader description portrays regulation as a constitutive element of medical activities and entities. Another way of expressing these two different understandings is by resorting to the distinction proposed by the French language between *réglementation* – the rules that must be followed by all those who want to take part in a given endeavor – and *régulation* that refers to the various kinds of interfaces that allow actors who do not necessarily share the same approach to cooperate while maintaining their autonomy. In short: regulation should not be reduced to a form of rationalization imposed upon medicine, so to speak, "from the outside," for instance by politicians and bureaucrats. While the activities of public health agencies are of course part of the story, we need to look in parallel at the "endogenous" regulatory work pursued by physicians and scientists *as part* of their biomedical (clinical and research) activities, which in turn establishes the conditions of possibility for other forms of regulation.[5]

To get a better sense of why regulation should be understood as constitutive of recent medicine, we can think of a common example of a regulatory interface, namely standards. According to Timmermans and Berg's typology,[6] standards come in different forms. Terminological standards fix the terms used to classify and describe phenomena (such as diseases) and substances (such as diagnostic or therapeutic compounds) in order to achieve comparability. Performance standards define outcomes or results acceptable in particular situations (e.g., response rates or side effects of a given therapy). Design standards define the characteristics of devices (from small, straight-forward tools to sophisticated medical equipment) that are part of complex systems of action, thus allowing them to function in daily routines. Finally, procedural standards specify the actions or protocols (ranging from relatively simple laboratory protocols to the complex, multi-faceted protocols at the core of clinical trials) that must be followed in given situations. This typology is to a large extent analytical, for the four kinds of standards obviously interact and, in a sense, presuppose each other, but it is useful insofar as each category implies a different form of coordination of work or, to put it otherwise, of interdependency. The fact that the relative weight accorded each of the categories has shifted with time[7] can thus be linked to the emergence of a distinctive configuration of *biomedical* work. This is not simply an organizational matter: the collective production of (different kinds of) standards co-produces the domains in which the standardized entities and practices are deployed. We have examined elsewhere how this is done, by using as examples the establishment of a nomenclature of novel reagents, the stabilization of the instrumental procedures with which these reagents are associated, and the clinical categories and interventions for which these tools should be used.[8] Regulation, furthermore, can be achieved by means other than standards and, in fact, regulatory decisions include decisions about what to standardize and what not to standardize. To put it differently, we can make sense of standards (and other regulatory interfaces) only by relating them to the whole set of explicit and tacit, formal and informal regulatory interventions that define a given domain.

Maybe readers, at this point, will start wondering again what all this has to do with the papers in this book. To clarify this point, it is useful to return to the work of Ludwik Fleck, considered by some as the founding father of the sociology of medical knowledge, and whose work was contemporaneous with several events discussed in this volume. In his seminal 1935 contribution *Entstehung und Entwicklung einer wissenschaftlichen Tatsache*,[9] Fleck built on the example of the production and standardization of the Wasserman test for the diagnosis of syphilis. In doing so, he also discussed the work of Paul Ehrlich, the German immunologist who is portrayed in several chapters of the present book as a central figure of *Wertbestimmung*. Instead of rehearsing the well-known content of Fleck's argument, we will focus on two passages that mention (in one case explicitly, in the other implicitly) a "civil servant" variety of science. In the first passage (pp. 15–17 of the English translation), discussing the discovery of *Spirochaeta pallida*, the causative agent of syphilis, Fleck claimed that "[the discovery] was the result of steady, systematic work by civil servants," adding that "the title 'discoverer of the syphilis agent' should properly be awarded" to this "team of civil servants." The second passage (p. 78 of the English translation) concerns the development of the Wasserman test: after providing an overview of the convoluted trajectory of this test, Fleck concluded that the "thought collective that made the Wasserman reaction usable and ... even practical ... standardized the technical process with genuinely social methods, at least by and large, through conferences, ordinances, and legislative measures." In both cases, Fleck's reference to civil servants, ordinances and legislative measures – all elements usually related to regulation – was *not* meant to denote second-rate science or to denounce the dubious influence of "external" elements on scientific activities. Rather, Fleck called attention to what we call, in today's Science and Technology Studies' (S&TS) jargon, the "heterogeneous collective" that established and managed the relation between syphilis, its causative agent and the diagnostic test.

To use, once again, today's S&TS terminology, Fleck's argument qualifies as a "co-productionist" account[10] of the development of the Wasserman test: instead of hypostasizing "Science," the "State" or "Industry" as independent, well-defined fields of activity influencing or constraining each other, he followed the meandering initiatives that, crisscrossing established lines of work, slowly and painstakingly led to the production and stabilization of the test. In turn, the outcomes of these heterogeneous processes redefined what it meant to engage in biomedical research, public health and the industrial production of diagnostic tests and therapeutic agents. There is some confusion surrounding this point. Take, for instance, Ilana Löwy's argument according to which, contrary to Fleck's belief "that the tumultuous early history of the Wassermann reaction ... [had been] successfully stabilized through the standardization of laboratory practices and thanks to the rise of a specific professional segment, the serologists", the meaning of the test was subsequently destabilized by regulatory measures "barely mentioned in [Fleck's] historical narrative"[11]. After claiming (it is not entirely clear whether approvingly or disapprovingly) that "Fleck argued that the 'external' and the 'internal' elements that shape scientific knowledge and its context cannot be separated," Löwy maintains that "the fate of the Wassermann reaction was affected not only by serologists' thoughts and deeds, but by material, economic and political considerations external to the scientific community," that "the majority of contemporary 'scientific facts' are produced by scientific communities that ... are open to external influence, pressures, and a critical appraisal" and that "the need for successful articulations between the world of medical research and external social worlds ... increases the complexity of interactions within 'thought collectives' of medicine and biomedical sciences, and between these collectives and other social groups." From a co-productionist point of view, it makes little sense to think in terms of "externality" and "social worlds" (remember that for Fleck a "thought collective" is not a fixed, substantial group or unit). The claim that "external" and "internal" elements cannot be separated (or, rather, that one could actually distinguish "external" from "internal" elements) and the claim that one should analyze the interactions between "external" and "internal" social worlds are obviously contradictory! These two categories can be examined as actors' categories, as Pauline Mazumdar does in her paper on the debate about the role of "technical" and "political" contributions to the League of Nations' standardization initiatives, but not used as analytical resources. In fact, as we have seen, and contrary to Löwy's claim that regulation had no place in Fleck's account, he was fully aware of what he called a "civil servant" science. From this point of view, the subsequent history of the Wasserman test, as summarized in Löwy's article[12] confirms, rather than belies, the heuristic value of Fleck's approach.

What lessons should we draw from this discussion? Although the papers in this volume differ from each other in a number of aspects, most of them share the explicit or implicit assumption that the development of *Wertbestimmung* at the beginning of the past century was intimately related to the growing role played by the state and industry in health care. Industry or, to be more precise, production modalities (such as quality control and standardization) that are held to define industrial practices, became core components of the development and production of new therapeutic agents, and, in turn, the mobilization of the novel agents as part of public health campaigns (e.g., vaccination) and in the fight against diseases taking a large toll on the child population (e.g., diphtheria) or on the adult population in wartime periods (e.g., tetanus) legitimized increased interventionist measures by state authorities. As pointed out by the four papers on the management of diphtheria serum in different countries (Anne I. Hardy on Germany, Gabriel Gachelin on France, Mariama Kaba on Switzerland, and Anne Hardy on the Danish case and, more generally, from a comparative perspective) the exact division of labor between public and private organizations in the development, production, evaluation and distribution of sera and vaccines differed from country to country, but even (semi) public organizations adopted (quasi) industrial production methods. As previously mentioned, this theme, so formulated, certainly captures important aspects of the transformations that took place during the first half of the twentieth century, but is at odds with the methodological principle that industry and state interventions should not be analyzed as *external* constraints on medical practices. Rather, as just argued, medical developments have acted as much as conditions of possibility for industrial and state initiatives as the latter have done for medical science. The "container metaphor,"[13] according to which scientific activities need to be recast within an allegedly "broader picture" is unable to capture the dynamics at work in the development of (bio)medical innovations.

If we take the specific example of the development and circulation of diphtheria antisera, while it is indeed possible to argue, to cite a passage from the Introduction to the present volume, that around that period governmental institutions "started taking an interest in the therapeutic value ... of new therapeutic agents," one should not forget that Ehrlich's development of his original, paradigmatic approach to *Wertbestimmung* (Ehrlich initially used the term *Wertbemessung*) was co-substantive with the development of his "side-chain theory" of antibody formation, first made public in a 1897 paper devoted, precisely, to the standardization of diphtheria antisera.[14] Not only it is impossible to dissociate Ehrlich's "fundamental" work on immunity, for which he shared the 1908 Nobel Prize for Physiology or Medicine with another immunologist (Elie Metchnikoff from the Pasteur Institute, an institution also directly involved in antisera production), from his more "practical" work on *Wertbestimmung*, but, as argued in Cay-Rüdiger Prüll's paper in this volume, these two elements of Ehrlich's work entertained a mutually generative relation. A similar argument could be made for the work carried out several decades later at the Danish State Serum Institute (DSSI) by the 1984 Nobel Prize winner Niels K. Jerne. As explained in Anne Hardy's paper, during the time span covered by the present volume the DSSI acted not only as a key *Wertbestimmung* institution but also became a "leading international scientific centre" under the direction of Thorvald Madsen, who entertained close relations not only with Ehrlich but also with yet another Nobel Prize winner (1903), Svante Arrhenius. In short, and to cite once again the Introduction to the present book, the new therapeutic agents produced at that time were simultaneously "novel objects of knowledge *and* therapeutic agents."

Such a dual status also applies to the regulatory arrangements that were put in place to turn the novel agents into clinical and public health tools. Hardy's paper on the DSSI convincingly shows that the standardization work carried out by that institution (and, by extension, by similar institutions in other countries) amounted to a full-fledged form of research both in itself and because research on standardization affected more traditional forms of research. Given that the novel therapeutic agents were based on the latest discoveries in emerging fields such as immunology, it is not surprising that pharmaceutical companies began looking for closer ties with academic researchers, but the arrow goes in both directions: indeed, as argued in Bonah's paper, clinicians and medical researchers called industrialists to their rescue in order to translate their discoveries into operational therapeutic tools, a process that implied, *inter alia*, the establishment of equivalents between different sera via such tools as proper production methods, common measurement techniques, and product standards.[15] Issues such as the dubious cleanliness if not outright filth of several leading laboratories (evoked in Hardy's paper on the DSSI), became legitimate topics of discussions aiming, for instance, at isolating possible factors leading to contingent, haphazard results as opposed to reproducible outcomes. The elaborate administrative techniques described, for instance, in Axel Hüntelmann's paper on the production of diphtheria antiserum are an example of the (administrative) solutions provided for the new quality requirements. In short, efforts to standardize operations led to the realization that medical research and practices were characterized by variation and differences and the effective management of these differences became an ongoing preoccupation for practitioners who began designing tools and creating institutions to reach this goal in concrete situations.

Taken alone, standards and standard measurements are examples of what Daston and Galison call "mechanical objectivity"[16] or, to use a more common term, "metrology"; in this sense, their roots reach back to the nineteenth century. I have argued, however, that from a retrospective vantage point *Wertbestimmung* draws the initial contours of a different form of (regulatory) objectivity that will come to characterize twentieth century biomedicine and that is less an epistemic virtue than a form of institutional action. While calling upon the metrology of standard substances, regulatory objectivity works by generating and enacting reflexive conventions: the tacit knowledge of thought collectives is replaced or, rather, complemented by the explicit, albeit not always transparent, decision-making procedures of consensus groups. To understand this transition it is useful, once again, to examine Fleck's 1935 book (p. 63 of the English translation) and, in particular, his claim that "no universally accepted system of measurement" was available in biology, and that "this [was] especially so in serology." Far from being swept away by subsequent technological developments, this issue grew larger with time, and especially with the larger space accorded to diagnostic technologies by contemporary biomedicine. Since no "true reference standards" exist for biological elements such as blood and blood cells, complex schemes for assessing the degree of consensus between diagnostic laboratories had to be put in place, vindicating Georges Canguilhem's claim that normalization does not presuppose a norm: instead, norms are the result of normalization processes.[17] *Wertbestimmung* signals the beginning of a process through which a growing number of "conventions"[18] came to define medical activities. While the term "conventions" refers to the existence of negotiations and compromise, it does not necessarily imply arbitrariness. Rather, in the case of medicine, conventions rely on the existence of material entities and protocols produced and maintained by heterogeneous (bio-clinical and industrial) collectives. They establish new regimes of coordination[19] that increasingly operate as conditions of possibility for the conduct of medical activities. Since the time of *Wertbestimmung* the number of interlocking conventions produced by increasingly complex networks and consortia has been steadily growing, producing an expanding array of embedded regulations that set the parameters of use of novel agents and generating layers of regulation that combine biomedical entities and protocols with clinical data also generated through coordination and conventions.

In conclusion, this collection of papers on the standardization of novel biological agents requires one to think about the meaning of standardization. The term can refer to a wide array of practices for, to use once again one of Krislov's elegant formulations, "there is no standard way of defining standards."[20] This is true not simply from a substantive point of view (the term can refer, for instance, to state, trade associations or third-party regulations) but also from a more analytical point of view, for far from being reducible to a single, pre-existing logic – an industrial frame or a rationalization movement – standardization practices, as already hinted, combine and promote different types of coordination (ranging, for instance, from close-contact collaboration to at-a-distance, impersonal interdependencies), some of which constitute an innovation in itself. Minimally, one should ask what is being standardized, who performs the standardization work, and for what purpose. In the present case, the answer to the "what" question seems quite easy: the entities being standardized are the novel biological therapeutic agents (sera, vaccines, etc.). But as soon as we look at *how* these standardization processes were carried out (and this is where the empirical richness of the contributions to this volume lie), we face a decidedly more complex situation, for standardizing a given substance meant not simply standardizing (certain aspects of) the production process, but also large sections of the network of heterogeneous actors defining the (therapeutic, diagnostic etc.) practices through which a given substance was enacted.[21] As for the carriers of the standardization process, the situation is similarly complex. Here again, while, as several papers in this volume have shown, established industrial organizations such as pharmaceutical companies can and did indeed play a role, they more often than not did so as part of multifaceted arrangements mobilizing other actors who sometimes used established industrial companies as *de facto* standardization devices. Similarly, and as mentioned in our discussion of Fleck's argument about civil servant science, hybrid collectives consisting of researchers, clinicians, public health and other public officials, rather than top down state agencies, carry out regulation. Finally, the question of the purpose of standardization once again seems to call for an easy answer: to ensure uniform quality and thus also safe and effective therapeutic use. But regulation increasingly appears to have become a condition of possibility for performing and coordinating innovation and routine biomedical tasks via the circulation of the complex biological entities that have since become essential components of the biomedical armamentarium.

My preceding remarks converge on the point that the analysis of the different forms of regulation should not be separated from the analysis of the constitution of the entities, processes and activities that are the subject of regulation. But if this is so, then the challenge for historians is to find a way of producing historical narratives that, far from relying on a common sense understanding of a small set of "usual suspects," each cleanly assigned to a preestablished, watertight domain (state, industry, science), and of the categories (political, economic, cultural, technical) within which their activities allegedly fall, will instead focus on the shifting composition and modalities of actions of the collectives that, as I have argued in this text, have produced and performed the network of conventions and inventions that has come to characterize twentieth-century biomedicine.

Notes

Acknowledgments: Research for this paper has been made possible by grants from the following agencies: the Canadian Institutes of Health Research (CIHR), the Social Sciences and Humanities Research Council of Canada (SSHRC) and the Fonds Québécois de recherches sur la société et la culture (FQRSC). I would like to thank Christoph Gradmann for inviting me to the conference where the initial version of this paper was presented and Peter Keating for his comments and suggestions on the final draft of the text.

1. This is already a theme of Cay-Rüdiger Prüll's contribution to the present volume.
2. Samuel Krislov, *How Nations Choose Product Standards and Standards Change Nations*. Pittsburgh: University of Pittsburgh Press, 1997, p. 9.
3. Peter Keating and Alberto Cambrosio, *Biomedical Platforms: Realigning the Normal and the Pathological in Late-Twentieth-Century Medicine*. Cambridge, MA: MIT Press, 2003.
4. Alberto Cambrosio, Peter Keating, Thomas Schlich and George Weisz, "Regulatory Objectivity and the Generation and Management of Evidence in Medicine". *Social Science and Medicine*, 2006 63: 189–99. For a more recent discussion, see the introduction to a special issue of *Social Studies of Science* on medical objectivity: Alberto Cambrosio, Peter Keating, Thomas Schlich and George Weisz, 'Biomedical Conventions and Regulatory Objectivity: A Few Introductory Remarks'. *Social Studies of Science*, 2009 39: 651–664.
5. I borrow the term "endogenous" (that should not be confused with "internal") from Michael E. Lynch, "Technical Work and Critical Inquiry: Investigations in a Scientific Laboratory". *Social Studies of Science*, 1982 12: 499–533.
6. Stefan Timmermans and Marc Berg, *The Gold Standard: The Challenge of Evidence-Based Medicine and Standardization in Health Care*. Philadelphia: Temple University Press, 2003.
7. George Weisz, Alberto Cambrosio, Peter Keating, Loes Knaapen, Thomas Schlich and Virginie Tournay, 'The Emergence of Clinical Practice Guidelines'. *The Milbank Quarterly*, 2007 85: 691–727.
8. Alberto Cambrosio, Peter Keating and Andrei Mogoutov, "Mapping Collaborative Work and Innovation in Biomedicine: A Computer-Assisted Analysis of Antibody Reagent Workshops". *Social Studies of Science*, 2004 34: 325–64. Peter Keating and Alberto Cambrosio, "Interlaboratory Life: Regulating Flow Cytometry". In: Jean-Paul Gaudillière and Ilana Löwy (eds) *The Invisible Industrialist: Manufacturers and the Construction of Scientific Knowledge*. London: Macmillan, 1998: 250–95. Peter Keating and Alberto Cambrosio, "Real Compared to What? Diagnosing Leukemias and Lymphomas". In: Margaret Lock, Allan Young and Alberto Cambrosio (eds) *Living and Working with the New Medical Technologies: Intersections of Inquiry*. Cambridge: Cambridge University Press, 2000: 103–34.
9. I will cite from the (not entirely satisfying) English translation of the book: Ludwik Fleck, *Genesis and Development of a Scientific Fact*. Chicago: The University of Chicago Press, 1979 (Original German Edition, 1935).

10. Andrew Barry, *Political Machines. Governing a Technological Society.* London: Athlone Press, 2001. Sheila Jasanoff (ed.), *States of Knowledge: The Co-Production of Science and Social Order.* London: Routledge, 2004.

11. Ilana Löwy, "'A River That Is Cutting Its Own Bed': The Serology of Syphilis Between Laboratory, Society and the Law". *Studies in History and Philosophy of Biological & Biomedical Sciences*, 2004 25: 509–24. I chose this example because it focuses directly on Fleck, but I could have picked many other examples.

12. See also Pauline M. H. Mazumdar, "'In the Silence of the Laboratory': The League of Nations Standardizes Syphilis Tests". *Social History of Medicine*, 2003 16: 437–59.

13. George Lakoff and Mark Johnson, *Metaphors We Live By.* Chicago: University of Chicago Press, 1980.

14. Paul Ehrlich, "Die Wertbemessung des Diphterieheilserums und deren theoretische Grundlagen". *Klinisches Jahrbuch*, 1897 6: 299–326.

15. Things have not changed in this respect: in the 1980s, scientists developing innovative applications using new reagents called monoclonal antibodies, explicitly relied on commercial firms to both standardize and circulate the new tools. See Alberto Cambrosio and Peter Keating, "Monoclonal Antibodies: From Local to Extended Networks". In: Arnold Thackray (ed.) *Private Science: Biotechnology and the Rise of the Molecular Sciences.* Philadelphia: University of Pennsylvania Press, 1998: 165–81.

16. Lorraine Daston and Peter Galison, *Objectivity.* New York: Zone Books, 2007.

17. Georges Canguilhem, *The Normal and the Pathological.* New York: Zone Books, 1989.

18. Nicolas Dodier, "The Conventional Foundations of Action: Elements of a Sociological Pragmatics". *Réseaux: The French Journal of Communication*, 1995 3(2): 147–66. Laurent Thévenot, "Un gouvernement par les normes: Pratiques et politiques des formats d'information". In: Bernard Conein and Laurent Thévenot (eds) *Cognition et information en société.* Paris: Éditions de l'École des Hautes Études en Sciences Sociales, 1997: 205–42.

19. Laurent Thévenot, "Organized Complexity: Conventions of Coordination and the Composition of Economic Arrangements". *European Journal of Social Theory*, 2001 4: 405–25.

20. Krislov, *How Nations Choose Product Standards*, 1997, p. 7.

21. On the notion of "enactment", see Annemarie Mol, *The Body Multiple: Ontology in Medical Practice.* Durham: Duke University Press, 2002. On the enactment of therapeutic substances, see Emilie Gomart, 'Methadone: Six Effects in Search of a Substance'. *Social Studies of Science*, 2002 32: 93–135. Alberto Cambrosio, Peter Keating and Andrei Mogoutov, "Protocols, Regimens and Substances: The Socio-Technical Space of Anti-Cancer Drugs". In: Jean-Paul Gaudillière and Volker Hess (eds) *Ways of Regulating: Therapeutic Agents Between Laboratories, Plants, Consulting Rooms.* Berlin: Max-Planck-Institute for the History of Science, 2008 (Preprint 363): 175–205.

Index